**Advances in
Electrochemical Science
and Engineering**

Volume 12
Photoelectrochemical
Materials and Energy
Conversion Processes

Advances in Electrochemical Science and Engineering

Advisory Board

Prof. Elton Cairns, University of California, Berkeley, California, USA
Prof. Adam Heller, University of Texas, Austin, Texas, USA
Prof. Dieter Landolt, Ecole Polytechnique Fédérale, Lausanne, Switzerland
Prof. Roger Parsons, University of Southampton, Southampton, UK
Prof. Laurie Peter, University of Bath, Bath, UK
Prof. Sergio Trasatti, Università di Milano, Milano, Italy
Prof. Lubomyr Romankiw, IBM Watson Research Center, Yorktown Heights, USA

In collaboration with the International Society of Electrochemistry

Advances in Electrochemical Science and Engineering

Volume 12
Photoelectrochemical Materials and
Energy Conversion Processes

Edited by
Richard C. Alkire, Dieter M. Kolb, Jacek Lipkowski,
and Philip N. Ross

WILEY-VCH Verlag GmbH & Co. KGaA

The Editors

Prof. Richard C. Alkire
University of Illinois
600 South Mathews Avenue
Urbana, IL 61801
USA

Prof. Dieter M. Kolb
University of Ulm
Institute of Electrochemistry
Albert-Einstein-Allee 47
89081 Ulm
Germany

Prof. Jacek Lipkowski
University of Guelph
Department of Chemistry
N1G 2W1 Guelph, Ontario
Canada

Prof. Philip N. Ross
Lawrence Berkeley National
  Laboratory
Materials Science Department
1 Cyclotron Road MS 2-100
Berkeley, CA 94720-0001
USA

All books published by Wiley-VCH are carefully produced. Nevertheless, authors, editors, and publisher do not warrant the information contained in these books, including this book, to be free of errors. Readers are advised to keep in mind that statements, data, illustrations, procedural details or other items may inadvertently be inaccurate.

Library of Congress Card No.: applied for

British Library Cataloguing-in-Publication Data
A catalogue record for this book is available from the British Library.

Bibliographic information published by the Deutsche Nationalbibliothek
The Deutsche Nationalbibliothek lists this publication in the Deutsche Nationalbibliografie; detailed bibliographic data are available on the Internet at http://dnb.d-nb.de.

© 2010 WILEY-VCH Verlag GmbH & Co. KGaA, Weinheim Germany

All rights reserved (including those of translation into other languages). No part of this book may be reproduced in any form – by photoprinting, microfilm, or any other means – nor transmitted or translated into a machine language without written permission from the publishers. Registered names, trademarks, etc. used in this book, even when not specifically marked as such, are not to be considered unprotected by law.

Typesetting Toppan Best-set Premedia Limited, Hong Kong
Printing and Binding betz-druck GmbH, Darmstadt
Cover Design Grafik-Design Schulz, Fußgöheim

Printed in the Federal Republic of Germany
Printed on acid-free paper

ISBN: 978-3-527-32859-8
ISSN: 0938-5193

Contents

Preface *IX*
List of Contributors *XIII*

1	**Applications of Electrochemistry in the Fabrication and Characterization of Thin-Film Solar Cells** *1*	
	Phillip Dale and Laurence Peter	
1.1	Introduction *1*	
1.2	Electrochemical Routes to Thin-Film Solar Cells *3*	
1.2.1	Basic Cell Configurations *3*	
1.2.2	Material Requirements for PV Applications *4*	
1.2.2.1	Implications of Materials Requirements for the Direct Synthesis of Absorber Layers by Electrodeposition *5*	
1.2.2.2	Synthetic Routes Involving Deposition and Annealing (EDA) *7*	
1.2.2.3	Summary of EDA Routes *11*	
1.2.3	EDA route to p-Type Semiconductors for Thin-Film Photovoltaics *13*	
1.2.3.1	Electrodeposition of CdTe for CdS|CdTe Solar Cells *13*	
1.2.3.2	Electrodeposition of CIGS for CIGS|CdS|ZnO Solar Cells *19*	
1.2.3.3	CZTS *30*	
1.2.4	Future *39*	
1.3	Characterization of Solar Cell Materials using Electrolyte Contacts *40*	
1.3.1	Overview *40*	
1.3.2	The Semiconductor–Electrolyte Junction *41*	
1.3.3	Photovoltammetry *42*	
1.3.4	External Quantum Efficiency (EQE) Spectra *43*	
1.3.5	Electrolyte Electroreflectance/Absorbance: EER/EEA *50*	
1.4	Conclusions *54*	
	Acknowledgments *55*	
	References *55*	

Advances in Electrochemical Science and Engineering. Edited by Richard C. Alkire, Dieter M. Kolb, Jacek Lipkowski, and Philip N. Ross
© 2010 WILEY-VCH Verlag GmbH & Co. KGaA, Weinheim
ISBN: 978-3-527-32859-8

2 Tailoring of Interfaces for the Photoelectrochemical Conversion of Solar Energy 61
Hans Joachim Lewerenz

2.1 Introduction 61
2.2 Operation Principles of Photoelectrochemical Devices 62
2.2.1 Currents, Excess Carrier Profiles, and Quasi-Fermi Levels 62
2.2.1.1 Dark Current and Photocurrent 62
2.2.1.2 Excess Minority Carrier Profiles 65
2.2.1.3 Quasi-Fermi Levels 69
2.2.2 Photovoltages and Stability Criteria 71
2.2.3 Photovoltaic and Photoelectrocatalytic Mode of Operation 77
2.2.3.1 Photovoltaic Photoelectrochemical Solar Cells 77
2.2.3.2 Photoelectrocatalytic Systems 78
2.2.4 Separation of Charge Transfer and Surface Recombination Rate 81
2.3 Surface and Interface Analysis Methods 83
2.3.1 *In Situ* Methods: I. Brewster Angle Analysis 84
2.3.2 *In Situ* Methods: II. Stationary Microwave Reflectivity 87
2.3.3 X-ray Emission and (Photo)Electron Spectroscopies 90
2.3.3.1 Selected X-ray Surface/Interface Analysis Methods 90
2.3.3.2 In-System Synchrotron Radiation Photoelectron Spectroscopy 94
2.3.3.3 High-Resolution Electron Energy Loss Spectroscopy 99
2.3.4 Tapping-Mode AFM and Scanning Tunneling Spectroscopy 99
2.3.4.1 Tapping-Mode AFM 100
2.3.4.2 Scanning Tunneling Spectroscopy 101
2.4 Case Studies: Interface Conditioning 104
2.4.1 Silicon Nanotopographies 107
2.4.1.1 Nanostructures by Divalent Dissolution 107
2.4.1.2 Step Bunched Surfaces 111
2.4.1.3 Oxide-Related Nanotopographies 121
2.4.2 Indium Phosphide 130
2.4.2.1 The InP(111) A-face 131
2.4.2.2 The In-Rich InP(100) (2×4) Surface 136
2.4.3 Copper Indium Dichalcogenides 137
2.4.3.1 $CuInSe_2$ 138
2.4.3.2 $CuInS_2$ 140
2.5 Photovoltaic, Photoelectrochemical Devices 143
2.5.1 Ternary Chalcopyrites 145
2.5.2 InP Solar Cells 146
2.5.3 Nanoemitter Structures with Silicon 147
2.5.3.1 Device Development 147
2.5.3.2 Surface Chemical Analysis of the Electrodeposition Process 154
2.6 Photoelectrocatalytic Devices 162
2.6.1 Nanoemitter Structures with p-Si 162
2.6.2 Thin-Film InP Metal–Interphase–Semiconductor Structure 165
2.6.2.1 Basic Considerations 165

2.6.2.2	Device Preparation and Properties *166*	
2.7	Synopsis *170*	
2.7.1	Summary *170*	
2.7.2	Reflections on Future Development Routes *171*	
	Acknowledgments *172*	
	Appendix 2.A *172*	
	Appendix 2.B *172*	
	Appendix 2.C *173*	
	References *173*	

3 **Printable Materials and Technologies for Dye-Sensitized Photovoltaic Cells with Flexible Substrates** *183*
Tsutomu Miyasaka

3.1	Introduction: Historical Background *183*
3.2	Low-Temperature Coating of Semiconductor Films *184*
3.3	Photoelectric Performance of Plastic Dye-Sensitized Photocells *186*
3.4	Polymer-Based Counter Electrodes with Printable Materials *190*
3.5	Investigation of High-Extinction Sensitizers and Co-adsorbents *197*
3.6	Durability Development for Plastic DSSCs *208*
3.7	Fabrication of Large-Area Plastic DSSC Modules *212*
3.8	Concluding Remarks *218*
	References *218*

4 **Electrodeposited Porous ZnO Sensitized by Organic Dyes – Promising Materials for Dye-Sensitized Solar Cells with Potential Application in Large-Scale Photovoltaics** *221*
Derck Schlettwein, Tsukasa Yoshida, and Daniel Lincot

4.1	Introduction *221*
4.2	Electrodeposition – A Well-Established Technology *225*
4.3	Electrodeposition of ZnO Thin Films *226*
4.4	Sensitization of ZnO *227*
4.5	Alternative Sensitizer Molecules *228*
4.5.1	Porphyrins and Phthalocyanines as Alternative Metal Complexes *230*
4.5.1.1	Frontier Orbital Positions *231*
4.5.1.2	Photosensitization by Porphyrins and Phthalocyanines *235*
4.5.2	Purely Organic Dyes *244*
4.6	Electrodeposition of Hybrid ZnO/Organic Thin Films *244*
4.7	Porous Crystalline Networks of ZnO as Starting Material for Dye-Sensitized Solar Cells *249*
4.8	Adaptation of Electrodeposition Towards Specific Demands of Alternative Substrate Materials *252*
4.8.1	Plastic Solar Cells *252*
4.8.2	Textile-Based Solar Cells *253*
4.9	State of the Art and Outlook *256*
	References *259*

5	**Thin-Film Semiconductors Deposited in Nanometric Scales by Electrochemical and Wet Chemical Methods for Photovoltaic Solar Cell Applications** *277*
	Oumarou Savadogo
5.1	Introduction *277*
5.2	Materials and Composite Materials Fabrication *279*
5.2.1	Fundamental Considerations *279*
5.2.1.1	Chemical Bath Deposition *279*
5.2.1.2	Electrodeposition *289*
5.2.1.3	Sol–Gel Method *295*
5.2.1.4	Other Wet Methods *299*
5.2.2	Preparation of Active Materials *307*
5.2.2.1	Preparation by Chemical Deposition *307*
5.2.2.2	Preparation by Electrochemical Deposition *325*
5.2.2.3	Preparation by the Sol–Gel Method *329*
5.2.2.4	Thin Films Deposited with Heteropolycompounds *330*
5.3	Systems Development *336*
5.3.1	State-of-the-Art Thin-Film Solar Technology using Chemical, Electrochemical, and/or Sol–Gel Fabrication Methods *336*
5.3.2	Toxicity and Sustainability Issues *338*
5.4	Conclusions and Perspectives *339*
	References *340*

Index *351*

Preface

The purpose of this series is to provide high-quality advanced reviews of topics of both fundamental and practical importance for the experienced reader. This volume focuses on photovoltaic materials for energy conversion processes with emphasis on electrochemical science aspects associated with phenomena, reactions, and materials; and engineering fundamentals associated with fabrication processes and functional capabilities. The chapters of this volume, along with more than 1100 references therein, illustrate the considerable potential of electrochemistry as a tool that can be used for the preparation and characterization of materials for solar cells. Electrochemical processes may soon take a central position as an enabling technology that will have an impact on the large-scale deployment of photovoltaic devices.

Lewerenz reviews the basic principles of the operation of photoelectrochemical devices, along with the physics of the underlying phenomena that occur. Experimental methods for the investigation of fundamental phenomena are described, along with results for major classes of materials including silicon, indium phosphide, and copper indium dichalcogenides. The role of these fundamentals in the operation of photovoltaic photoelectrochemical devices is discussed with emphasis on ternary chalcopyrites, indium phosphide solar cells, and nanostructured silicon. In addition, the role of photocatalysis in tandem structures to capture the broad spectral range of photonic energy is discussed for p-Si nanoemitter configurations, as well as thin expitaxial p-InP films.

Dale and Peter review material requirements for inorganic thin-film solar cells, the development of new and sustainable materials, and prospects for fabrication technologies based on the direct synthesis of solar cells by electrochemical processing routes. The electrodeposition of cadmium telluride is reported with special emphasis on reaction mechanisms as well as the resulting structural properties that are critically important for solar cell applications. Various methods for preparing of copper–indium–gallium–selenium (CIGS) compounds are described along with experience in achieving high-efficiency performance. Extensive coverage is devoted to the preparation of copper–zinc–tin–sulfur compounds, which offer the attractive benefit of sustainability owing to the widespread occurrence of these elements. Also discussed are the advantages offered by room-temperature ionic liquids. Issues associated with the semiconductor–electrolyte junction are presented in the context of characterizing solar cell device components.

Schlettwein, Yoshida, and Lincot provide a critical review of the research literature on porous zinc oxide sensitized by organic dyes, and the potential use of these materials for the development of large-scale photovoltaic solar cells. An introduction is provided to the electrodeposition of zinc oxide thin films, their means of sensitization by a variety of organic molecules, and their preparation by one-step deposition from solutions containing water-soluble dyes to produce hybrid materials suitable for solar cell. The advantages of various substrates for large-scale cells are discussed for plastic and textile materials.

Miyasaka reviews the photoelectrochemistry of dye-sensitized mesoporous semiconductor electrodes with emphasis on application to the fabrication of thin, flexible photovoltaic cells. The discussion focuses on low-temperature preparation methods for electrode materials and the use of these methods in rapid, printable processes that enable manufacture of low-cost, lightweight integrated modules on plastic substrates. A wide variety of novel procedures are described for achieving high levels of light-harvesting performance, as well as bifacial photovoltaic capability. The complex interplay between various system components is discussed in the content of optimizing spectral sensitizers and developing printable materials for catalyzed electrochemical processes on the counterelectrode.

Savadogo presents the fundamental thermodynamic and kinetic considerations for fabricating inorganic thin-film materials with special emphasis on chemical deposition from mixed solutions, electrodeposition, and sol–gel methods. The exceptional range of materials that have been prepared by these methods is noteworthy, as is the low cost of the processing methods. Treated in detail are the syntheses of the component chemicals, as well as the mechanistic and synergistic role of the components including the important role played by additives.

The field of semiconductor electrochemistry cannot be discussed without recognizing the central contributions of Prof. H. Gerischer, who also served as co-editor of this monograph series from 1977 to 1994. Stimulated by work in the mid-1950s on the germanium–electrolyte interface by Brattain and Garrett at the Bell Telephone Laboratories, Prof. Gerischer soon contributed a series of pioneering publications that clarified many of the underlying principles on which the field rests today. These include establishing criteria for the stability of semiconductor electrodes under illumination, developing experimental techniques for study of photosensitization, elucidating the principles behind electrochemical photo and solar cells, and photoassisted oxidation of organic molecules on semiconductor particles. He continued to publish on semiconductor electrochemistry for the remainder of his life. This body of literature has, in turn, served to inspire others throughout the world to contribute to the field of semiconductor electrochemistry and engineering.

Prof. H. Gerischer, 1919–1994

Urbana, Illinois, March 2010   *Richard Alkire*

List of Contributors

Phillip Dale
Université du Luxembourg
Laboratoire Photovoltaïque
41 rue du Brill
4422 Belvaux
Luxembourg

Hans Joachim Lewerenz
Institute of Solar Fuels and Energy
Storage Materials
Division of Solar Energy
Helmholtz Center Berlin for
Materials and Energy
Lise-Meitner-Campus,
Hahn-Meitner-Platz 1
14109 Berlin
Germany

Daniel Lincot
IRDEP, Institut de Recherche et
Developpement sur l'Energie
Photovoltaique
UMR 7174, EDF-CNRS-ENSCP, 6
quai Watier
78401 Chatou Cedex
France

Tsutomu Miyasaka
Toin University of Yokohama
Graduate School of Engineering
1614 Kurogane-cho, Aoba-ku, Yokohama
Kanagawa 225-8502
Japan

Laurence Peter
University of Bath
Department of Chemistry
Bath BA2 7AY
UK

Oumarou Savadogo
École Polytechnique de Montréal
Laboratory of New Materials for
Electrochemistry and Energy
(LaNoMat)
CP 6079 Succ. Centre-ville
Montréal, Qc, H3C 3A7
Canada

Derck Schlettwein
Justus-Liebig-Universität Gießen
Institut für Angewandte Physik
35394 Gießen
Germany

Tsukasa Yoshida
Gifu University
Graduate School of Engineering
Environmental and Renewable Energy
Systems (ERES) Division
Yanagido 1-1
Gifu 501-1193
Japan

1
Applications of Electrochemistry in the Fabrication and Characterization of Thin-Film Solar Cells

Phillip Dale and Laurence Peter

1.1
Introduction

Tackling climate change has become a priority for the scientific as well as the political community. The European Union has set a target for photovoltaic (PV)-generated electricity to become competitive with conventional electricity generation by 2020–2030, and the Council of the European Union aims to reduce greenhouse gas emission by 60–80% by 2050. The UK's Stern Review [1], which deals with the economic impact of global climate change, states that "the benefits of strong, early action on climate change outweigh the costs" and goes on to identify "development of a range of low-carbon and high-efficiency technologies on an urgent timescale" as essential to any strategy that aims to address the problems of climate change. These targets have implications for the science and technology of PV. If PV is to make a major contribution to a low-carbon energy economy, issues of materials and manufacturing costs [2] as well as materials sustainability need to be considered. Current non-silicon PV technologies are based on cadmium telluride and the chalcopyrite materials $Cu(In,Ga)Se_2$ (CIGS) and $CuInS_2$. Both of these materials are unlikely to be sustainable in the long term for terawatt deployment [3] of terrestrial PV, so that a search for alternative materials has assumed considerable importance.

Figure 1.1 illustrates the nature of the problem by summarizing the natural abundance and raw elemental costs for materials that are used in thin-film solar cells. It is important to note that a logarithmic scale has been used for both the abundance and the raw materials cost. The most expensive material by far is indium, and the availability and cost of indium have become geopolitical issues in recent months since it is used in the manufacture of liquid crystal displays and touch screens.

The rarest element shown in Figure 1.1 is tellurium, and it is reasonable to suppose that this has implications for the long-term sustainability of cadmium telluride solar cell technology. Sustainability issues of this kind provide the rationale for expansion of the range materials that deserve study for PV applications. A promising new candidate for sustainable PV is Cu_2ZnSnS_4

Advances in Electrochemical Science and Engineering. Edited by Richard C. Alkire, Dieter M. Kolb, Jacek Lipkowski, and Philip N. Ross
© 2010 WILEY-VCH Verlag GmbH & Co. KGaA, Weinheim
ISBN: 978-3-527-32859-8

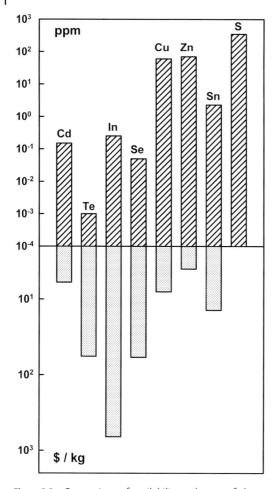

Figure 1.1 Comparison of availability and costs of elements used for the fabrication of solar cells. Note the logarithmic scales.

(CZTS), a compound with electronic properties similar to those of CIGS that contains cheap and plentiful non-toxic elements. However, remarkably little work has been carried out to characterize the potential of this material for PV applications.

Sustainability and environmental issues are also associated with the fabrication processes used in PV. The authors of this chapter believe that electrochemical methods have the potential for large-scale low-cost preparation of PV materials [4], and, in addition, electrochemical methods are powerful tools for the characterization of PV materials and device components [5]. These two topics are explored in the present chapter.

1.2
Electrochemical Routes to Thin-Film Solar Cells

1.2.1
Basic Cell Configurations

Examples of thin-film PV devices based on compound semiconductor absorber films are shown in Figure 1.2. The n-CdS|p-CdTe solar cell (Figure 1.2a) is an example of a superstrate cell. The thin CdS layer is grown on glass coated with a transparent conducting layer (e.g., fluorine-doped tin oxide, FTO). CdTe is then deposited onto the CdS layer and, following thermal treatment, the device is completed by the addition of ohmic contacts to the CdTe. CIGS solar cells are mostly made in the substrate configuration (Figure 1.2b) where the CIGS is deposited onto a metallic back contact, for example Mo-coated glass (although efficient CIGS superstrate cells have also been fabricated). The CIGS is coated with a thin CdS layer followed by a layer of intrinsic ZnO and a layer of Al-doped ZnO, which acts as a conducting transparent top contact. An Al grid is used to collect the current.

CdTe devices have achieved 16.7% efficiency in the laboratory [6], whereas CIGS cells have reached 20.0% [7]. The most efficient CIGS devices contain around 30% Ga distributed non-uniformly through the film, with higher concentrations preferred at the front and back of the layer. Kesterites are emerging as suitable In-free materials for absorber layers, but current understanding of the factors that

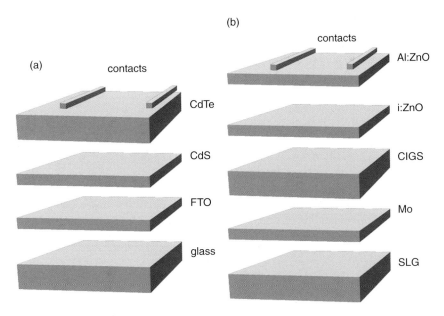

Figure 1.2 Comparison of superstrate and substrate solar cell configurations. (a) Superstrate configuration used for CdS|CdTe solar cells. (b) Substrate configuration commonly used for CIGS solar cells.

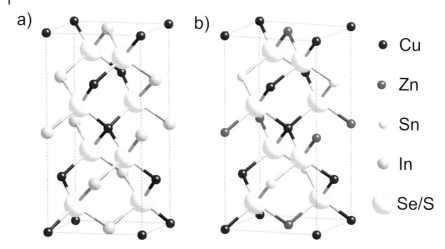

Figure 1.3 Crystal structures of (a) chalcopyrite and (b) kesterite [12].

determine the optoelectronic properties of these materials is limited. The best kesterite-based devices, like CIGS devices, have absorber layers that are Cu deficient [8, 9]. The best published efficiency of 6.7% has been reported by Katagiri et al. [10] for a device that is Zn rich and Sn deficient. Our research has shown that CZTS device performance improves if the surface of the absorber layer is etched in KCN which is known to remove CuS phases [11].

The kesterite structure Cu_2ZnSnS_4 is isoelectronic with chalcopyrite $CuInS_2$. Half of the In(III) atoms are replaced with Zn(II) atoms, and the other half are replaced with Sn(IV) atoms. The crystal structures of the chalcopyrite and kesterite are shown in Figure 1.3.

The methods available for preparation of the different layers in thin-film solar cells include physical methods such as vacuum sputtering, vapor-phase deposition, and molecular beam epitaxy as well as chemical methods such as chemical vapor-phase deposition, metal organic vapor-phase epitaxy, chemical bath deposition (CBD), and electrochemical deposition (ED). This chapter explores the potential of electrodeposition as a route to the fabrication of absorber layers such as CdTe, CIGS, and CZTS for thin-film solar cells. Electrochemistry may also be useful for the preparation of transparent layers such as ZnO; this topic has been reviewed by Pauporte and Lincot [13].

1.2.2
Material Requirements for PV Applications

In this section we discuss first the general materials requirements for an absorber layer. Specific issues for particular materials are then considered in subsequent sections. The specific optoelectronic properties of each absorber layer will not be discussed in detail here, but rather the structural, chemical, and morphological properties of the layer that are required to minimize electrical losses. For specifics, the interested reader is referred to reviews of each material: CdTe [14, 15], $Cu(In,Ga)(S,Se)_2$ [8, 15], and kesterite [16]. For the fundamentals of semiconductor physics, the reader is referred to standard textbooks, for example Sze [17].

The electron–hole pairs that are created in the absorber layer when light is absorbed are driven to the contacts by the Fermi level gradients in the device [18]. Loss of these carriers is due either to recombination or to electrical shorting. In order to minimize recombination, the absorber layer should be crystalline with as few lateral grain boundaries and chemical impurities as possible. The material should be single phase without parasitic secondary phases, which in the worst case act as recombination centers and in the best case just reduce the photogeneration volume. If secondary phases extend through the absorber layer, they may give rise to electrical shorting. Similarly, if there are pinholes between the individual grains of the film, subsequent deposition of the next layer may lead to shorting.

For use in PV devices, the absorber layer must be uniform laterally on both the micrometer and centimeter length scales with respect to its composition, and vertically on the micrometer length scale with respect to its thickness. Lateral uniformity on the centimeter scale is a prerequisite for fabrication of larger area PV devices, and the vertical thickness needs to be uniform in order to avoid stress in subsequent deposited layers and to avoid pinhole formation. A final criterion to consider is that the absorber layer must adhere well to the substrate so that it can withstand subsequent thermal treatment.

1.2.2.1 Implications of Materials Requirements for the Direct Synthesis of Absorber Layers by Electrodeposition

Compound semiconductor absorber layers for PV applications are complex multi-element materials with stringent requirements regarding structure, composition, and morphology. Meeting these requirements presents formidable challenges for any electrodeposition process. Even in those cases where the desired material can be obtained directly by electrodeposition, additional treatments are required before electrodeposited layers can be used in working PV devices. In other cases, a two-step electrodeposition and annealing process is required to obtain PV-grade material. For work prior to 1995, the reader is referred to Pandey [19].

The absorber layer in the thin-film solar cells considered here is typically a polycrystalline p-type compound semiconductor with grain sizes of the order of a micrometer. In order to electrodeposit such a thin film directly, a number of criteria have to be fulfilled. The nucleation density must not be too high so that large grains can be grown uniformly, and secondary nucleation should be avoided if possible. At the same time, overlap of the grains should not leave pinholes that lower the shunt resistance of the device. The layer should also be reasonably flat to minimize recombination losses, and its composition should be as ideal as possible, with low impurity concentrations.

The requirement for strict control of stoichiometry has three consequences. Firstly, electrons need to be readily available for the reduction of ionic precursor species. This can be a problem if the deposit is p-type. Secondly, deposited atoms must be incorporated into the correct positions in the growing semiconductor lattice. However, these may not be the lowest energy positions on the surface, and if atoms are deposited elsewhere, there may be insufficient thermal energy to move them to minimum energy positions. Thirdly, the deposition rates of the atomic constituents have to be controlled so that the correct atoms are deposited in the correct order. While it may be possible to satisfy some of the requirements regarding morphology and stoichiometry in an ED process, it is virtually impossible to satisfy all of them at the same time. Possibly the closet approach to ideal deposition has been achieved in electrochemical atomic layer epitaxy (ECALE) [20–22]. This method involves layer-by-layer growth, where a solution of single component is flowed over the substrate and one monolayer is deposited and then the solution is removed and replaced by another with the next species to be deposited. This process is repeated many times to form the semiconductor of interest. This approach works particularly well on single-crystal substrates [21], but it is not suitable for a large-area commercial fabrication process.

ED often produces compact pinhole-free compound semiconductor films with the correct global stoichiometry, but with small grain sizes of between 50 and 100 nm. Figures 1.4a and 1.4b compare an as-deposited $CuInSe_2$ precursor film with an annealed film. The crystalline quality of the films can be improved by annealing them at high temperature. The optoelectronic properties also improve, and the films have been used to fabricate devices of 6.7% efficiency [23], in spite of the fact that only small grain growth is observed in the SEM image of Figure 1.4b.

The XRD pattern for the as-deposited film depicted in Figure 1.5a shows the main characteristic reflexes of the chalcopyrite phase, but the peaks are broad, indicating poor crystallinity. The optoelectronic properties of the as-deposited films are also poor (see Section 1.3). A narrowing of the reflexes in the XRD pattern is observed after annealing the film (Figure 1.5b), confirming the improvement in crystallinity.

 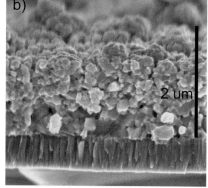

Figure 1.4 SEM images of co-deposited $CuInSe_2$: (a) precursor; (b) annealed.

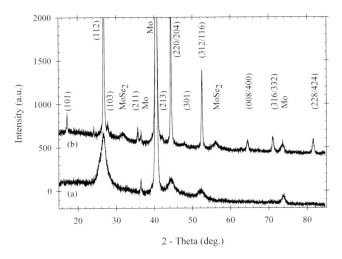

Figure 1.5 X-ray diffractogram of co-deposited CuInSe$_2$: (a) precursor; (b) annealed.

It is clear from this discussion that electrodeposition alone does not give high-quality p-type compound semiconductor layers with optoelectronic properties that are suitable for PV applications. ED therefore generally forms part of a multistage fabrication process. In the first step some or all of the elements of the semiconductor are deposited electrochemically to form a thin film, and in the subsequent steps the film is thermally annealed in an appropriate chemical environment to give it the desired properties.

1.2.2.2 Synthetic Routes Involving Deposition and Annealing (EDA)

The electrodeposition and annealing (EDA) route is based on recognition of the fact that ED alone has not so far proved suitable for preparing good-quality p-type compound semiconductors for PV. In the EDA process, the primary purpose of the ED step is to form a reactive precursor film that can be converted to a compound semiconductor with the desired optoelectronic properties by a process such as reactive annealing. The EDA approach increases the number of different synthetic routes to form the semiconductor. The main ED routes are (i) deposition of metal elements only, (ii) deposition of all constituent elements simultaneously in one layer, (iii) deposition of all constituent elements in two or more layers, and (iv) other hybrid methods including an ED stage. The main requirement for the electrodeposition step is to ensure that the lateral distribution of the elements is uniform. Whichever route is taken, the as-deposited films generally require annealing. The objective of the annealing step depends on the nature of the as-deposited film. For example, in the case where the electrodeposition step is used to form an alloy layer or a stack of metal layers, reactive annealing is necessary to introduce the chalcogen so as to form the desired semiconductor compound. At the other extreme, in the case where a stoichiometric semiconductor film such as CdTe has been formed by electrodeposition, annealing may be used to control the conductivity type (air anneal) or to improve crystallinity (CdCl$_2$ anneal).

1.2.2.2.1 Electrodeposition

The aim of the ED step is to provide a compact, adherent, laterally uniform precursor film with the desired stoichiometry: for commercial applications, lateral compositional uniformity needs to be guaranteed over large areas. The desired properties can only be obtained if the ED process is properly controlled. The nature of deposited layers depends on electrode material, precursor species in solution, applied potential program, hydrodynamic conditions, and temperature. For single metal systems, these factors are well understood. For deposition of multiple elements including a chalcogen, the ED process becomes substantially more complex.

The key challenges facing attempts to electrodeposit chalcogen-containing compounds are summarized below.

(i) **Substrate electrode.** The substrate electrode must be clean, preferably free of passivating oxides, and able to sustain the annealing conditions. The substrate also needs to be sufficiently conducting to minimize the potential drop so that the thickness and composition of the electrodeposit do not vary from one area to another due to the potential difference between the two areas. This is a particular problem with transparent conducting oxide (TCO) layers.

(ii) **Precursor species in solution.** The chalcogenide compounds used for absorber layers can contain from two to five elements. Metal precursor ions may be available in one or more oxidation states, each of which has different solubility and reduction potentials. The metal ion can be introduced as a salt or supplied *in situ* by anodic dissolution of a sacrificial anode. The metal ions can also be complexed with ligands to increase their solubility and prevent precipitation, particularly in alkaline solutions. Soluble chalcogen sources are more limited. H_2SeO_3 and $HTeO_2^+$ [24] are commonly used for selenium and tellurium and $S_2O_3^{2-}$ [25] for sulfur. However $S_2O_3^{2-}$ solutions tend to be rather unstable, slowly precipitating colloidal sulfur. In this case, an alternative strategy is to use organic solvents such as polyethylene glycol [26] or room temperature ionic liquids [27], which can dissolve elemental sulfur. Other components of the electroplating bath include inert supporting electrolytes, acids, bases, buffers, and organic additives. Supporting electrolytes are used to increase the conductivity of the solution to give a more uniform current distribution. An acid, base, or pH buffer is frequently added to the plating solution to control the solubility of the ionic species. Finally organic additives are used either to control reaction rates or to improve the morphology of the deposits.

(iii) **Applied potential.** Each species of ion in solution has a different reduction potential. At first sight, the simplest way to incorporate all desired species into the deposit would be to apply a sufficiently negative potential to reduce all the species. The deposits may contain the species in elemental form, but more likely in the form of binary or ternary compounds depending on the number of species in solution. The applied potential does not need to be constant: it has been shown that use of a voltage ramp can assist the incorporation of electronegative elements and reduce dendritic growth [28]. Whatever the potential program used, the question arises—which compounds will be formed? This question is related to the mechanism of deposition, which in turn depends on the applied potential and the availability of ionic species for reduction. For the CdTe system, it is possible to electrodeposit the thermodynamically favored compound uniformly with the correct stoichiometry, but for the $CuInSe_2$ system, it is common to electrodeposit several different phases simultaneously, rather than the thermodynamically favored semiconductor, from a bath containing copper, indium, and selenium ions.

One approach to the growth of the thermodynamically favored semiconductor is based on the work of Kroger [29], who showed for chalcogenides that the reduction potential of the more electronegative element in the compound could be moved positive by an amount equivalent to the free energy of formation of the compound. This mechanism—which is equivalent to underpotential deposition—is self-limiting because the more electronegative element cannot be reduced without the presence of the chalcogen ion or chalcogen-containing compound at the electrode surface.

A common problem encountered during electrodeposition of ternary chalcogen compounds is the excess of the most noble metal (generally copper) in the deposit. To solve this problem, the concentration of the metallic precursor is either lowered in order to reduce its flux to the surface, or an organic ligand is added to complex the metal ion in order to shift its reduction potential to more negative values. Figure 1.6 illustrates this for Cu(I) ions depositing on a carbon electrode with the reduction potential shifted more negative with increasing concentration of added SCN^- ligand.

(iv) **Mass transport conditions.** The rate of electrodeposition of ions from solution is limited either by kinetics or mass transport. Kinetically limited deposition occurs when the electron transfer process from the electrode to the adsorbed ion is the rate-determining step, and mass transport-limited deposition occurs when the flux of ions to the surface is the rate-determining step. In order to obtain a uniform deposit under mass transport-limited conditions, the flux of ions to the surface must also be uniform over all parts of the electrode surface. This can be achieved in the laboratory for small areas using a rotating disk electrode (RDE), but larger areas require engineering solutions to the problem. To illustrate the problem of non-uniformity we carried out a simple copper deposition onto two molybdenum electrodes: one a stationary vertically held electrode, and the other an electrode rotated at 100 rpm. Energy dispersive x-ray (EDX) analysis at 20 keV was used to measure an approximate composition of the deposit. If the deposit has a constant thickness then the EDX measurement should return a constant compositional value for copper. Figure 1.7 illustrates the locations at which the copper composition was measured, with points 0 to 12 going horizontally and points 13 to 25 going vertically. The stationary vertical electrode has thicker layers of copper at the edges and at the bottom, consistent with an uncontrolled convection at the surface [19]. By contrast, the copper film formed on the rotated electrode shows little variation in thickness.

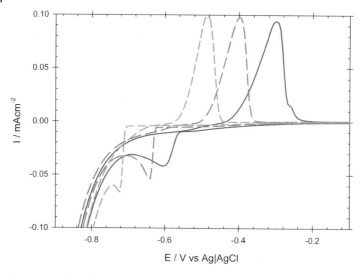

Figure 1.6 Effect of SCN⁻ ligand on the Cu deposition potential on carbon surface at pH 2. Solid line, 1 M KSCN; dashed line, 2 M KSCN; short-dashed line, 4 M KSCN.

(v) **Temperature.** Changes in temperature influence all of the kinetic and thermodynamic factors involved in electrodeposition. The deposition of CdTe, for example, is carried out at 85–90 °C. The higher temperature is necessary to obtain stoichiometric CdTe films; deposition at room temperature generally gives films that are poorly crystallized and contain excess Te. It seems probable that elevated temperatures enhance surface diffusion of adatoms so that the crystalline phase can grow properly. Metal plating is often carried out at elevated temperatures too. Generally, the optimum temperature is found empirically since the plating systems are complex with many temperature-dependent variables.

1.2.2.2.2 Annealing

The annealing step in the EDA process can take different forms with different purposes: (i) increasing crystallinity of the as-deposited film by allowing atoms to relax to their equilibrium positions, (ii) increasing grain size, (iii) driving solid-state reactions between two or more phases, (iv) adding or exchanging the chalcogen component, and (v) type conversion. A number of technical and safety issues have to be considered for annealing. In particular, annealing should be carried out in a closed environment to prevent release of toxic materials (Cd, Te, Se) to the environment. The heating rate, maximum temperature, and cooling rate all need to be controlled and optimized. A fast heating rate may be desirable in order to prevent formation of unwanted intermediate phases. The temperature must be sufficiently high to achieve its purpose, but it may be limited by (i) degradation of the underlying substrate and (ii) loss of volatile elements or compounds from the semiconductor film. An example of the second of these limitations is selenium loss during thermal annealing of CIGS precursor films [30]. To prevent this,

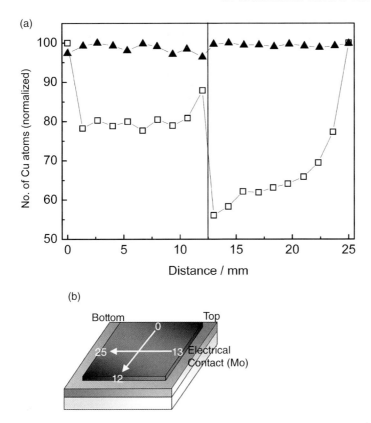

Figure 1.7 (a) Comparison of normalized copper EDX signal for electrodes with deposited copper layer. Electrodes were stationary vertical (squares) or rotated (triangles). (b) Sketch of mapped points on electrode surface for EDX measurements.

annealing is carried out in the presence of excess selenium. More generally for chalcopyrite and kesterite absorber layers, a source of chalcogen is present during the annealing process. The chalcogen can be added as an elemental solid (S, Se) which sublimes at elevated temperatures or added as the hydrogen chalcogenide (H_2S, H_2Se) in the gas phase. The atmosphere in the annealing chamber may also contain an inert gas such as argon or nitrogen, or a reducing atmosphere such as forming gas. The hydrogen in the forming gas has two roles: firstly to remove oxides, and secondly it can form the hydrogen chalcogenides from the elements *in situ*. For example Verma *et al.* showed that the formation reaction of hydrogen and selenium must be considered at 450 °C to explain the concentrations of H_2Se observed during experiments to observe its decomposition [31]. For CdTe films, a $CdCl_2$ flux can be used during annealing to induce recrystallization.

1.2.2.3 Summary of EDA Routes

Four distinct types of precursors that can be used for the EDA process are illustrated in Figure 1.8: (i) a stack of metal layers, (ii) an alloy of the metals, (iii) a stack of binary chalcogenides, and (iv) a layer containing all of the constituent elements

Figure 1.8 Common electrodeposited precursor types for annealing. Mn = metal and X = chalcogen.

including the chalcogen. Approaches (i) and (ii) have been used for fabricating CIGS [32–34] and CZTS [11, 35–37, 39]. Both approaches allow for high deposition rates; for example Voss et al. [34] electroplated a Cu–In precursor in 3.5 minutes compared to a Cu–In–Se precursor that takes 60 minutes [23]. The stacked metal layer approach is the easiest in terms of plating bath chemistry but adds multiple steps and requires that each layer deposits satisfactorily on the preceding one. Some of the problems encountered in fabrication of multilayer metal stacks for kesterite synthesis are discussed in Section 1.2.3.3. The alloy approach (ii) is generally difficult since the composition is determined not only by electron transfer kinetics and mass transport but also the nature of the stable alloy phases.

The chalcogen reacts with the metal precursors during the annealing step, and the resulting semiconductors usually have large grains [34]. The disadvantages of this method are that adhesion of the semiconductor layer may be poor and that the film may contain cracks, blisters, and pinholes resulting from the substantial volume increase during the solid-state reaction.

Type (iii) precursors consist of two or more layered stacks containing binary chalcogenides. As far as we are aware, this approach has only been used for CuInSe$_2$ [28, 41, 42], although there is no reason to suppose that it would not work for other systems. The deposition of both Cu–Se and In–Se compounds is mass transport limited, and the process is complicated by the fact that several phases with different stoichiometries can be formed within a narrow range of experimental conditions (see Section 1.2.3.2.2). The binary stacks react together during annealing, inducing a phase transformation that gives rise to large grains of the ternary semiconductor compound. One advantage of this method is that there are no large changes in volume during the formation of the ternary semiconductor, but the chalcogenide deposition steps generally take longer, and the thermodynamic driving force for the reaction is not so high as in cases (i) and (ii).

Type (iv) precursors consist of a single layer containing all the necessary metals and the chalcogen(s) electrodeposited simultaneously from a single bath. CdTe [24, 43], CIS [23, 30], and CIGS [44] have all been deposited by this method. Typically the deposition of the layer is much slower than for the metal depositions due to the lower concentrations in solution. An extreme case is the deposition of CdTe, where the concentration of $HTeO_2^+$ is sub-millimolar. The deposition process is difficult to control in the case of ternary or quaternary chalcogenides, but the subsequent annealing step in the presence of excess chalcogen is used to react secondary phases and to create larger crystals of the desired absorber material.

1.2.3
EDA route to p-Type Semiconductors for Thin-Film Photovoltaics

1.2.3.1 Electrodeposition of CdTe for CdS|CdTe Solar Cells

1.2.3.1.1 Overview

II–VI (or more correctly 12,16) compound semiconductors can be synthesized electrochemically in a number of ways. In the case of sulfides, one of the simplest methods is to anodize the parent metal in an alkaline sulfide solution. This method has been used to grow films of CdS on cadmium as well as Bi_2S_3 on bismuth [45–47] for application in photoelectrochemical solar cells [46, 48–50]. Generally, however, this method is not useful for solid-state PV devices, where CdS layers are commonly deposited from a chemical bath. By contrast, cathodic synthesis of II–VI semiconductors from aqueous solutions has been developed extensively. The metal precursor is generally the M^{2+} ion, whereas the chalcogenide precursor can be thiosulfate ($S_2O_3^{2-}$) [25], selenosulfate ($SSeO_3^{2-}$) [51], or $HTeO_2^+$ [24] for sulfides, selenides, and tellurides, respectively. In the case of oxides such as ZnO, the chalcogen precursor can be molecular oxygen or hydrogen peroxide [13]. II–VI compounds have also been synthesized electrochemically from non-aqueous organic solvents [52, 53] and from ionic liquids [54]. II–VI compounds have also been synthesized by ECALE [22, 55], but this process is not suited for large-scale fabrication.

1.2.3.1.2 Historical Development

From the point of view of II–VI PV devices, by far the most important electrochemical process is the electrodeposition of cadmium telluride for CdS|CdTe heterojunction solar cells. This was the first electrodeposition process to be used in large-scale manufacture of PV devices. BP Solar's Apollo® modules, which were fabricated via an electrodeposition route, reached an advanced stage of manufacture, with a module plant coming on line in 1998. However, manufacture of Apollo modules was discontinued later, possibly as a consequence of environmental concerns. Since then, First Solar's cadmium telluride modules, which are fabricated by close space sublimation rather than electrodeposition, have achieved efficiencies of over 12%, with costs poised to fall below $1 per watt peak.

Interest in electrodeposition of CdTe goes back 30 years to the work of Panicker *et al.* [24] who established the feasibility of a process based on using an aqueous solution containing large excess of Cd^{2+} ions and Te in the form of TeO_2 dissolved in acidic solution as $HTeO_2^+$. These authors deposited CdTe on nickel and on glass coated with antimony-doped tin oxide and found that the material was n-type if pure solutions were used, whereas the presence of low-level copper impurities gave p-type layers. By 1984, Basol *et al.* [56] had reported a 9.35% efficient CdS|CdTe heterojunction cell fabricated by electrodeposition of a 60 nm CdS film onto indium tin oxide-coated glass followed by electrodeposition of the CdTe layer (1.2–1.5 μm thick). The structure was then annealed at 400 °C to convert the as-deposited n-CdTe to p-CdTe, forming the required heterojunction.

The experimental conditions used by Basol *et al.* for CdTe deposition have been followed closely in subsequent studies. The cadmium precursor consisted of 0.5 M $CdSO_4$ and the Te precursor was TeO_2 dissolved in a pH 2 electrolyte as $HTeO_2^+$ at a concentration in the range 24–40 ppm. CdTe deposition on conducting glass substrates was also studied at around the same time by Rajeshwar and co-workers [57–60]. The subsequent large-scale implementation of the electrodeposition route to CdS|CdTe solar cells was carried out by researchers at BP Solar [61, 62], culminating in the production of large-area monolithic Apollo modules with efficiencies over 10%. At this point, the development was halted, and, as a consequence, interest in electrochemical routes to CdTe solar cells has declined sharply in spite of the advantages of electrodeposition as a low-cost, low-temperature processing route. Since the electrodeposition of CdTe serves as a well-documented illustration of transferring an electrochemical process from the laboratory to a large-scale manufacturing operation, it will be examined in the following sections.

1.2.3.1.3 Mechanism of CdTe Electrodeposition

Thermodynamic aspects of cathodic electrodeposition have been discussed by Kroger [29]. The potentials for the deposition of elemental Te and Cd can be obtained from the equilibrium potentials

$$HTeO_2^+ + 3H^+ + 4e^- = Te + H_2O \quad E = 0.559 + 0.0148\log[HTeO_2^+] - 0.0443\,pH \tag{1.1}$$

$$Cd^{2+} + 2e^- = Cd \quad E = -0.4025 + 0.0295\log[Cd^{2+}] \tag{1.2}$$

Te can be reduced at more cathodic potentials to form H_2Te, and the corresponding equilibrium potential for this process is

$$Te + 2H^+ + 2e^- = H_2Te \quad E = -0.740 - 0.0295\log[H_2Te] - 0.059\,pH \tag{1.3}$$

The molar Gibbs energy of formation of CdTe is $-92\,kJ\,mol^{-1}$, so that CdTe can be deposited at less cathodic potentials than Cd. The relevant equilibrium potential is

$$Te + Cd^{2+} + 2e^- = CdTe \quad E = 0.074 + 0.0295\log[Cd^{2+}] \tag{1.4}$$

At more cathodic potentials, CdTe can be reduced to Cd and H_2Te. The equilibrium potential for this process is

$$CdTe + 2H^+ + 2e^- = Cd + H_2Te \quad E = -1.217 - 0.0295\log[H_2Te] - 0.059\,pH \tag{1.5}$$

These potentials effectively define the regions in which CdTe is stable to both anodic and cathodic decomposition. It can be seen that the CdTe deposition window is limited by the anodic limit of CdTe stability and by the cathodic deposition potential of bulk Cd.

The regions of stability derived from the thermodynamic analysis are shown in Figure 1.9 below a cyclic voltammogram obtained using a vitreous carbon

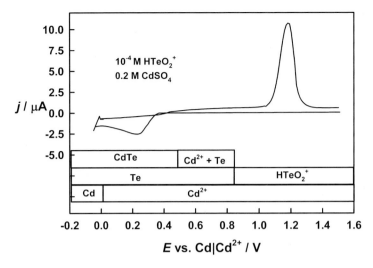

Figure 1.9 Cyclic voltammogram for deposition and stripping of CdTe at a vitreous carbon electrode in unstirred solution (pH 2, 0.2 M CdSO$_4$, 10^{-4} M HTeO$_2^+$, sweep rate 10 mV s^{-1}). The figure also shows the regions of stability of various phases. Data taken from reference [63].

electrode in an unstirred solution containing 0.5 M CdSO$_4$ and 10^{-4} M HTeO$_2^+$ at pH 2 [63]. In this case, the Cd^{2+}/Cd equilibrium potential has been chosen as a convenient zero point on the potential scale.

It can be seen from the forward sweep of the voltammogram that deposition begins around 400 mV positive of the reversible Cd^{2+}/Cd potential. This corresponds to a region where CdTe is expected to be the stable product. The beginning of Cd deposition is just evident at the cathodic limit of the voltammogram before the sweep is reversed and a well-defined anodic peak due to oxidation of the CdTe deposit is seen.

Several different mechanisms have been proposed for the electrodeposition of CdTe. Many authors have assumed that the first step is the four-electron reduction of HTeO$_2^+$ to Te followed by the reaction [64]

$$Te + Cd^{2+} + 2e^- = CdTe \qquad (1.6)$$

Danaher and Lyons [65], on the other hand, assumed that the first step in CdTe deposition involves the six-electron reduction of HTeO$_2^+$ to H$_2$Te, which then reacts with Cd^{2+} to form CdTe, whereas Engelken and Vandoren [66] proposed that CdTe is formed as a consequence of a solid-state reaction between Cd and Te. A more detailed model involving free and occupied adsorption sites on the growing CdTe deposit was developed by Sella *et al.* [67]. The model is based on the competition for active sites between HTeO$_2^+$, Cd^{2+}, Cd, and Te. The importance of substrate–adsorbate interactions was also recognized by Saraby-Reintjes *et al.* [63], who identified slow relaxation processes during deposition of CdTe on a RDE that reflect the dynamics of adsorption processes. These relaxation processes give rise

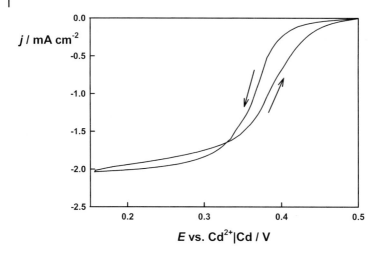

Figure 1.10 Cyclic voltammogram for deposition of CdTe at a stainless steel rotating disc electrode. Rotation rate 16 Hz, sweep rate 1 mV s^{-1}. Solution 10^{-4} M HTeO$_2^+$, 0.1 M CdSO$_4$ and 0.1 M K$_2$SO$_4$. Note the crossover between the forward and reverse sweeps that is attributed to relaxation of the surface coverages with Cd and Te species. Data taken from reference [63].

to transient behavior when the system is perturbed by applying small potential steps or steps in the rotation rate. The relaxation effects are also evident in the cyclic voltammogram for CdTe deposition at a RDE shown in Figure 1.10. The most striking feature of the voltammogram is the crossover of the cathodic and anodic sweeps which clearly has nothing to do with nucleation phenomena. Saraby-Reintjes *et al.* were able to show that this crossover is characteristic of a system in which deposition is controlled by the surface coverage of active sites. The model developed by these authors is discussed in some detail below.

A good starting point for the discussion of the mechanisms giving rise to the crossover in the voltammogram is the work of Stickney *et al.*, who developed the method known as ECALE [20–22, 68, 69]. The ECALE method relies on the fact that underpotential deposition of Cd occurs at Te surface sites, whereas Te is deposited preferentially on Cd sites. Saraby-Reintjes *et al.* assumed that stoichiometric CdTe is deposited so that, under steady-state conditions, the rate of two-electron reduction of Cd^{2+} ions at Te sites is equal to the rate of four-electron reduction of HTeO$_2^+$ at Cd sites. According to this mechanism, the Cd and Te surface coverage values self-adjust so as to maintain the 1 : 1 stoichiometric balance. The detailed model developed by Saraby-Reintjes *et al.* on the basis of these concepts predicts not only the steady-state behavior of the system but also the response of the system to perturbations in potential or mass transport (i.e., rotation rate of the RDE). The interested reader is referred to the original paper for full details. The key expression that predicts the stationary and dynamic current response relates the time dependence of the Cd coverage (θ_{Cd}) to the rates of deposition and removal ($\vec{v}_{Cd}$ and $\bar{v}_{Cd}$, respectively) of Cd at Te sites and the rate of deposition of Te (v_{Te}) on Cd sites (it is assumed that $\theta_{Cd} + \theta_{Te} = 1$):

$$M\frac{d\theta_{Cd}}{dt} = \vec{v}_{Cd}(1-\theta_{Cd}) - \bar{v}_{Cd}\theta_{Cd} - v_{Te}\theta_{Cd} \tag{1.7}$$

where M is a proportionality factor (mol cm^{-2}) corresponding to monolayer coverage. Solution of this differential equation shows that any step perturbation of the system will give rise to transient currents associated with readjustment of the surface coverages to their new steady-state values. The corresponding current transient takes the form

$$\Delta j = 2F[2v_{Te} - (\vec{v}_{Cd} - \bar{v}_{Cd})](\theta_{Cd}^0 - \theta_{Cd}^\infty)e^{t/\tau} \tag{1.8}$$

where the time constant τ that characterizes the relaxation process is given by

$$\tau = \frac{M}{\vec{v}_{Cd} + \bar{v}_{Cd} + v_{Te}} \tag{1.9}$$

Although this model is clearly simplified, it correctly predicts the signs of the current transients observed when the potential is stepped towards or away from the reversible Cd potential, mainly changing v_{Cd}. The model also correctly predicts the sign of the transient current response to changes in the rotation rate of the RDE, which mainly perturbs v_{Te}. Most convincingly of all, it predicts that there will be a crossover in the cyclic voltammogram when the term in square brackets is equal to zero. The main limitation of the model is that it does not take into account changes in surface area of the CdTe film during deposition.

1.2.3.1.4 Structural Properties of Electrodeposited CdTe Layers

A number of factors control the structural properties of electrodeposited CdTe. Good-quality near-stoichiometric crystalline material is obtained by deposition at 85 °C onto CBD CdS films on FTO-coated glass (CdTe films can also be grown on FTO-coated glass that has been "sensitized" by previous cathodic reduction [70], but generally they are of lower quality). The deposits exhibit preferential orientation along the <111> axis, with the strongest texturing observed for films deposited close to the reversible Cd^{2+}|Cd potential [70]. The strong preferential orientation and the quality of the deposit can be seen in Figure 1.11 which shows a high-resolution transmission micrograph of a CdTe deposit on a CdS substrate. The stacking faults that are evident as dark/light bands are also seen in high-quality epitaxial layers of CdTe grown from the vapor phase.

The very high quality that can be achieved for the cathodic deposition of CdTe is illustrated by the work of Lincot *et al.* [71], who showed that epitaxial layers of CdTe can be grown electrochemically on the ($\bar{1}\bar{1}\bar{1}$) surface of single-crystal InP. Perfectly oriented films could be produced by pre-depositing a thin (20–30 nm) buffer layer of epitaxial CdS on the InP. Figure 1.12 illustrates the RHEED pattern observed for a 0.12 μm CdTe film on InP, demonstrating a high degree of epitaxy, with the CdTe growing with no rotation of the crystallographic directions in the basal (111) plane.

Figure 1.11 High-resolution TEM image of electrodeposited CdTe film showing columnar structure and stacking faults. Daniel Lincot.

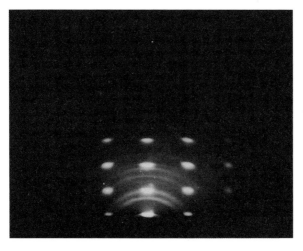

Figure 1.12 RHEED of electrodeposited CdTe on InP/CdS. The substrate is oriented in the [112] azimuth. Daniel Lincot.

1.2.3.2 Electrodeposition of CIGS for CIGS|CdS|ZnO Solar Cells

1.2.3.2.1 Historical Perspective

The first report of a $CuInSe_2$/CdS heterojunction solar cell – based on single-crystal $CuInSe_2$ – appeared in 1974 [72]. The cell achieved an efficiency of 5%. In 1983 the first electrochemical approach to produce a polycrystalline thin film of $CuInSe_2$ was reported by Bhattacharya [73] in which Cu, In, and Se were deposited simultaneously from an acidic solution. Other approaches to the synthesis of polycrystalline $CuIn(S,Se)_2$ thin films appeared quickly after the first report: for example, Hodes *et al.* annealed an electrodeposited Cu–In alloy [74], Kapur *et al.* annealed stacked Cu–In layers [75], and Bhattacharya *et al.* annealed In–Se Cu–Se stacks [28]. The EDA routes which currently produce the most efficient $CuIn(S,Se)_2$ solar cells involve co-deposition of all of the constituent elements [76] followed by annealing in a sulfur atmosphere, and the electrodeposition of a Cu–In alloy with a small amount of Se, again followed by annealing in sulfur [34]. Both routes produce devices with around 11% power conversion efficiency.

Over a series of many articles stretching from 1994 (Guillemoles, Lincot and Vedel [77]) through to the present day (Chassaing *et al.* [78]), the ENSCP group in Paris have studied the mechanisms for the simultaneous three-element electrodeposition of $CuInSe_2$. These detailed investigations have shown that the deposition mechanism is complicated and time dependent, but current understanding is sufficiently good to be able to control the process so as to yield solar cells with 11.5% efficiency via the EDA route [76].

In order to compete with the high efficiencies of cells fabricated using physical vapor deposition [7], it is necessary to optimize the bandgap of the materials by incorporating Ga with a controlled concentration profile. Ga incorporation in ED films was first achieved by Calixto *et al.* in a co-deposition of all elements [79]. The best electrochemically incorporated Ga-containing devices have been reported by Kampmann *et al.* [33] who achieved a device efficiency of over 10% by using a stacked layered approach.

1.2.3.2.2 Cu(In,Ga)Se$_2$ via Stacked Metal Layers

As described in Section 1.2.3, CIGS can be formed by annealing thin films of the metal elements in a chalcogen atmosphere: the Showa–Shell–Siemens industrial process is based on this route [80, 81]. Whether a stack of metals or an alloy layer is used depends in part on the annealing process. Single-element depositions are easier to control, but an alloy deposition reduces the number of deposition steps.

Copper salts are available in the +1 or +2 oxidation state, while the stable In and Ga oxidation states are +3. The relevant standard reduction potentials are given below.

$$Cu^{2+} + 2e^- = Cu \quad E = 0.34 - RT/2F \ln[aCu^{2+}] \tag{1.10}$$

$$In^{3+} + 3e^- = In \quad E = -0.34 - RT/3F \ln[aIn^{3+}] \tag{1.11}$$

$$Ga^{3+} + 3e^- = Ga \quad E = -0.53 - RT/3F \ln[aGa^{3+}] \tag{1.12}$$

Copper salts are reasonably soluble over a wide pH range, but In and Ga salts only exhibit solubilities above 1 mM at pH < 3 for In^{3+} and at pH < 1.5 for Ga^{3+} [82]. Rotating disk electrode experiments show that Cu deposits under mass transfer-controlled conditions on Mo, whereas In deposits under kinetic control at room temperature. There are a few reports on electrodeposition of Cu/In or In/Cu stacks for the realization of CIS semiconductor compounds [83, 84]. One technologically interesting report from Penndorf et al. [83] describes the use of copper tape as both the substrate and source of Cu for the formation of $CuInS_2$. Indium was electrodeposited on the copper tape in a roll-to-roll process with remarkably high current densities of 150–200 mA cm^{-2} with the help of thiourea. Cell efficiencies of up to 6% were reported with this approach.

One recent route taken in our laboratories is the electrodeposition of stacked Cu/In/Ga layers on Mo followed by annealing in a Se/forming gas atmosphere. Each layer is deposited from a separate plating bath: details of the baths and deposition conditions are given in Table 1.1. The basic Cu plating bath was adapted from Barbosa et al. [84] with the addition of a zwitterionic surfactant that reduces grain size and surface roughness [11]. A basic Cu bath is used because the Mo substrate is passivated by a surface oxide in acidic conditions, leading to poorly adherent films. The indium plating bath is adapted from the review paper of Walsh and Gabe [85]. Different Ga plating baths have been tried with different organic ligands to enhance the solubility of Ga(III) at reasonable pH values. The ligands glycine, KSCN, and ethylenediaminetetraacetic acid have been tested in both basic and acidic media. The Ga plating bath with glycine at pH 13 is found to give reasonable deposits of Ga on Cu surfaces with 57% plating efficiency (measured by integrating the charge of a forward and reverse cyclic voltammogram), but on an In surface the plating efficiency is found to be less than 10%, as measured by inductive coupled plasma mass spectrometry.

Table 1.1 Typical plating recipes for Cu/In/Ga stacked layer deposition.

Element	Solution	E_{dep} vs. Ag|AgCl (V)	Rotation speed (rpm)
Cu	0.1 M $CuCl_2$ 3 M NaOH 0.2 M sorbitol 0.25 µg l^{-1} Empigen BB	−1.20	150
In	0.261 M $InCl_3$ 2.3 M H_3NO_3S 2.0 M NaOH 8.0 g l^{-1} dextrose, 2.29 g l^{-1} triethanolamine	−0.95	0
Ga	0.1 M $Ga_2(SO_4)_3$ 3 M NaOH 0.5 M glycine	−1.70	100

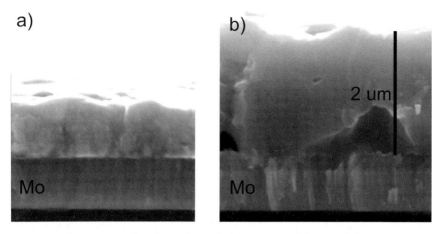

Figure 1.13 SEM images of (a) Cu/In/Ga stacked precursor and (b) annealed precursor.

The substrates used for the deposition were 1 inch square Mo-coated glass pieces held in a RDE with contacts at the corners. Rotation was used to ensure lateral thickness uniformity across the entire sample. The separate layers of a deposited Cu/In/Ga stack are not visible in the SEM image of Figure 1.13a, possibly due to alloying. The stack was annealed in a mixture of forming gas and selenium vapor, and the resulting crystalline semiconductor layer is shown in Figure 1.13b. The annealing process has formed large crystals of CIGSe, and some voids can be observed at the interface.

The annealed $CuInSe_2$ films are p-type and give reasonable photocurrent responses when tested in an electrolyte (see Section 1.3). However, to date device performance has been poor, about 4%, which we attribute to large pinholes in the films, and some blistering. Investigations are under way to determine if this is a function of the annealing or of the electrodeposition.

Kampmann *et al.* [86] have successfully deposited layers in this manner, albeit with a thermally evaporated Se capping layer on the Cu/In/Ga stack. In their case no details are given due to the industrial nature of the work, but device efficiencies of 10.4% are reported [86].

To successfully deposit elemental Ga with high rates the hydrogen evolution reaction must be suppressed by using organic additives. Some suitable additives and conditions for suppressing hydrogen evolution are discussed in the following sections.

1.2.3.2.3 Cu(In,Ga)Se$_2$ via Metal Alloys

Hodes *et al.* [74] plated Cu–In alloys on titanium substrates at pH 2 using aqueous $CuSO_4$, $In_2(SO_4)_3$ (or $InCl_3$), and H_2SO_4 at a potential of -1.0 V vs. standard calomel electrode (SCE). These alloy films were annealed in a 1:1 H_2S:argon mix for 30 minutes. Annealing at 400 °C produced only sphalerite (cubic) phases while at 500–550 °C the chalcopyrite (tetragonal) phase dominated and grain size increased.

Hodes *et al.* suggested that this was due to quasi-rheotaxy type of crystalline growth. The films annealed at the higher temperatures also showed higher photoactivity.

The most significant work on the alloy route has been published by the Attotech–Erlangen collaboration in a series of papers [34, 87–90]. Attotech – a company specializing in electroplating – developed a Cu–In alloy plating bath that allowed deposition of precursor layers in 3.5 minutes [34]. Unfortunately no details of the chemical composition of the plating bath are available, but it evidently contains complexants and brightening and wetting agents. Additionally the plating bath contained a very small amount of Se, equivalent to less than 5% of the necessary chalcogen – presumably to promote adhesion to the Mo substrate. Voss *et al.* [34] demonstrated that the Cu/In ratio was constant when the hydrodynamic conditions were varied from 3 to 10 rad s^{-1} and the applied potential was varied by 300 mV. This relative insensitivity to experimental conditions is important, since there can be a substantial lateral variation of potential across large-area substrates due to the iR drop. Furthermore Voss *et al.* found that the Cu/In ratio could be varied from 0.8 to 1.6 by adjusting the pH from 2.8 to 3.5. Uniform electrodeposits on areas of 30 × 30 cm were achieved using a spray nozzle array to control the hydrodynamic conditions. XRD analysis of the deposited precursors showed the expected $Cu_{11}In_9$ and $CuIn_2$ phases, but no selenide phases were observable due to the very small amount of Se in the precursor. When annealed in a mixed S and Se atmosphere, the precursor converted into the semiconductor $CuIn(S,Se)_2$; no additional phases were observed by XRD. The best final cells fabricated by Voss *et al.* had a power conversion efficiency of 11.4% equaling the efficiencies achieved by vacuum deposition techniques for $CuIn(S,Se)_2$ [91].

Friedfeld *et al.* deposited a $CuGa_2$ alloy phase consisting of 0.5 μm isolated grains from a 5 M NaOH metal sulfate bath using a RDE [92]. A layer of $CuInSe_2$ was then added in a one-step electrodeposition, and the combined layer was annealed to form CIGS. The In/Ga ratio could be varied by varying the relative thicknesses of the two layers, as evidenced by the shifting cell lattice parameters obtained by XRD.

Zank *et al.* co-deposited In and Ga potentiostatically onto 10 cm^2 Cu-coated Mo using a 2.1 M KCN bath in 0.5 M KOH with 62 mM dextrose [32]. Cyanide is a very strong complexing agent for cations in solution, but also appears to bind to the electrode surface to suppress the hydrogen reaction and reduce the electron transfer rate. As a consequence, the deposition is kinetically limited, which eliminates the need to control the mass transport. Films containing approximately 60% Ga could be deposited at very negative potentials, and the Ga content increased with increasing negative potential up to the limit of solvent breakdown. Interestingly the morphology of the deposit was found to depend on the Ga content. When a Ga content of 35.7% was deposited, the layer was in a liquid phase, as the alloy is a liquid below at the deposition temperature. The Ga reacts with the Cu phase underneath and forms the $CuGa_2$ phase.

Kois *et al.* took this approach one step further and co-deposited Cu–In–Ga simultaneously from a 0.4 M acetate bath (pH 5) containing 50 mM metal ions and

2 M KSCN [93]. With these conditions, the bath was stable and the reduction of the SCN ligand was not observed. Cu and In deposition were found to be hydrodynamically controlled, while Ga deposition was found to be unaffected by stirring. Metal precursors containing variable amounts (from 2 to 30 at%) of Ga could be made by altering the stirring in the bath. Precursors annealed in Se vapor exhibited a broadened (112) peak demonstrating mixed CuInSe$_2$ and CuGaSe$_2$ phases rather than just one phase.

1.2.3.2.4 Cu(In,Ga)Se$_2$ via Stacked Binary Layers

The deposition of the binary selenide phases is discussed in this section in the order Cu–Se, In–Se, and Ga–Se followed by a summary of some recent results from our laboratory on the stacking and annealing of Cu–Se/In–Se stacks.

Massaccesi et al. [94] studied in detail the room temperature deposition of copper selenide films onto rotating disk tin oxide electrodes from deoxygenated solutions at pH 2.45 containing 1 mM CuSO$_4$, 0.1 M K$_2$SO$_4$, and between 0 and 10 mM H$_2$SeO$_3$. Two Se(IV) species are present at this pH, and their reduction potentials are given below.

$$H_2SeO_3 + 4H^+ + 4e^- = Se + 3H_2O \quad E = 0.741 - 0.0391 pH + 0.0148 \log(H_2SeO_3) \tag{1.13}$$

$$HSeO_3^- + 5H^+ + 4e^- = Se + 3H_2O \quad E = 0.778 - 0.0739 pH + 0.0148 \log(HSeO_3^-) \tag{1.14}$$

Massaccesi et al. studied the deposition of copper and selenium separately and then co-deposition of the two elements. Copper deposition on tin oxide was found to be facile, but selenium appeared to passivate the electrode surface, and no limiting current was observed in a linear voltammogram. However, limiting currents for selenium could be observed when the tin oxide electrode was replaced with a copper electrode. Co-deposition on tin oxide was studied using voltammetry in solutions containing 1 mM Cu and variable amounts of H$_2$SeO$_3$. It was observed that copper deposits at less negative potentials than selenium, which is the reverse of what their standard potentials suggest (see Equations 1.10, 1.13, and 1.14). Furthermore the addition of selenium to the copper solution did not move the copper reduction potential more positive, as would be expected for the Kroger underpotential deposition mechanism [29]. To explain the observed voltammograms and the compositional stoichiometry of the deposits, Massaccesi et al. defined the ratio of Cu to Se fluxes (r_J) at the surface of the electrode as

$$r_J = \frac{J_{Cu}}{J_{Se}} = \left(\frac{D_{Cu}^n}{D_{Se}^n}\right)\frac{[Cu]^{sol}}{[Se]^{sol}} \tag{1.15}$$

where $J_{[X]}$ is the limiting current, $[X]^{sol}$ is the concentration of ions in the bulk, and n is the exponent of the diffusion coefficient which is equal to 2/3. For films deposited between potentials of −0.5 and −0.8 V vs. a mercury sulfate electrode (MSE, +0.64 V vs. SHE), it was found that the composition of the deposit was directly proportional to the flux ratio r_J, suggesting that the copper and selenium reduce independently of voltage and one another. At potentials more negative than −0.8 V vs. MSE, only Cu_2Se was formed. Massaccesi et al. found that if the flux of Se(IV) ions is higher than that needed for the formation of Cu_2Se, Se(I) and Se(II) species were formed which diffuse back into solution and reduce the H_2SeO_3 to form red colloidal Se(0):

$$H_2SeO_3 + 2H_2Se \rightarrow 3Se + 3H_2O \tag{1.16}$$

Massaccesi et al. calculated phase maps showing the species present as a function of r_J and applied voltage. The complexity of the deposition process is illustrated by the calculations for $r_J = 0.8$, when the following species are formed at the electrode surface going from positive to negative potentials: Cu, Cu_2Se, Cu_3Se_2, CuSe, Se(0), Se_2^{2-}, Se^{2-}.

Lippkow and Strehblow [95] studied the effect of temperature on copper selenide deposition from an electrolyte similar to that used by Vedel et al. and found that the Cu_2Se phase was preferred when deposition was carried out at 80 °C, whereas a mixed phase of Cu_2Se and $CuSe_2$ was formed at lower temperatures.

The first detailed study of the electrodeposition of In_2Se_3, onto tin oxide RDE, was reported by Massaccesi et al. [96], who used a bath containing 2 mM $In_2(SO_4)_3$, 1 mM H_2SeO_3, with 100 mM K_2SO_4 as background electrolyte adjusted to pH 3.5 with 10 mM H_2SO_4. Again, as for the Cu–Se case, the authors examined the deposition of the elements individually first and the co-deposition second. At room temperature H_2SeO_3 was reduced to Se(0), passivating the surface of the electrode producing a red deposit, while at 82 °C a gray deposit was formed that was consistent with the gray conducting allotrope of selenium. On the cathodic sweep, indium deposition could not be readily distinguished from the reduction of the tin oxide electrode. Co-deposition of both elements at room temperature yielded a passivated electrode at potentials equal to or more positive than −1.05 V vs. MSE and a gray conducting In–Se phase at more negative potentials. At potentials more negative than −1.05 V vs. MSE, H_2SeO_3 is reduced to H_2Se when no indium is present in the solution. Therefore it appears at room temperature that In(III) only deposits in the presence of H_2Se. When insufficient In(III) is available at the surface, the H_2Se reacts with H_2SeO_3 to form Se(0) (1.16). However, at 82 °C, strongly adherent gray films containing a Se/In ratio similar to that of In_2Se_3 were successfully deposited at −1.05 V vs. MSE. Annealing the films in vacuum at 390 °C crystallized the films into photoactive n-type β-In_2Se_3.

Bhattacharya et al. [28] successfully deposited In_2Se_3 at room temperature using much higher concentration baths than Massaccesi et al.: the baths contained 25 mM $In_2(SO_4)_3$ and 25 mM H_2SeO_3 at pH 1.5. These authors recommended plating a 50 nm thin Cu film on the Mo substrate to improve adhesion and observed that an In-rich film can be obtained by replacing $In_2(SO_4)_3$ by $InCl_3$ in the deposition bath.

Aksu et al. [97] have deposited In–Se compounds from basic solutions containing complexing agents such as tartrate and citrate. The deposition solutions consisted of $InCl_3$, H_2SeO_3, and 0.7 M complexing agent, and the pH was adjusted with NaOH. For basic solutions, the predominant Se(IV) ion in solution is SeO_3^{2-}, with the reduction potential given below:

$$SeO_3^{2-} + 6H^+ + 4e^- = Se + 3H_2O \quad E = 0.875 - 0.0886\,pH + 0.0148\log[SeO_3^{2-}] \tag{1.17}$$

Thus, the SeO_3^{2-} reduction potential is more negative in basic solutions and therefore closer to the reduction potential of In(III) (1.11). Aksu *et al.* showed that they could control the In/Se ratio from 4/7 to 9/1. Furthermore using this approach of complexing agents, and exchanging $InCl_3$ for $GaCl_3$, they also managed to deposit Ga–Se compounds, but only in a narrow range of pH between 7 and 8.5.

In our laboratory we have recently prepared In_2Se_3 films at 82 °C for use in In_2Se_3/Cu–Se precursor stacks for converting into $CuInSe_2$. Initially the recipe of Massaccesi *et al.* [96] was employed, but later the deposition bath was modified by inclusion of a pH 2 buffer, and deposition was carried out under controlled hydrodynamic conditions. The buffer was introduced to control pH better at the electrode surface, and also a thin underlayer of copper was deposited onto the molybdenum substrate to reduce the initial hydrogen evolution. The cyclic voltammograms for a stationary and a rotated electrode are shown in Figure 1.14. For the stationary working electrode, a H_2SeO_3 to Se(0) reduction peak is observed on the forward scan at −0.4 V vs. Ag/AgCl followed by a poorly defined peak due to the deposition of an indium selenide phase. Hydrogen evolution is observed at potentials more negative than −0.6 V vs. Ag/AgCl. On the return sweep a small anodic stripping peak is observed at −0.6 V vs. Ag/AgCl. For the RDE at 250 rpm, the initial current density is more than double the stationary current density, and again the H_2SeO_3 to Se(0) reduction peak is observed at −0.4 V vs. Ag/AgCl. In this case no distinct In–Se reduction peak is evident. On the reverse sweep, the current remains similar to the forward sweep, decreasing at about −0.5 V vs.

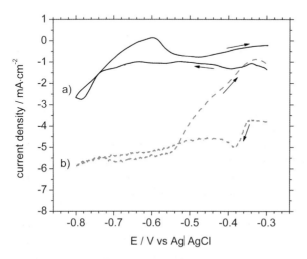

Figure 1.14 Cyclic voltammogramms for 1 mM H_2SeO_3 and 2 mM $In_2(SO_4)_3$ buffered at pH 2 at 82 °C. Scan rate 10 mV s^{-1}. RDE with a 100 nm Cu layer plated on Mo substrate. (a) 0 rpm, (b) 250 rpm.

Ag/AgCl, to indicate where In–Se deposition ceases to be possible. It was found that stoichiometric In_2Se_3 could be deposited at fixed potentials more negative than −0.6 V vs. Ag/AgCl and selenium-rich layers at potentials more positive than this.

Deposition on 1 inch square substrates without hydrodynamic control was found to give non-homogeneous films with visible variation in the color and light-scattering properties. EDX measurements confirmed differences in composition and layer thickness across the film. By contrast, smooth and uniform reflective gray films were obtained when the deposition was performed under controlled hydrodynamic conditions, and EDX measurements confirmed uniform thickness and stoichiometry. SEM images comparing the deposits formed on the vertical stationary deposit and on the rotated electrode are shown in Figure 1.15. The rotated substrates produce very compact uniform layers of indium selenide, while the non-rotated substrates produce a nodular type of growth.

It is known that $CuInSe_2$ can be formed by reacting Cu–Se and In–Se phases [98, 99]. The Gibbs free energy change for this reaction is −49 to −120 kJ mol^{-1} depending on the stoichiometry of the two binary compounds [100]. The potential benefits of this reaction as a route to the final product include recrystallization of the layers in the absence of large lattice changes since the chalcogen is already present. Guillen and Herrero [41] electrodeposited Cu/In_2Se_3 stacks and observed better recrystallization during the annealing of the stacks (compared to co-deposited elements) as shown by the narrowing of the main $CuInSe_2$ XRD peaks.

In_2Se_3/Cu_xSe_y stacks have been electrodeposited and annealed by Hermann et al. [101]. Additionally this group has managed to incorporate gallium into an indium gallium selenide layer by using solutions saturated in chloride ions to stabilize the gallium [102].

In our own laboratory, we have prepared laterally uniform In_2Se_3/Cu–Se stacks and annealed them in a forming gas/elemental selenium atmosphere. Firstly In_2Se_3 layers were deposited at 400 rpm on 10 nm thick copper underlayers (see above). The copper selenide phase was then deposited on top of the In_2Se_3 at room temperature from an electrolyte containing 5.5 mM H_2SeO_3, 2.6 mM $CuCl_2$, 236 mM LiCl, and a pH 3 buffer comprising sulfamic acid and potassium

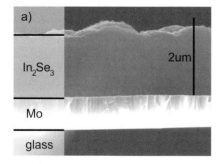 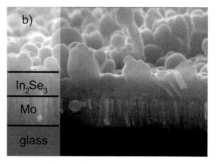

Figure 1.15 SEM images of In_2Se_3 deposits on thin Cu-coated Mo substrates rotated at (a) 250 and (b) 0 rpm.

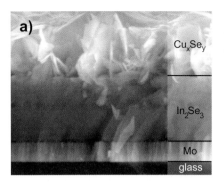 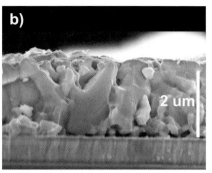

Figure 1.16 SEM images of binary selenide stack: (a) precursor; (b) annealed.

biphthalate. The deposition was carried out at −0.30 V vs. Ag/AgCl on a RDE at 100 rpm. The resulting Cu/Se ratio of the layer was 0.75 as measured by ICPMS, indicating a mixture of CuSe and Se(0) phases were present, in agreement with Thouin et al. [103]. Figure 1.16a shows an SEM image of the precursor and Figure 1.16b of the annealed sample. The two distinct metal selenide phases are observable in the precursor, and the recrystallization of both the layers is clearly evident in the annealed layer: large grains extending from molybdenum surface to the top of the layer can be seen in the SEM image. XRD clearly shows that only the $CuInSe_2$ phase is present. Using the photoelectrochemical tests described in Section 1.3, it was found that the layer is p-type doped with only a small photoresponse.

1.2.3.2.5 Cu(In,Ga)Se$_2$ via Simultaneous Deposition of All Elements

Simultaneous co-deposition of all elements is the widely reported route for the formation of CIGS semiconductor layers or their precursors. Simultaneous electrodeposition of three or four elements is more complicated than deposition of the binary selenides. The main problems are (i) the large differences in standard reduction potentials and (ii) the fact that copper deposition is diffusion controlled. Three main approaches have been taken to tackle these problems, namely use of controlled hydrodynamic conditions, the use of organic additives to change reduction potentials, and pulse plating. All three approaches are discussed in this section, starting with $CuInSe_2$ for simplicity and then considering CIGS.

One of the first detailed investigations into the mechanism for co-depositing $CuInSe_2$ was conducted by Mishra and Rajeshwar, who used cyclic photovoltammetry and a RRDE as investigative tools [104]. They used a simple metal sulfate bath containing SeO_2, with the pH adjusted to 1.0 using sulfuric acid. Comparing Cu–Se and In–Se systems at room temperature, they observed that the In–Se system was kinetically much slower than the Cu–Se system, even though the concentration of In(III) was an order of magnitude larger than that of Cu(II). They found that indium incorporates into the copper selenide phase at potentials lower than the In–Se deposition potential. Cyclic photovoltammetry was carried out using chopped white light. On the forward scan in the three-element solution, the first two peaks were attributed to the formation of $Cu_{2-x}Se$, and no photoresponse was observed. Between the third and fourth reduction peaks, a cathodic p-type photoresponse, attributed to the formation of $CuInSe_2$, was observed. Mishra and Rajeshwar proposed the following mechanism for the incorporation of indium into the deposit:

$$Cu_{2-x}Se.H_2Se_{(adsorbed)} + In^{3+} = CuInSe_2 + 2H^+ + (1-x)Cu^+ \quad (1.18)$$

The key point in this proposed mechanism is that $Cu_{2-x}Se$ needs to be present on the surface together with the adsorbed H_2Se.

The Vedel group studied the co-deposition under three different conditions [105, 106]. They used solutions similar to those employed for their study of the Cu–Se binary system [94, 103] with the addition of 1 mM In(III). Again the key parameter of importance is the ratio of the fluxes of copper and selenium in solution. Vedel et al. define α which is just the reciprocal of r_j (Section 1.2.3.2.4) or the flux ratio of Se(IV) ions to Cu(II) ions incident on the surface [105]. They found that when α was less than two, it was possible to deposit mixtures of $CuInSe_2 + Cu_2Se$ or of $CuInSe_2 + Cu$. When α was greater than two, $CuInSe_2 + In_2Se_3$ was obtained. It was proposed that in the presence of In(III), Se(IV) could be more easily reduced to Se(II). In further work, the same authors found that having an excess of In(III) in solution allowed them to control the compounds deposited on the electrode surface. Furthermore they developed a phase map which is shown in Figure 1.17 [106].

The latest work from the Paris laboratory is that of Chassaing and co-workers [78, 107–109], who investigated the deposition of $CuInSe_2$ onto a Mo RDE from a

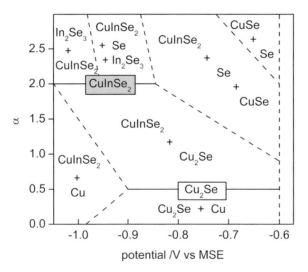

Figure 1.17 Electrochemical phase diagram showing the composition of co-deposited films. Parameter α is the flux ratio of Se(IV) and CU(II) ionic species to the surface. MSE is a mercury sulfate electrode (MSE, +0.64 V vs. NHE) [106]. Reproduced by permission of The Electrochemical Society.

sulfate solution containing 1 mM Cu, 1.7 mM selenous acid, and 6 mM In(III) adjusted to pH 2. They investigated the deposition process *in situ* as a function of time by polarization and impedance spectroscopy, and *ex situ* by chemical, XPS, and XRF analysis. Using these values of concentration and the diffusion coefficients from [94] a value for α of around 2 or slightly higher was obtained depending on the value of n used, and single phase CuInSe$_2$ could be obtained with the correct applied potential. Depositions were carried out at various potentials for fixed amounts of time. At -0.9 V vs. MSE after 3 seconds of deposition there were 10^{11} cm^{-2} nuclei approximately 10 nm in diameter formed on the molybdenum surface, consisting mainly of copper and selenium. After the first 10 seconds, the Se/Cu surface ratio tended to a value of 2, while at longer times the ratio tended to 3. At deposition times longer than 10 seconds the In/Cu ratio at the surface was just 0.5 while in the bulk it was 1.0. XPS data showed that the selenium on the surface of the films was a mixture of Se(0) and Se(II) oxidation states. Furthermore when depositing below potentials of -0.6 V vs. MSE Chassaing et al. [78] observed deposits of binary Cu–Se phases as well as elemental Se(0). For deposition at potentials more negative than -0.6 V vs. MSE, no elemental Se(0) was seen in the deposits, but incorporation of indium into the Cu–Se phases occurred – a process that became more pronounced at more negative potentials. Impedance data showed that at potentials just positive of -0.6 V vs. MSE, the deposited layer was highly capacitive due to the presence of Se(0) on the surface, whereas the capacitance fell at more negative potentials due to the formation of the semiconducting CuInSe$_2$ phase. Chassaing et al. [107] proposed the following reaction scheme based on their results:

$$2Cu^{2+} + H_2SeO_3 + 8e^- + 4H^+ = Cu_2Se + 3H_2O \qquad (1.19)$$

$$H_2SeO_3 + 4e^- + 4H^+ = Se(0)_{(adsorbed)} + 3H_2O \qquad (1.20)$$

$$Cu_2Se + 3Se(0)_{(adsorbed)} = 2CuSe_{2(surface)} \qquad (1.21)$$

$$CuSe_{2(surface)} + In^{3+} + 3e^- = CuInSe_2 \qquad (1.22)$$

Equation 1.20 requires that Se(0) is absorbed onto a deposition surface containing Cu$_2$Se. When sufficient Se(0) is available it is then consumed to form CuSe$_2$ at the surface which has been shown to be the site where In reacts to form the CuInSe$_2$ chalcopyrite compound [109]. This proposed mechanism requires, of course, that the electrode surface should be continually refreshed in Cu$_2$Se and Se.

One of the more successful co-depositions of CIGS was reported by Calixto et al. [110], who used a ratio of selenium and copper ions in the deposition bath similar to that employed by Chassaing et al., but chose a chloride-based supporting electrolyte to stabilize the Ga(III) ions. These authors achieved a Ga/(Ga + In) ratio of 0.2 in the precursor film as indicated in the XRD analysis of the annealed semiconductor by the shift of the (112) peak from $d = 3.348$ to 3.314 Å. The best device had an efficiency of 6.2%. Bhattacharya et al. also used similar routes to co-deposit CIGS and incorporated additional Ga into the deposit by non-electrochemical means [111].

Ligands or complexing agents have been used to try and reduce the dependence of the electrodeposited precursor stoichiometry on applied potential and solution composition. Kemell et al. [112, 113] showed that a constant composition close to the required 1:1:2 for $CuInSe_2$ was achieved over a 300 mV potential range using a SCN ligand. The deposition bath contained 0.01 M CuCl and $InCl_3$, 1 mM SeO_2, and 2 M KSCN. Higher concentrations of KSCN resulted in the deposited films containing traces of the SCN ligand. Using this approach, photoactive p-type $CuInSe_2$ was obtained after annealing. Other popular ligands include citric acid, chloride ions, triethanolamine, and ethylenediaminetetraacetic acid – see Lincot et al. [30] and references therein.

Pulse plating, the application of a cyclic potential program to the work electrode, has been used to deposit $CuInSe_2$ precursor films from a pH 2 chloride bath with a ratio of Cu to Se ions of 1:2 in an unstirred solution [114]. Nomura et al. found that smooth stoichiometric precursors could be obtained by the application of two potentials of −0.7 and 0 V vs. SCE with a duty cycle of 33%.

1.2.3.3 CZTS

The ternary compound CZTS is currently being investigated as an indium-free alternative to CIGS in thin-film solar cells. Bandgaps between 1.36 and 1.62 eV have been reported for CZTS, making the material suitable for a single-junction solar cell [10, 115]. CZTS films show p-type behavior with doping levels that are suitable for thin-film devices [35, 116, 117]. Photovoltaic devices are generally fabricated by the same methods used for CIGS cells, except that the CIGS layer is replaced by CZTS. The current record device, with a reported AM 1.5 efficiency of 6.7%, was prepared by a two-stage process involving preparation of a co-sputtered precursor using Cu, ZnS, and SnS targets followed by annealing in H_2S [10]. Single-stage co-evaporation methods, which produce the best devices for CIGS, have yielded efficiencies up to 4.2% [118].

There are currently two basic approaches to fabricate CZTS layers by electrodeposition. In the stacked elemental (SEL) approach, the parent metals are electrodeposited sequentially to form a stack [35, 119] which is then annealed in a sulfur atmosphere to from CZTS. In an alternative approach, the parent metals are co-deposited to form a mixed metallic or alloy layer which is then annealed in the presence of sulfur [120, 121]. Both approaches have yielded cells with AM 1.5 efficiencies of over 3%.

1.2.3.3.1 Fabrication of Thin-Film Solar Cells via the SEL Method

The SEL approach has some advantages over co-deposition, because it allows simpler coulometric control of the precursor stoichiometry. The SEL method may also be more suitable for large-scale fabrication since there is no requirement to balance the deposition rates of several different metals, so higher current densities can be used. If an alloyed precursor is preferred for the sulfur annealing step, a short (<5 min) heat treatment at 200–350 °C is sufficient to completely alloy the stacked precursor. In this section, we highlight some of our recent work on the fabrication of CZTS solar cells by the SEL route [11].

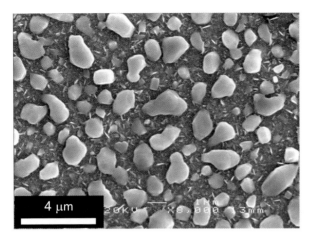

Figure 1.18 SEM image of Sn deposited on Cu using sorbitol bath [123]. Reproduced from [11].

Earlier work in our laboratory encountered problems with non-uniform thickness and incomplete coverage [35–37]. The morphology of the Sn layers in particular was poor with large islands as shown in Figure 1.18 [36, 122].

Figure 1.19a illustrates the problem of poor microscopic morphology in the case of a CZTS film prepared from a Cu/Sn/Zn precursor with the island structure shown in Figure 1.18. The sulfurized CZTS film contains small grains, and the morphology is uneven. Figures 1.19b and 1.19c show elemental distribution maps for Sn and Zn over the same area. The contrast variations in the two elemental maps are not coincident, suggesting that the film contains other phases besides CZTS. These variations probably arise from the island morphology of the Sn layer in the precursor seen in Figure 1.18, which was not laterally homogenized during sulfurization.

In later work in our laboratory, thickness variations in the precursor layers were eliminated by carrying out electrodeposition under hydrodynamic control using a RDE [122]. In addition the electrodeposition processes and stacking order were modified to improve the precursor morphology. The improved Cu/Sn/Cu/Zn stacks were sulfurized and converted into devices, the best of which showed an efficiency similar to that reported by Araki *et al.* [120] and Ennaoui *et al.* [121] for CZTS devices fabricated using metal co-deposition routes. Electrodeposition was carried out potentiostatically in three-electrode mode onto square 25×25 mm Mo-coated soda-lime glass substrates mounted in the face of a cylindrical polypropylene RDE which was placed face-down in the electrodeposition solution and rotated during deposition. The deposition conditions used for growth of the sequential metal layers are given in Table 1.2. Deposition efficiencies were obtained from ICP-MS measurements of dissolved films. The precursors were placed with 100 mg of sulfur inside a graphite container in a tube furnace filled to 500 mbar with 10% H_2 in N_2 and heated to 575 °C for 2 hours. Prior to device fabrication and materials characterization, the sulfurized films were etched in aqueous KCN

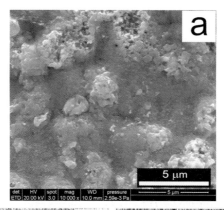

Figure 1.19 (a) SEM image of a sulfurized Cu/Sn/Zn precursor stack and EDX element distribution maps of (b) Sn and (c) Zn from the same area as (a). Reproduced from [11].

Table 1.2 Chemical composition and electrodeposition parameters for each layer of the Cu/Sn/Cu/Zn precursor. Potentials are given with respect to the Ag|AgCl reference electrode.

Layer	Composition of electrolyte	E (V)	Rotation (rpm)
Cu	3 M NaOH, 0.1 M CuSO$_4$, 0.2 M Sorbitol, 0.9 mM Empigen BB (surfactant)	−1.20	150
Sn	50 mM Sn(SO$_3$CH$_3$)$_2$, 1 M CH$_3$SO$_3$H, 3.6 mM Empigen BB	−0.72	100
Zn	0.1 M ZnSO$_4$.7H$_2$O, pH3 hydrion buffer, 0.5 M K$_2$SO$_4$	−1.20	150

to remove copper sulfide phases. Devices were fabricated by deposition of a CdS layer by CBD followed by deposition of i-ZnO and Al:ZnO layers by sputtering. Contact to the cells was made by a nickel grid. Following fabrication, devices were heated in air at 200 °C for up to 5 minutes.

Growth of Copper Layers Nucleation and growth of metal layers on the Mo-coated soda-lime glass substrates is not without problems. In acidic conditions, the untreated Mo surface is passivated by an oxide layer [82] that prevents effective nucleation of Cu, resulting in non-adherent, powdery deposits. Adhesion can be improved by sensitizing the substrate with Pd prior to electrodeposition [119]. Since the oxide layer on Mo is soluble in alkaline media, surface passivation can be avoided, and excellent Cu layer deposits can be obtained using an RDE in an alkaline Cu electrolyte adapted from the work of Barbosa et al. [84]. Introduction of a quaternary ammonium surfactant (Empigen BB) to the electrolyte improves the morphology of the Cu deposit.

Growth of Tin Layers Tin layers deposited in our previous work [122] from the alkaline sorbitol electrolyte consisted of large crystals with low nucleation density, giving incomplete coverage. Much improved flat layers showing specular reflectivity were obtained using a methanesulfonic acid electrolyte containing Empigen BB. The film morphology of the tin layer produced in this electrolyte is shown in Figure 1.20, where it is the underlying layer.

Growth of Zinc Layers The morphology of the Zn layer is strongly influenced by the structure of the underlying Sn layer. Although excellent Zn films could be grown on Cu layers, nucleation of Zn on Sn layers was found to be very uneven. A comparison of the voltammetric behavior of $Zn^{2+/0}$ at Sn and Cu electrodes (Figure 1.21) shows that underpotential deposition (UPD) of Zn on Cu occurs due

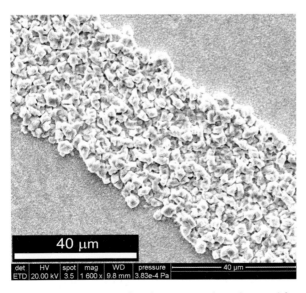

Figure 1.20 SEM image of Zn deposit on Sn layer deposited from a methanesulfonic acid electrolyte. Reproduced from reference [11].

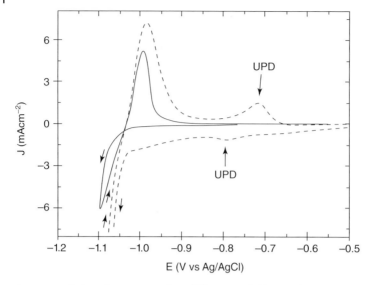

Figure 1.21 Cyclic voltammogram for $Zn^{2+/0}$ at Sn (solid line) and Cu (dashed line) surfaces. Reproduced from reference [11].

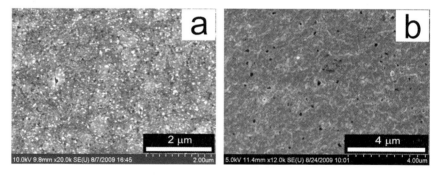

Figure 1.22 SEM images of (a) Cu layer deposited onto Sn and (b) Zn deposited onto Cu. Reproduced from reference [11].

to the energetically favorable formation of a Cu–Zn alloy. This UPD layer completely covers the Cu surface in a thin layer of Zn, strongly inhibiting hydrogen evolution. By contrast, no UPD of Zn on Sn was observed and hydrogen evolution interferes with Zn deposition and nucleation is non-uniform.

This problem was solved by depositing a second Cu layer on top of the Sn film, using a negative potential to inhibit displacement of Sn^{2+} from the film. It was found that good Zn films could grown on top of this intermediate Cu layer to give a Cu/Sn/Cu/Zn stack. Figure 1.22 shows the morphology of the second Cu layer and the final Zn layer. This arrangement of the layers may be advantageous since Araki *et al.* reported that a precursor stack in which Cu and Sn were neighbors gave the best devices after sulfurization [124]. This observation is consistent with the reaction scheme proposed by Hergert and Hock [125], which describes the growth of CZTS as proceeding via intermediate formation of Cu_2SnS_3 and ZnS [126].

Characteristics of Stacked Precursor Films The new Cu/Sn/Cu/Zn precursors have good lateral uniformity, and the layers consist of small compact grains. The EDX maps in Figure 1.23 show the macroscopic uniformity of the precursor stacks over a $1.4\,cm^2$ area, in terms of the compositional ratios Cu/(Zn + Sn) and Zn/Sn. Both show only small variations, significantly better than those achieved previously [122]. In the example shown, the Cu/(Zn + Sn) ratio was 0.80 ± 0.02 and the Zn/Sn ratio was 1.37 ± 0.05.

The XRD diffractogram (Figure 1.24) measured 14 days after preparation of the stacked layer precursor indicates that the metal layers have alloyed partly at room temperature, even though there are no visible changes in color or texture. The only elemental phase observed is Cu, with binary Cu–Sn and Cu–Zn phases accounting for the remaining peaks. No ternary phase is observed, in agreement with the findings of Chou and Chen [127], whose study of the Cu–Sn–Zn system at 250 °C showed only binary phases.

SIMS measurements on the precursor confirm intermixing of the layers occurs at room temperature (Figure 1.25). In particular, the Cu signal is clearly seen in both the Sn and the Zn phases. The majority of the Zn is still confined to the top of the precursor, while the Sn is distributed towards the bottom, suggesting it is mainly diffusion of Cu that is responsible for alloy formation.

Characterization of Annealed CZTS Films Sulfurization of the Cu/Sn/Cu/Zn precursor stacks gave better microscopic uniformity than the previous Cu/Sn/Zn stacks. An SEM image and EDX maps of the Sn and Zn distributions in a film produced from the Cu/Sn/Cu/Zn precursor stack is shown in Figure 1.26. A comparison with Figure 1.23 shows a much more uniform distribution of the elements for the film formed from the four-layer stack precursor. The high degree

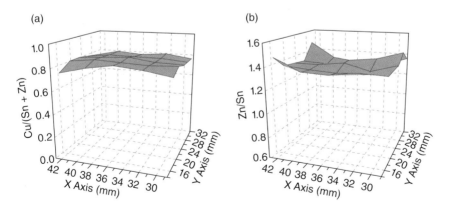

Figure 1.23 Large-area EDX maps of a Cu/Sn/Cu/Zn precursor stack in terms of the ratios (a) Cu/(Zn + Sn) and (b) Zn/Sn. Reproduced from reference [11].

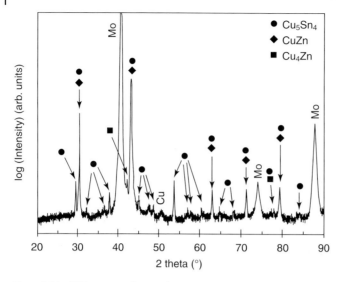

Figure 1.24 XRD spectra of a stoichiometric Cu/Sn/Cu/Zn precursor stack deposited on Mo. Reproduced from reference [11].

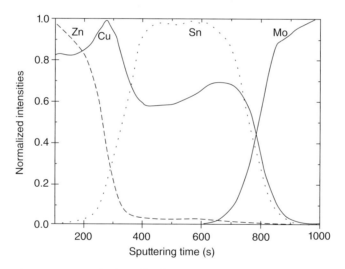

Figure 1.25 SIMS depth profile of Cu/Sn/Cu/Zn precursor layer after four weeks at room temperature. Each element has been normalized to its highest value. Reproduced from reference [11].

of correlation between the two elements suggests that the films are single phase. It appears that the only published study of the phase diagram of the Cu_2S–ZnS–SnS_2 system is by Olekseyuk et al. [128], who have reported that Cu_2ZnSnS_4 only exists as a single phase over a very small range of composition varying by less than ±1.5% absolute from the stoichiometric values for each element. The CZTS film in Figure 1.26 has a Cu/(Zn + Sn) ratio of 0.70 ± 0.07 and a Zn/Sn ratio of

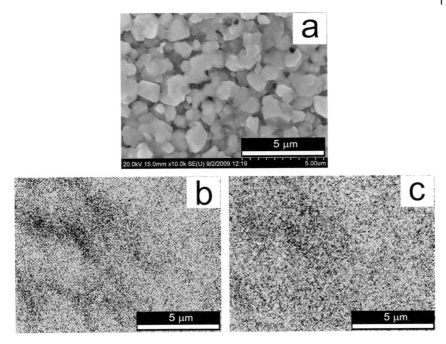

Figure 1.26 (a) SEM image of a sulfurized Cu/Sn/Cu/Zn precursor stack. (b) Sn and (c) Zn elemental EDX maps corresponding to (a). Reproduced from reference [11].

1.06 ± 0.12. Both values lie outside the single-phase region suggested by Olekseyuk et al., but no significant secondary phases could be distinguished within the layer with the resolution of mapping used here.

A grazing incidence XRD spectrum of a sulfurized film is shown in Figure 1.27. All reflexes can be assigned to the kesterite structure using ICSD card 01-075-4122 (inorganic crystal structure database) except for two peaks (labeled with asterisks) that are assigned to a copper sulfide phase. This assignment is supported by the fact that the two peaks disappear when the film is etched in aqueous KCN. The figure also shows the largest reflexes for the most probable secondary phases, ZnS and Cu_2SnS_3. It is important to note that these phases have very similar unit cells to CZTS, and are essentially indistinguishable from it using XRD.

1.2.3.3.2 CZTS Solar Cells

Several cells were fabricated using the CZTS layers. Figure 1.28 shows a cross-sectional image of the best cell showing that the CZTS layer near the Mo substrate consists of small grains with some voids, whereas the upper part is formed of large close-packed grains of around a micron in size. This kind of structure is also observed for $CuInSe_2$ films prepared by a similar electrodeposition–annealing approach [87].

SIMS measurements were made on the best cell, after removal of the ZnO and CdS layers: the results are shown in Figure 1.29. The concentration of all three

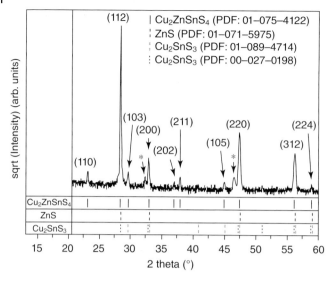

Figure 1.27 Grazing incidence XRD pattern of sulfurized Cu/Sn/Cu/Zn precursor layer. The asterisks indicate a secondary phase assigned to copper sulfide. Cu/(Zn + Sn) = 0.70 ± 0.07, Zn/Sn = 1.06 ± 0.12. Reproduced from reference [11].

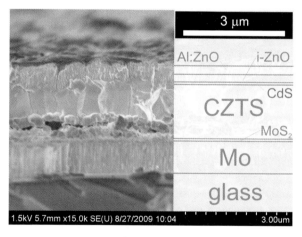

Figure 1.28 SEM image of a cross section through the best performing cell. Reproduced from reference [11].

metallic elements is constant throughout the bulk of the film. The CZTS film appears to be slightly Cu and Zn deficient at the front interface. The S concentration is also uniform throughout the depth of the film, increasing slightly at the front of the film, possibly due to the presence of residual CdS.

The best device gave an open circuit voltage (V_{oc}) of 480 mV, a short circuit current density (J_{sc}) of 15.3 mA cm^{-2}, and a fill factor of 45%. The corresponding power conversion efficiency was 3.2%, which is similar to the values reported by Araki *et al.* (3.16% [120]) and Ennaoui *et al.* (3.4% [121]).

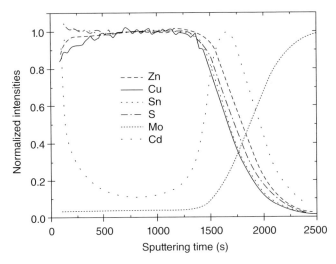

Figure 1.29 SIMS profile of the best performing cell after chemical removal of the ZnO and CdS layers. Each element has been normalized to its highest value. Reproduced from reference [11].

This section has shown that improvements in the microscopic and macroscopic uniformity of electrodeposited stacked elemental layer precursors and the use of a four-layer stack–Cu/Sn/Cu/Zn–result in better solar cells. Clearly CZTS has some way to go before it can compete with CIGS, but these early results are certainly encouraging.

1.2.4
Future

Electrodeposition of CIS and CIGS absorber films has been realized from aqueous solution, but all of the precursor films suffer from a number of problems. The incorporation of In and more especially Ga is not trivial. Due to their limited solubility and large negative reduction potentials close to the hydrogen evolution reaction of water, it is necessary to use strong complexation chemistry, involving toxic ligands such as cyanide and to work under highly alkaline conditions to avoid hydrogen evolution which causes uncontrolled stirring and poor morphology of the deposits. Moreover, In and Ga are often deposited together with their oxide or hydroxide forms under these conditions. Nevertheless, the electrodeposited precursor layers are transformed into device-quality CIGS during the highly energy consuming post-annealing step under Se atmosphere at high temperature.

A possible solution to these problems could be to replace the aqueous deposition media by room temperature ionic liquids (RTILs). These are defined as liquids consisting of molecular cations and anions that have a melting point below 100 °C. Due to the size of the molecular ions and the fact that the charge is delocalized through the molecule, these liquids have large electrochemical windows. Another potential benefit is that RTILs have high degradation temperatures and low vapor pressures, allowing them to be used at temperatures up to 250 °C. These wide electrochemical and thermal windows allow processes which are impossible in aqueous or organic solvents.

The electrochemical window of water varies between 1 and 2 V, while those of most common pyrrolidinium-based ionic liquids are between 5 and 6 V [129]. Shivagan et al. [130] have shown that In and Ga can be reversibly electrodeposited from a eutectic-based ionic liquid composed of a 1:2 choline chloride:urea mixture. El Abedin et al. [131] have investigated the electrochemistry of Cu, In, and Se individually in the ionic liquid 1-butyl-1-methylpyrrolidinium-bis(trifluoromethylsulfonyl)amide. More recently, the same group showed that Ga can be deposited electrochemically and electroless on Au(111) electrodes [132]. The very low vapor pressure of ionic liquids allows them to be used as electrodeposition media at elevated temperatures, thus being potentially interesting to improve the crystallinity of the deposited thin films. Two reports from the literature mention the effect of temperature on the deposition of compound semiconductors from ionic liquids. Yang et al. [133] showed that for the electrodeposition of InSb from 1-ethyl-3-methylimidazolium chloride/tetrafluoroborate, at temperatures up to 120 °C, the crystallinity of the deposit increased with the deposition temperature. The as-deposited films were p-type and photoactive. Hsiu et al. [134] deposited CdTe thin films at constant potential at temperatures up to 140 °C onto glassy carbon, titanium, and tungsten substrates from 1-ethyl-3-methylimidazolium chloride/tetrafluoroborate.

From the very limited literature describing ED of compound semiconductors from ionic liquids it can be seen that more research is needed to understand the fundamental aspects of ED from these liquids. However it is reasonable to conclude that due to their unique electrochemical and thermal properties, ionic liquids are likely to play an important role in the field of compound semiconductor electrodeposition. Ongoing work in our laboratories is investigating controlled gallium incorporation and high-temperature electrodeposition of CIGS thin films from ionic liquids.

1.3
Characterization of Solar Cell Materials using Electrolyte Contacts

1.3.1
Overview

The electrochemical and photoelectrochemical properties of semiconductors in contact with electrolytes have been widely studied using single-crystal samples as well as polycrystalline thin films, and several excellent general books are available in the literature [135–138]. By contrast, the scope of this section is relatively narrow since it focuses specifically on practical applications of electrolyte contacts for the purpose of characterizing thin-film solar cell layers and partially completed PV devices of various kinds.

The main advantage of using electrolyte contacts to characterize the semiconducting layers that are used in PV devices is that liquid contacts are transparent and easily removed without damaging the material. Furthermore, provided that the semiconductor is stable in the electrolyte used, liquid contacts tend to be rather ideal since they avoid the lattice mismatch problems and chemical interactions that are often encountered with metal contacts. Formation of a Schottky barrier at the semiconductor–electrolyte junction allows convenient variation of the band bending in the space charge region by controlling the electrode potential with respect to a reference electrode potential. For measurements under illumination, where minority carriers are driven to the interface, the electrolyte generally contains a suitable scavenger species to prevent anodic or cathodic photodecomposition of the semiconductor. This section describes the main semiconductor characterization techniques that employ electrolyte junctions. A recent comprehensive survey of experimental techniques in semiconductor electrochemistry by Peter and Tributsch [5] contains useful background information about the techniques described here.

1.3.2
The Semiconductor–Electrolyte Junction

Basic aspects of the semiconductor–electrolyte junction are discussed in the references given in the previous section. Here we summarize the main points, placing particular emphasis on their relevance to the characterization of thin-film PV materials. For further details on semiconductor characterization in general, the reader is referred to the excellent book by Schroder [139].

If an inert electrolyte (i.e., one not containing a redox system) is used to form a semiconductor–electrolyte contact, the potential distribution at the interface is determined by the applied electrode potential. When the semiconductor is held at the *flatband potential*, there is no space charge in the semiconductor, and therefore the conduction and valence bands are flat. For characterization purposes, the junction is usually polarized so as to create depletion conditions where a space charge region is formed in the semiconductor as majority carriers are withdrawn from the junction. This corresponds to reverse biasing a conventional Schottky junction. For n-type materials, polarization to more positive electrode potentials corresponds to reverse bias, whereas the polarization direction is the opposite for p-type materials. Provided that the material is not highly doped, the current flowing under depletion conditions in the dark is small, although in practice pinholes and defects in polycrystalline layers give rise to higher currents than those observed with single crystals. Polarization of the semiconductor–electrolyte junction into accumulation is best avoided since this is likely to lead to decomposition reactions involving majority carriers: n-type materials will be reduced and p-type materials will be oxidized. These decomposition reactions can modify the semiconductor surface substantially, complicating characterization measurements.

When a semiconductor is illuminated under depletion conditions with light of sufficient energy ($h\nu > E_g$, where E_g is the bandgap), electron–hole pairs are created and minority carriers (electrons for p-type and holes for n-type material) move to the interface where they can take part in electron transfer reactions, generating a photocurrent. In the case of n-type materials, the photocurrent is anodic, whereas for p-type materials it is cathodic.

If a redox electrolyte is used, equilibration of the electron free energies in the semiconductor and electrolyte phases at open circuit results in the alignment of the electron Fermi levels, E_F, so that E_F (semiconductor) = E_F (redox). If, for example, the redox Fermi level before contact is lower than the Fermi level of an n-type semiconductor, a depletion layer will be formed as electrons are extracted by the redox system until equilibration of the Fermi levels is achieved, and the bands will be bent upwards at the interface. For a p-type semiconductor sample, a depletion layer is formed when it is brought into contact with a redox system that has a higher Fermi energy. In this case the bands are bent downwards at the interface. The formation of semiconductor–electrolyte junctions by choosing suitable redox couples is analogous to the formation of metal–semiconductor Schottky junctions by choosing metals with appropriate work functions. This principle is the basis for the design of regenerative photoelectrochemical solar cells, which were widely investigated in the 1980s and achieved remarkably high efficiencies [140]. In practice, adequate stabilization of semiconductor layers for characterization purposes can only be achieved if decomposition reactions are suppressed by choosing electrolytes which capture carriers before they can react with the semiconductor lattice. Usually it is best to choose redox electrolytes that form Schottky barriers since this ensures that minority carriers will be scavenged. Examples include the use of Eu^{3+}/Eu^{2+} couple for p-type materials such as CIS. However, in some cases it may be preferable to use an electrolyte in which the electron transfer process is not only fast but also irreversible. For example Na_2SO_3 is often used with n-type materials like CdS since the SO_3^{2-} ion is an excellent hole scavenger, reacting rapidly and irreversibly to form sulfate.

1.3.3
Photovoltammetry

A rapid assessment of photoactivity and majority carrier type is often required for screening semiconductor layers that have been synthesized for PV applications. This can be achieved (without the need to fabricate an entire device) by using an electrolyte contact and carrying out a cyclic potential sweep while illuminating the electrode with chopped light from a light-emitting diode (LED). The wavelength is usually chosen to lie in the region where the material absorbs strongly, although white (red/green/blue) LEDs are also useful. The sign of the photocurrent can be used to identify the carrier type, and the magnitude gives an indication of the external quantum efficiency (EQE), which is also referred to as the IPCE (incident photon conversion efficiency). The scan also gives information about the behavior of the sample–electrolyte junction in the dark. The presence of pinholes in the film that allow contact between the electrolyte and the substrate can often be inferred from higher dark currents and the appearance of voltammetric

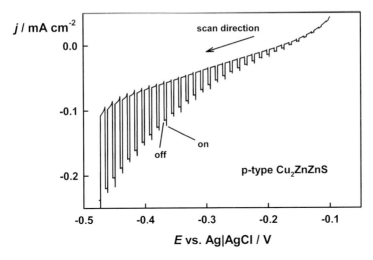

Figure 1.30 Photovoltammogram of a Cu_2ZnSnS_4 layer measured in 0.2 M $Eu(NO_3)_3$ under pulsed illumination from a white LED. The cathodic photoresponse shows that the sample is p-type.

waves associated with oxidation and reduction of the redox couple. Similar effects can be seen if the grain boundaries are quasi-metallic due to impurities or high doping. The transition from depletion to accumulation conditions as a function of applied potential is generally marked by a steep increase in dark current, but this region is best avoided in order to prevent damage to the semiconductor surface.

Photovoltammetry is used routinely in our laboratories to assess new and established materials. Figure 1.30 shows an example of the photovoltammogram recorded for a sample of Cu_2ZnSnS_4 prepared by sulfurization of a stack of three layers of the parent metals [35, 36, 141]. In this case, the electrolyte used contained Eu(III) as an electron scavenger. The cathodic photoresponse confirms that the sample is p-type, and the dark current seen at negative potentials indicates that the film has pinholes that expose the underlying molybdenum film. The photocurrent response to chopped illumination can also be detected using a lock-in amplifier, allowing a more accurate determination of the photocurrent onset potential, which provides an estimate of the flatband potential (the photocurrent for p-type semiconductors onset is usually 200–300 mV more negative than the flatband potential as a consequence of recombination via surface states).

1.3.4
External Quantum Efficiency (EQE) Spectra

The semiconductor–electrolyte junction is essentially an analog of the junction that is formed in a solid-state device such as a solar cell. For this reason, measurements of the EQE as a function of photon energy can be used to obtain information

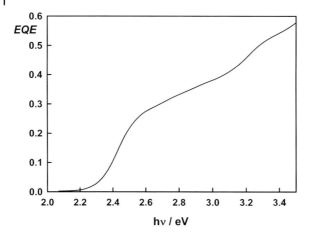

Figure 1.31 EQE spectrum of n-CdS film on FTO glass measured in 0.1 M Na$_2$SO$_3$.

about the quality of materials as well as, in some cases, doping densities and minority carrier diffusion lengths.

EQE spectra can be recorded for a range of layer structures that represent different stages in the fabrication of a PV device. In the case of fabrication of CdS/CdTe heterojunction cells by electrodeposition of CdTe onto a thin CdS layer on FTO glass, the EQE response has been used to characterize the CdS films, which are prepared by CBD [142]. Figure 1.31 shows an EQE spectrum measured for an 80 nm CdS film on FTO glass following an annealing in CdCl$_2$ at 415 °C. Analysis of the spectrum using absorbance data obtained by transmission shows that the *internal* quantum efficiency approaches unity over the entire spectral range where the CdS absorbs. As-deposited and air-annealed CdS films have different bandgaps, but they are both photoactive with internal quantum efficiencies approaching unity. Interestingly, the CBD CdS appears to lose its photoactivity entirely when the final structure is annealed to form a CdS/CdTe heterojunction solar cell (this point is discussed in more detail below).

Another interesting example of the application of EQE spectroscopy is its use to follow type conversion in CdS|CdTe heterojunction solar cells fabricated by electrodeposition (see Section 1.2). The method was used originally by Basol on completed solid-state CdS|CdTe cells [143] and later extended by others to characterize partially completed structures using electrolyte contacts [144–147]. The location of the photoactive regions is probed by comparing the EQE spectra for substrate side (SS) and electrolyte side (ES) illumination of the samples as illustrated schematically in Figure 1.32.

As-deposited CdTe films are n-type, so that no heterojunction is formed between the CdTe and the underlying n-type CdS. When the as-deposited CdTe film on CdS is contacted with an electrolyte and is polarized so that a depletion layer forms at the n-CdTe–electrolyte junction, electron–hole pairs generated by illumination are separated, with the holes moving towards the electrolyte interface where they

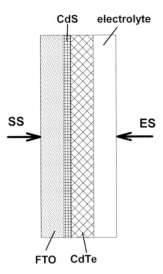

Figure 1.32 Illumination geometry used to study type conversion of CdTe layers on CdS. SS, substrate side illumination; ES, electrolyte side illumination.

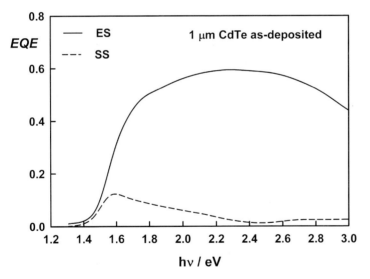

Figure 1.33 EQE spectra for SE and EE illumination of an as-deposited 1 μm CdTe film on CdS/FTO. Comparison of the SE and EE spectra indicates that the CdTe film is n-type.

are scavenged by ions in the electrolyte (0.1 M Na_2SO_3). As Figure 1.33 shows, the EQE spectrum for ES illumination of a 1 μm thick CdTe layer on CdS exhibits an onset at the CdTe bandgap and a fairly flat response at higher photon energies with EQE values of around 0.6. When the illumination is switched to the SS, the

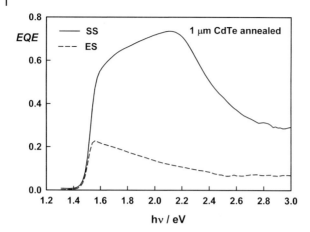

Figure 1.34 SS and ES EQE spectra for CdTe layer on CdS/FTO following type conversion (compare with Figure 1.4).

EQE spectrum shows a maximum near the band edge, but almost no response at higher energies. This indicates that the active junction is indeed located at the CdTe–electrolyte interface, since light can only penetrate through to the CdTe–electrolyte interface from the substrate side when the absorption coefficient of the CdTe is sufficiently low, as is the case near the band edge.

Heating the electrodeposited CdS|CdTe structure in air for 15 minutes at 430 °C converts the CdTe to p-type, forming a heterojunction with the underlying n-CdS. The type conversion process can be followed by measuring the EQE spectra. As Figure 1.34 shows, illumination of the type-converted structure through the glass substrate produces a photocurrent over the entire spectral region above the bandgap of CdTe except for the high-energy region where the CdS absorbs. For SS illumination, the incident light is absorbed close to the n-CdS|p-CdTe junction, and electrons (minority carriers) from the p-CdTe are transferred to the FTO contact via the n-CdS. The "blue" defect seen in the EQE spectrum in the CdS absorption region indicates that the CdS layer is evidently no longer photoactive after the annealing process. This may indicate that the active junction is in fact not at the CdS|CdTe interface but is located inside the CdTe–effectively an n–p homojunction, with no electrical field in the CdS film. The CdS therefore acts as a filter, reducing the response in the blue part of the EQE spectrum. By contrast, after annealing, the EQE spectrum for SE illumination now only exhibits a response near the band edge of the CdTe, since light must be able to penetrate from the electrolyte side through to the internal p–n hetero- or homojunction, which is located close to the substrate contact.

Type conversion has also been followed as a function of annealing time by measuring EQE spectra. Figure 1.35 illustrates how the SS EQE spectra of a 0.5 μm CdTe film on CdS/FTO evolve as the film is annealed. The results show that internal p–n junction is formed after 5 minutes and improves further with

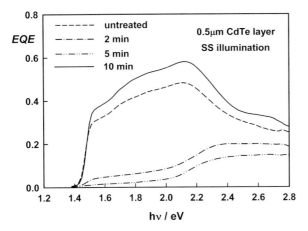

Figure 1.35 EQE spectra for substrate side illumination for CdTe films on CdS/FTO that have been annealed for different times. The spectra show that formation of the p–n junction requires at least 5 minutes annealing time.

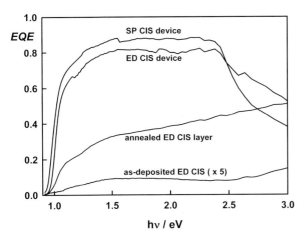

Figure 1.36 EQE spectra of as-deposited and annealed CIS layers compared with the EQE spectra of completed devices based on electrodeposited (ED) and sputtered (SP) CIS layers.

subsequent annealing. Interestingly the spectra also reveal that the CdS layer is photoactive initially but loses its activity when the junction is formed.

Another more recent example of the application of EQE spectra to characterize solar cell absorber layers is shown in Figure 1.36 for different stages in the fabrication of $CuInSe_2$ solar cells. The ED CIS precursor layer was prepared by electrodeposition onto Mo-coated glass, followed by annealing in Se vapor and subsequent etching in KCN to obtain the photoactive material [23] (see Section 1.2). The sputtered CIS device was prepared by magnetron sputtering of a Cu–In precursor layer followed by annealing in a selenium atmosphere, KCN etching, and deposition of CBD CdS and sputtering of transparent contact layers.

It can be seen that the photoresponse of the as-deposited ED film is very weak – it has been magnified five times in Figure 1.36 to make it visible. The substantial improvement brought about by annealing the ED precursor film in selenium and then etching it in KCN to remove copper selenides is evident from the greatly enhanced EQE response. Interestingly, however, the subsequent fabrication of a device by deposition of CdS and ZnO layers brings about a further improvement, with the finished ED device exhibiting a performance that is almost comparable to that of the sputtered device. It appears that the main reason for this additional improvement is that the CBD of CdS on the CIS film reduces recombination losses substantially. Both devices were fabricated without an antireflection coating, so it can be concluded that the internal quantum efficiency probably approaches unity. The AM 1.5 efficiencies of the finished devices in this study were 6.7% for the ED cell and 8.3% for the sputtered cell. The CdS layer in the finished devices also gives rise to the same "blue" defect in the EQE spectra – the decrease in EQE in the spectral region above 2.4 eV where CdS absorbs the incident light – that is seen in CdTe|CdS devices. The band tailing evident below in the EQE spectra for both ED and sputtered materials probably indicates a high density of sub-bandgap states.

EQE spectra measurements have been used recently as a tool for compositional optimization of Cu_2ZnSnS_4 films prepared by sulfurization of stacked metal layers [35, 36, 141] (see Section 1.2). The spectra shown in Figure 1.37 illustrate the sensitivity of the photoresponse to the Cu/(Zn + Sn) ratio. The highest EQE value (up to 0.4) was obtained for a mean Cu/(Zn + Sn) ratio of 0.86. Recent work in our laboratory has shown that the EQE can be mapped across the surface and correlated with the local composition determined by EDX [36].

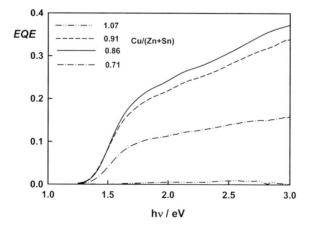

Figure 1.37 EQE spectra of Cu_2ZnSnS_4 layers with different Cu/(Zn + Sn) ratios measured in Eu(III) electrolyte at −0.3 V vs. Ag|AgCl.

EQE spectra can be analyzed to obtain the bandgap of absorber materials. The photocurrent response (j_{photo}) of the semiconductor electrolyte junction is described by the Gärtner equation [148] as

$$\text{EQE} = \frac{j_{photo}}{qI_0} = \frac{1-e^{-\alpha W}}{1+\alpha L} \qquad (1.23)$$

where I_0 is the incident photon flux corrected for reflection losses, α is the absorption coefficient of the absorber at the wavelength of interest, W is the width of the space charge region, and L is the minority carrier diffusion length. For small values of α and L, expansion of the exponential term shows that the EQE becomes linearly proportional to α. For a direct optical transition, the energy dependence of the absorption coefficient is given by

$$\alpha h\nu \propto (h\nu - E_g)^{1/2} \qquad (1.24)$$

so that a plot of $(\text{EQE} \times h\nu)^2$ vs. photon energy should be linear with an intercept at the bandgap energy E_g. This method is illustrated in Figure 1.38 for CuInSe$_2$ absorber layers prepared by ED and by sputtering (SP) [23]. It can be seen that the two films give slightly different values of E_g. This observation, which appears to be related to the Cu/In ratio in the layers, is discussed in the next section, which deals with the (more precise) determination of bandgaps by electrolyte electroreflectance (EER) spectroscopy.

Doping densities in single-crystal semiconductor samples can usually be determined from capacitance and voltage measurements uing suitable electrolyte contacts. However, in the case thin polycrystalline layers fabricated for PV devices, dark currents arising from cracks and pinholes often make interpretation of impedance data difficult. In this case a different approach based on analysis of the EQE response may be useful. If the minority diffusion length in the sample is sufficiently small to fulfill the condition $\alpha L \ll 1$ – as is usually the case for

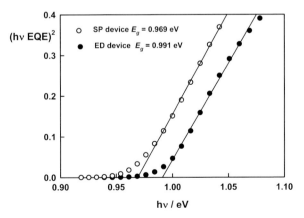

Figure 1.38 Plots used to determine the bandgap of CIS absorber layers from EQE spectra. SP, sputtered film; ED, electrodeposited film.

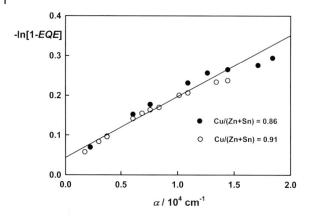

Figure 1.39 Plot used to estimate W, the width of the space charge region, and hence the acceptor density for two Cu$_2$ZnSnS$_4$ layers with different Cu/(Zn + Sn) ratios.

polycrystalline compound semiconductors – Equation 1.23 can be written in the form

$$-\ln(1-EQE) = \alpha W \tag{1.25}$$

It follows that a plot of $-\ln(1-EQE)$ vs. α should be a linear with a slope equal to the width of the space charge region, W, which is given for a p-type semiconductor by

$$W = \left(\frac{2(E_{fb}-E)\varepsilon_0\varepsilon_r}{qN_a}\right)^{1/2} \tag{1.26}$$

where $(E - E_{fb})$ is the potential relative to the flatband potential E_{fb}, ε_r is the relative permittivity of the material, and N_a is the acceptor density. Figure 1.39 illustrates the application of this method to estimate the doping density in p-type Cu$_2$ZnSnS$_4$ samples. The required α values as a function of photon energy were taken from work by Tanaka et al. [149], and the flatband potential was estimated from the photocurrent onset (cf. Figure 1.1). The acceptor density obtained from the plot is 3×10^{16} cm^{-3}, a value similar to those reported in the literature for CZTS films prepared by various routes [149, 150].

1.3.5
Electrolyte Electroreflectance/Absorbance: EER/EEA

The technique of EER (or electrolyte electroabsorbance (EEA) in the case of thin films) is a powerful way of determining bandgaps and also of assessing compositional and structural defects. The method is based on perturbing the optical properties of a semiconductor by application of an electric field [151, 152]. The method involves applying an AC modulation to the applied potential when the sample is the working electrode held in depletion conditions in an electrochemical cell to generate a field in the space charge region. The synchronous modulation of the reflectance or transmittance of the sample is detected using a lock-in amplifier [5]. The changes in reflectance (or transmittance) of the sample brought about by this perturbation generally give rise to sharp third-derivative structures located at critical points in the joint density of states function. A detailed theoretical treatment is outside the scope of this chapter – the interested reader is referred to reviews by Aspnes [151, 152]. For the present purposes, the most important critical point is located at the bandgap energy.

The dependences of EER and EEA spectra on photon energy, E, are generally fitted to a generalized third-derivative fit function given by Aspnes:

$$\frac{\Delta R}{R} \text{ or } \frac{\Delta T}{T} = \text{Re}[Ce i^{\theta}(E - E_g + i\Gamma)] \tag{1.27}$$

where C, θ and Γ are amplitude, phase, and broadening parameters, respectively, and m is a half-integer value that depends on the type of optical transition.

Fitting EER/EEA spectra not only gives precise values of the bandgap but also provides information about the quality of samples. A large value of the broadening parameter Γ can point towards structural or compositional disorder in the sample. An example of the effects of crystallinity on EEA spectra is shown in Figure 1.40. The broad spectrum measured for the air-annealed CBD CdS film on FTO is consistent with the fact that the film is made up of nanocrystallites with sizes of the order of 10 nm [142]. When the film is recrystallized by heating with $CdCl_2$, the spectrum becomes much sharper and the bandgap moves to higher energies. This is consistent with a color change from orange to yellow. In fact in the example shown, the value of Γ for the recrystallized film is actually smaller than the spectral resolution of the monochromator. The EEA spectra show clearly that the recrystallization produces material of high quality.

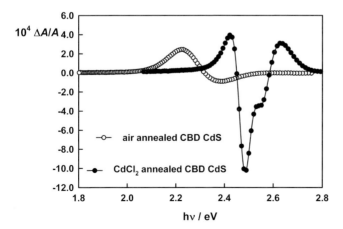

Figure 1.40 Comparison of EEA spectra for air-annealed CdS film on FTO and $CdCl_2$-annealed CdS film showing the effect of recrystallization on the broadening parameter.

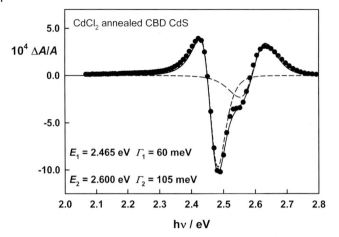

Figure 1.41 Fit of the EEA spectrum of CdCl$_2$-annealed CdS layer on FTO glass to two transitions with the bandgap and broadening parameters shown.

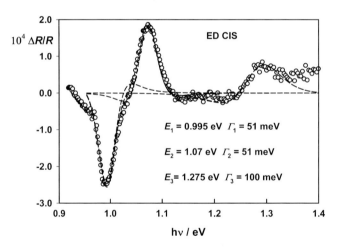

Figure 1.42 Fit of the EER spectrum of an annealed electrodeposited CuInSe$_2$ film.

The fitting of EEA spectra is illustrated in Figure 1.41 for the case of a CdCl$_2$-annealed CdS film. The figure shows that the more complex line shape can be fitted to two transitions at 2.465 and 2.600 eV. Figure 1.42 shows a similar fit of the EER spectrum of a CIS layer, in this case for three optical transitions.

The determination of bandgaps by EER or EEA is illustrated in Figure 1.43, which contrasts the EER spectra measured for ED and SP CIS samples (the spectrum for the sputtered layer is more noisy since the film is rough, scattering a considerable faction of the incident light). The fit for the electrodeposited sample gives a bandgap of 0.99 eV and a broadening parameter of 65 meV. This contrasts with the fit for the sputtered layer, which gives E_g = 1.00 eV and Γ = 80 meV.

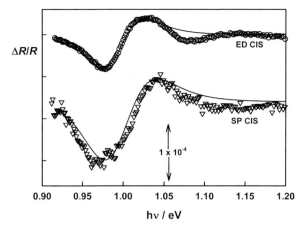

Figure 1.43 EER spectra of electrodeposited (ED) and sputtered (SP) CIS films.

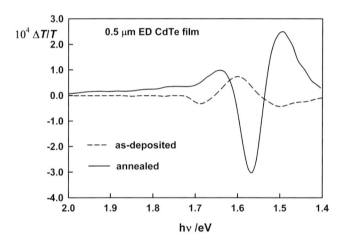

Figure 1.44 EEA spectra of as-deposited and air-annealed CdTe layers on CdS/FTO substrates. The inversion in the peaks is due to conversion of the CdTe from n-type to p-type during annealing.

Shirakata et al. [153] have reported EER spectra for single-crystal CIS, and their work indicates that the bandgap depends on the Cu/In ratio in the material, so the difference in E_g values between the ED and SP samples may be significant.

EEA has also been used to follow type conversion in electrodeposited CdTe layers [147]. The change from n-type to p-type is evident in Figure 1.44 as an inversion in the peaks located at the bandgap energy, and a more detailed analysis has shown that the bandgap of the CdTe decreases slightly as a consequence of diffusion of sulfur into the CdTe from the CdS layer to form $CdTe_{1-x}S_x$, where $x = 0.05–0.07$.

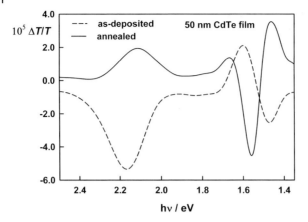

Figure 1.45 EEA spectra of 50 nm CdTe layer on 80 nm CdS film showing effect of air annealing. The structure in the 2.0–2.4 eV range arises from the CdS layer. The lower energy structure is from the CdTe. Note the inversion of both regions brought about by type conversion.

EEA measurements on ultra-thin (50 nm) CdTe layers of electrodeposited CdTe on CdS/FTO show a well-defined response from the CdS layer as well as the CdTe [147], as shown in Figure 1.45. In this case the CdS response is retained after air annealing, indicating that the electric field penetrates into the CdS. This contrasts with the lack of response seen when thicker CdTe films are grown on top of the CdS and air annealed. This suggests that a very thin CdTe film does not supply sufficient Te to diffuse into the CdS during annealing to make it photoinactive.

1.4
Conclusions

This chapter aims to illustrate the considerable potential of electrochemistry as a tool that can be used for the preparation and characterization of materials for thin-film solar cells. Electrochemical methods are not as well established in semiconductor science as they could be. Compared with the industrial importance of electrodeposition of metals for magnetic storage devices and device contacts, the electrodeposition of semiconductors and semiconductor precursors is still underdeveloped. However, the range of relatively sophisticated experimental techniques available to the electrochemist provides a sound basis for further progress. The authors believe that the next few years could see electrochemistry taking a more central position as an enabling technology that will impact on large-scale deployment of photovoltaics as a response to the looming challenges posed by global climate change. The search for new and sustainable materials and fabrication technologies is particularly relevant in this context, and the authors hope that this chapter will stimulate speculation about further applications of electrochemistry to address real-world problems.

Acknowledgments

The authors would like to thank Florencia Balbastro, Dominik Berg, Johannes Fischer, Masato Kurihara, Jonathan Scragg, and Marc Steichen for their help with the preparation of the manuscript.

References

1. The Economics of Climate Change. Nicholas Stern. Cabinet Office – HM Treasury 2007. ISBN-13: 9780521700801.
2. Beach, J.D. and McCandless, B.E. (2007) *MRS Bull.*, **32**, 225.
3. Feltrin, A. and Freundlich, A. (2008) *Renewable Energy*, **33**, 180.
4. Lincot, D. (2005) *Thin Solid Films*, **487**, 40.
5. Peter, L.M. and Tributsch, H. (2008) In *Nanostructured and Photoeletrochemical Systems for Solar Photon Conversion*. Editors M.D. Archer and A.J. Nozik. Imperial College Press, London.
6. Green, M.A., Emery, K., Hishikawa, Y., and Warta, W. (2009) *Prog. Photovoltaics*, **17**, 320.
7. Bar, M., Repins, I., Contreras, M.A., Weinhardt, L., Noufi, R., and Heske, C. (2009) *Appl. Phys. Lett.*, **95**, 3.
8. Kemell, M., Ritala, M., and Leskela, M. (2005) *Crit. Rev. Solid State Mater. Sci.*, **30**, 1.
9. Katagiri, H., Jimbo, K., Tahara, M., Araki, H., and Oishi, K. (2009) Proceedings of MRS 2009 Spring Meeting, San Francisco.
10. Katagiri, H., Jimbo, K., Yamada, S., Kamimura, T., Maw, W.S., Fukano, T., Ito, T., and Motohiro, T. (2008) *Appl. Phys. Express*, **1**, 041201.
11. Scragg, J.J., Berg, D.M., and Dale, P.J. (2010) *J. Electroanal. Chem.*, **646**, 52.
12. Schorr, S. (2007) *Thin Solid Films*, **515**, 5985.
13. Pauporte, T. and Lincot, D. (2000) *Electrochim. Acta*, **45**, 3345.
14. Miles, R.W., Zoppi, G., and Forbes, I. (2007) *Mater. Today*, **10**, 20.
15. Romeo, A., Terheggen, A., Abou-Ras, D., Batzner, D.L., Haug, F.J., Kalin, M., Rudmann, D., and Tiwari, A.N. (2004) *Prog. Photovoltaics*, **12**, 93.
16. Dale, P.J., Hoenes, K., Scragg, J., and Siebentritt, S. (2009) 34th IEEE Photovoltaic Specialists Conference, vols 1–3, 1956–61, Philadelphia.
17. Sze, S.M. (1985) *Semiconductor Devices Physics and Technology*, John Wiley & Sons, Inc., New York.
18. Wurfel, P. (2005) *Physics of Solar Cells*, Wiley-VCH Verlag GmbH, Weinheim.
19. Pandey, R.K. (1996) *Handbook of Semiconductor Electrodeposition*, 1st edn, CRC.
20. Gregory, B.W. and Stickney, J.L. (1991) *J. Electroanal. Chem.*, **300**, 543.
21. Muthuvel, M. and Stickney, J.L. (2006) *Langmuir*, **22**, 5504.
22. Suggs, D.W., Villegas, I., Gregory, B.W., and Stickney, J.L. (1992) *J. Vac. Sci. Technol. A*, **10**, 886.
23. Dale, P.J., Samantilleke, A.P., Zoppi, G., Forbes, I., and Peter, L.M. (2008) *J. Phys. D*, **41**, 085105.
24. Panicker, M.P.R., Knaster, M., and Kroger, F.A. (1978) *J. Electrochem. Soc.*, **125**, 566.
25. Dennison, S. (1993) *Electrochim. Acta*, **38**, 2395.
26. Kashyout, A.B., Arico, A.S., Monforte, G., Crea, F., Antonucci, V., and Giordano, N. (1995) *Solar Energy Mater. Solar Cells*, **37**, 43.

27 Dale, P.J., Samantilleke, A.P., Shivagan, D.D., and Peter, L.M. (2007) *Thin Solid Films*, **515**, 5751.
28 Bhattacharya, R.N., Fernandez, A., Contreras, M., Keane, J., Tennant, A.L., Ramanathan, K., Tuttle, J.R., Noufi, R., and Hermann, A.M. (1996) *J. Electrochem. Soc.*, **143**, 854.
29 Kroger, F.A. (1978) *J. Electrochem. Soc.*, **125**, 2028.
30 Lincot, D., Guillemoles, J.F., Taunier, S., Guimard, D., Sicx-Kurdi, J., Chaumont, A., Roussel, O., Ramdani, O., Hubert, C., Fauvarque, J.P., Bodereau, N., Parissi, L., Panheleux, P., Fanouillere, P., Naghavi, N., Grand, P.P., Benfarah, M., Mogensen, P., and Kerrec, O. (2004) *Solar Energy*, **77**, 725.
31 Verma, S., Varrin, R.D., Birkmire, R.W., and Russell, T.W.F. (1991) Conference Record of the 22nd IEE Photovoltaic Specialists Conference, Vols 1 and 2.
32 Zank, J., Mehlin, M., and Fritz, H.P. (1996) *Thin Solid Films*, **286**, 259.
33 Kampmann, A., Rechid, J., Raitzig, A., Wulff, S., Mihhailova, R., Thyen, R., and Kalberlah, K. (2003) Proceeding of the MRS Spring Meeting, San Francisco.
34 Voss, T., Schulze, J., Kirbs, A., Palm, J., Probst, V., Jost, S., Hock, R., and Purwins, M. (2007) 22nd European Photovoltaic Solar Energy Conference, 3–7 September 2007, Milan.
35 Scragg, J.J., Dale, P.J., and Peter, L.M. (2008) *Electrochem. Commun.*, **10**, 639.
36 Scragg, J.J., Dale, P.J., and Peter, L.M. (2009) *Thin Solid Films*, **517**, 2481.
37 Scragg, J.J., Dale, P.J., Peter, L.M., Zoppi, G., and Forbes, I. (2008) *Phys. Stat. Sol. (b)*, **245**, 1772.
38 Reference deleted.
39 Araki, H., Kubo, Y., Mikaduki, A., Jimbo, K., Maw, W.S., Katagiri, H., Yamazaki, M., Oishi, K., and Takeuchi, A. (2009) *Solar Energy Mater. Solar Cells*, **93**, 996.
40 Reference deleted.
41 Guillen, C. and Herrero, J. (1996) *J. Electrochem. Soc.*, **143**, 493–8.
42 Fischer, J., Dale, P.J., and Siebentritt, S. (2009) 216th Electrochemical Society, Vienna.
43 Basol, B.M. and Tseng, E.S.F. (1983) *J. Electrochem. Soc.*, **130**, C243.
44 Calixto, M.E., Dobson, K.D., McCandless, B.E., and Birkmire, R.W. (2005) Conference Record of the 31st IEEE Photovoltaic Specialists Conference, p. 378.
45 Miller, B., Menezes, S., and Heller, A. (1978) *J. Electroanal. Chem.*, **94**, 85.
46 Menezes, S., Miller, B., and Heller, A. (1979) *J. Electrochem. Soc.*, **126**, C376.
47 Peter, L.M. (1979) *J. Electroanal. Chem.*, **98**, 49.
48 Miller, B. and Heller, A. (1976) *Nature*, **262**, 680.
49 Peter, L.M. (1978) *Electrochim. Acta*, **23**, 1073.
50 Peter, L.M. (1978) *Electrochim. Acta*, **23**, 165.
51 Riveros, G., Guillemoles, J.F., Lincot, D., Meier, H.G., Froment, M., Bernard, M.C., and Cortes, R. (2002) *Adv. Mater.*, **14**, 1286.
52 Baranski, A.S., Bennett, M., and Fawcett, W.R. (1983) *J. Electrochem. Soc.*, **130**, C110.
53 Fatas, E., Duo, R., Herrasti, P., Arjona, F., and Garciacamarero, E. (1984) *J. Electrochem. Soc.*, **131**, 2243.
54 Freyland, W., Zell, C.A., Abedin, S.Z.E., and Endres, F. (2003) *Electrochim. Acta*, **48**, 3053.
55 Gregory, B.W., Suggs, D.W., and Stickney, J.L. (1991) *J. Electrochem. Soc.*, **138**, 1279.
56 Basol, B.M., Ou, S.S., and Stafsudd, O.M. (1984) *J. Appl. Phys.*, **58**, 3809.
57 Rajeshwar, K., and Bhattacharya, R.N. (1984) *J. Electrochem. Soc.*, **131**, C314.
58 Rajeshwar, K., Bhattacharya, R.N., and Ho, S.I. (1984) *J. Electrochem. Soc.*, **131**, C313.
59 Bhattacharya, R.N. and Rajeshwar, K. (1985) *J. Appl. Phys.*, **58**, 3590.
60 Bhattacharya, R.N., Rajeshwar, K., and Noufi, R.N. (1985) *J. Electrochem. Soc.*, **132**, 732.
61 Turner, A.K., Woodcock, J.M., Ozsan, M.E., Cunningham, D.W., Johnson, D.R., Marshall, R.J., Mason, N.B., Oktik, S., Patterson, M.H., Ransome,

S.J., Roberts, S., Sadeghi, M., Sherborne, J.M., Sivapathasundaram, D., and Walls, I.A. (1994) *Solar Energy Mater. Solar Cells*, **35**, 263.

62. Cunningham, D., Rubcich, M., and Skinner, D. (2002) *Prog. Photovoltaics*, **10**, 159.

63. Saraby-Reintjes, A., Peter, L.M., Ozsan, M.E., Dennison, S., and Webster, S. (1993) *J. Electrochem. Soc.*, **140**, 2880.

64. Mori, E. and Rajeshwar, K. (1989) *J. Electroanal. Chem.*, **258**, 415.

65. Danaher, W.J. and Lyons, L.E. (1984) *Aust. J. Chem.*, **37**, 689.

66. Engelken, R.D. and Vandoren, T.P. (1985) *J. Electrochem. Soc.*, **132**, 2904.

67. Sella, C., Boncorps, P., and Vedel, J. (1986) *J. Electrochem. Soc.*, **133**, 2043.

68. Colletti, L.P. and Stickney, J.L. (1998) *J. Electrochem. Soc.*, **145**, 3594.

69. Stickney, J.L. (1999) *Electroanal. Chem.*, **21**, 75.

70. Cowache, P., Lincot, D., and Vedel, J. (1989) *J. Electrochem. Soc.*, **136**, 1646.

71. Lincot, D., Kampmann, A., Mokili, B., Vedel, J., Cortes, R., and Froment, M. (1995) *Appl. Phys. Lett.*, **67**, 2355.

72. Wagner, S., Shay, J.L., Migliorato, P., and Kasper, H.M. (1974) *Appl. Phys. Lett.*, **25**, 434.

73. Bhattacharya, R.N. (1983) *J. Electrochem. Soc.*, **130**, 2040.

74. Hodes, G., Engelhard, T., Cahen, D., Kazmerski, L.L., and Herrington, C.R. (1985) *Thin Solid Films*, **128**, 93.

75. Kapur, V.K., Basol, B.M., and Tseng, E.S. (1987) *Solar Cells*, **21**, 65–72.

76. Kessler, J., Sicx-Kurdi, J., Naghavi, N., Guillemoles, J.F., Lincot, D., Kerrec, O., Lamirand, M., Legras, L., and Mogensen, P. (2005) 20th European Photovoltaic Solar Energy Conference, 6–10 June 2005, Barcelona.

77. Guillemoles, J.F., Lusson, A., Cowache, P., Massaccesi, S., Vedel, J. and Lincot, D. (1994) *Adv. Mater.*, **6**, 376–9.

78. Chassaing, E., Grand, P.-P., Etcheberry, A., and Lincot, D. (2009) *ECS Trans.*, **19**, 149.

79. Calixto, M.E., Sebastian, P.J., Bhattacharya, R.N., and Noufi, R. (1999) *Solar Energy Mater. Solar Cells*, **59**, 75.

80. Probst, V., Palm, J., Visbeck, S., Niesen, T., Tolle, R., Lerchenberger, A., Wendl, M., Vogt, H., Calwer, H., Stetter, W., and Karg, F. (2006) *Solar Energy Mater. Solar Cells*, **90**, 3115.

81. Palm, J., Probst, V., Brummer, A., Stetter, W., Tolle, R., Niesen, T.P., Visbeck, S., Hernandez, O., Wendl, M., Vogt, H., Calwer, H., Freienstein, B., and Karg, F. (2003) *Thin Solid Films*, **431**, 514.

82. Pourbaix, M. (1974) *Atlas of Electrochemical Equilibria in Aqueous Solutions*, NACE International Cebelcor.

83. Penndorf, J., Winkler, M., Tober, O., Roser, D., and Jacobs, K. (1998) *Solar Energy Mater. Solar Cells*, **53**, 285.

84. Barbosa, L.L., de Almeida, M.R.H., Carlos, R.M., Yonashiro, M., Oliveira, G.M., and Carlos, I.A. (2005) *Surf. Coat. Technol.*, **192**, 145.

85. Walsh, F.C. and Gabe, D.R. (1979) *Surface Technology*, **8**, 87–99.

86. Kampmann, A., Rechid, J., Wulff, S., Mihhailova, R., Thyen, R., and Lossin, A. (2004) 19th European Photovoltaic Solar Energy Conference, Paris, p. 1806.

87. Jost, S., Hergert, F., Hock, R., Schulze, J., Kirbs, A., Voss, T., and Purwins, M. (2007) *Solar Energy Mater. Solar Cells*, **91**, 1669.

88. Jost, S., Hergert, F., Hock, R., Schulze, J., Kirbs, A., Voss, T., Purwins, M., and Schmid, M. (2007) *Solar Energy Mater. Solar Cells*, **91**, 636.

89. Jost, S., Hergert, F., Hock, R., Voss, T., Schulze, J., Kirbs, A., Purwins, M., Probst, V., and Palm, J. (2007) *Thin-Film Compound Semiconductor Photovoltaics–2007*, ed T. Gessert, et al. (Warrendale: Materials Research Society) pp 175–80.

90. Jost, S., Schurr, R., Hergert, F., Hock, R., Schulze, J., Kirbs, A., Voss, T., Purwins, M., Palm, J., and Mys, I. (2008) *Solar Energy Mater. Solar Cells*, **92**, 410.

91. Siemer, K., Klaer, J., Luck, I., Bruns, J., Klenk, R., and Braunig, D. (2001) *Solar Energy Mater. Solar Cells*, **67**, 159.

92. Friedfeld, R., Raffaelle, R.P., and Mantovani, J.G. (1999) *Solar Energy Mater. Solar Cells*, **58**, 375.

93. Kois, J., Ganchev, M., Kaelin, M., Bereznev, S., Tzvetkova, E., Volobujeva,

O., Stratieva, N., and Tiwari, A.N. (2008) *Thin Solid Films*, **516**, 5948.
94 Massaccesi, S., Sanchez, S., and Vedel, J. (1993) *J. Electrochem. Soc.*, **140**, 2540–6.
95 Lippkow, D. and Strehblow, H.H. (1998) *Electrochim. Acta*, **43**, 2131.
96 Massaccesi, S., Sanchez, S., and Vedel, J. (1996) *J. Electroanal. Chem.*, **412**, 95.
97 Aksu, S., Wang, J.X., and Basol, B.M. (2009) *Electrochem. Solid State Lett.*, **12**, D33.
98 Kim, S., Kim, W.K., Kaczynski, R.M., Acher, R.D., Yoon, S., Anderson, T.J., Crisalle, O.D., Payzant, E.A., and Li, S.S. (2005) *J. Vac. Sci. Technol. A*, **23**, 310.
99 Kim, W.K., Kim, S., Payzant, E.A., Speakman, S.A., Yoon, S., Kaczynski, R.M., Acher, R.D., Anderson, T.J., Crisalle, O.D., Li, S.S., and Craciun, V. (2005) *J. Phys. Chem. Solids*, **66**, 1915.
100 Cahen, D. and Noufi, R. (1992) *J. Phys. Chem. Solids*, **53**, 991.
101 Hermann, A.M., Mansour, M., Badri, V., Pinkhasov, B., Gonzales, C., Fickett, F., Calixto, M.E., Sebastian, P.J., Marshall, C.H., and Gillespie, T.J. (2000) *Thin Solid Films*, **361**, 74.
102 Hermann, A.M., Westfall, R., and Wind, R. (1998) *Solar Energy Mater. Solar Cells*, **52**, 355.
103 Thouin, L., Sanchez, S., and Vedel, J. (1993) *Electrochim. Acta*, **38**, 2387.
104 Mishra, K.K. and Rajeshwar, K. (1989) *J. Electroanal. Chem.*, **271**, 279.
105 Thouin, L., Massaccesi, S., Sanchez, S., and Vedel, J. (1994) *J. Electroanal. Chem.*, **374**, 81.
106 Thouin, L. and Vedel, J. (1995) *J. Electrochem. Soc.*, **142**, 2996.
107 Chassaing, E., Grand, P.-P., Saucedo, E., Etcheberry, A., and Lincot, D. (2009) *ECS Trans.*, **19**, 189.
108 Chassaing, E., Guillemoles, J.-F., and Lincot, D. (2009) *ECS Trans.*, **19**, 1.
109 Chassaing, E., Ramdani, O., Grand, P.P., Guillemoles, J.F., and Lincot, D. (2008) *Phys. Stat. Sol. (c)*, **5**, 3445.
110 Calixto, M.E., Dobson, K.D., McCandless, B.E., and Birkmire, R.W. (2006) *J. Electrochem. Soc.*, **153**, G521.
111 Bhattacharya, R.N., Batchelor, W., Granata, J.E., Hasoon, F., Wiesner, H., Ramanathan, K., Keane, J., and Noufi, R.N. (1998) *Solar Energy Mater. Solar Cells*, **55**, 83.
112 Kemell, M., Ritala, M., and Leskela, M. (2001) *J. Mater. Chem.*, **11**, 668.
113 Kemell, M., Saloniemi, H., Ritala, M., and Leskela, M. (2001) *J. Electrochem. Soc.*, **148**, C110.
114 Nomura, S., Nishiyama, K., Tanaka, K., Sakakibara, M., Ohtsubo, M., Furutani, N., and Endo, S. (1998) *Jpn. J. Appl. Phys. Part 1*, **37**, 3232.
115 Ito, K. and Nakazawa, T. (1988) *Jpn. J. Appl. Phys. Part 1*, **27**, 2094.
116 Dale, P.J., Hoenes, K., Scragg, J., and Siebentritt, S. (2009) 34th IEEE Photovoltaic Specialists Conference, Vols 1–3, (New York: IEEE) pp 1956–61.
117 Katagiri, H. (2005) *Thin Solid Films*, **480–481**, 426.
118 Schubert, B., Marsen, B., Cinque, S., Unold, T., Klenk, R., Schorr, S., and Schock, H.W. Progress in Photovoltaics: Research and Applications DOI: 10.1002/pip.976
119 Araki, H., Kubo, Y., Mikaduki, A., Jimbo, K., Maw, W.S., Katagiri, H., Yamazaki, M., Oishi, K., and Takeuchi, A. (2007) 17th International Photovoltaic Science and Engineering Conference, Fukuoka, p. 996.
120 Araki, H., Kubo, Y., Jimbo, K., Maw, W.S., Katagiri, H., Yamazaki, M., Oishi, K., and Takeuchi, A. (2009) *Physica Status Solidi (C) Current Topics in Solid State Physics*, **6**, 1266–8.
121 Ennaoui, A., Lux-Steiner, M., Weber, A., Abou-Ras, D., Kötschau, I., Schock, H.W., Schurr, R., Hölzing, A., Jost, S., Hock, R., Voß, T., Schulze, J., and Kirbs, A. (2009) *Thin Solid Films*, **517**, 2511.
122 Kurihara, M., Berg, D., Fischer, J., Siebentritt, S., and Dale, P.J. (2009) *Phys. Stat. Sol. (c)*, **6**, 1241.
123 Broggi, R.L., Oliveira, G.M.D., Barbosa, L.L., Pallone, E., and Carlos, I.A. (2006) *J. Appl. Electrochem.*, **36**, 403.
124 Araki, H., Mikaduki, A., Kubo, Y., Sato, T., Jimbo, K., Maw, W.S., Katagiri, H., Yamazaki, M., Oishi, K., and Takeuchi, A. (2008) *Thin Solid Films*, **517**, 1457.

125 Hergert, F. and Hock, R. (2007) *Thin Solid Films*, **515**, 5953.
126 Schurr, R., Holzing, A., Jost, S., Hock, R., Voss, T., Schulze, J., Kirbs, A., Ennaoui, A., Lux-Steiner, M., Weber, A., Kotschau, I., and Schock, H.W. (2008) Symposium on Thin Film Chalcogenide Photovoltaic Materials held at the EMRS 2008 Spring Conference, Strasbourg, p. 2465.
127 Chou, C.Y. and Chen, S.W. (2006) *Acta Mater.*, **54**, 2393.
128 Olekseyuk, I.D., Dudchak, I.V., and Piskach, L.V. (2004) *J. Alloys Compd*, **368**, 135.
129 Abedin, S.Z.E. and Endres, F. (2006) *ChemPhysChem*, **7**, 58.
130 Shivagan, D.D., Dale, P.J., Samantilleke, A.P., and Peter, L.M. (2007) *Thin Solid Films*, **515**, 5899.
131 Abedin, S.Z.E., Saad, A.Y., Farag, H.K., Borisenko, N., Liu, Q.X., and Endres, F. (2007) *Electrochim. Acta*, **52**, 2746.
132 Gasparotto, L.H.S., Borisenko, N., Hofft, O., Al-Salman, R., Maus-Friedrichs, W., Bocchi, N., Abedin, S.Z.E., and Endres, F. (2009) *Electrochim. Acta*, **55**, 218.
133 Yang, M.H., Yang, M.C., and Sun, I.W. (2003) *J. Electrochem. Soc.*, **150**, C544.
134 Hsiu, S.I. and Sun, I.W. (2004) *J. Appl. Electrochem.*, **34**, 1057.
135 Pleskov, Y.V. and Gurevich, Y.Y. (1985) *Semiconductor Photoelectrochemistry*, Consultants Bureau, New York.
136 Sato, N. (1998) *Electrochemistry at Metal and Semiconductor Electrodes*, Elsevier, Amsterdam.
137 Memming, R. (2001) *Semiconductor Electrochemistry*, Wiley-VCH Verlag GmbH, Weinheim.
138 Licht, S., Bard, A.J., and Stratmann, M. (2002) *Encyclopedia of Electrochemistry, Volume 6: Semiconductor Electrodes and Photoelectrochemistry*, Wiley-VCH Verlag GmbH, Weinheim.
139 Schroder, D.K. (1998) *Semiconductor Material and Device Characterization*, 2nd edn, John Wiley & Sons, Inc., New York.
140 Pleskov, Y. (1990) *Solar Energy Conversion. A Photoelectrochemical Approach*, Springer, Berlin.
141 Scragg, J.J., Dale, P.J., Peter, L.M., Zoppi, G., and Forbes, I. (2008) *Phys. Stat. Sol. (b)*, **245**, 1772.
142 Ozsan, M.E., Johnson, D.R., Sadeghi, M., Sivapathasundaram, D., Goodlet, G., Furlong, M.J., Peter, L.M., and Shingleton, A.A. (1996) *J. Mater. Sci. Mater. Electron.*, **7**, 119.
143 Basol, B.M. (1984) *J. Appl. Phys.*, **55**, 601.
144 Kampmann, A., Cowache, P., Vedel, J., and Lincot, D. (1995) *J. Electroanal. Chem.*, **387**, 53.
145 Duffy, N.W., Lane, D., Ozsan, M.E., Peter, L.M., Rogers, K.D., and Wang, R.L. (2000) *Thin Solid Films*, **36**, 314.
146 Duffy, N.W., Peter, L.M., Wang, R.L., Lane, D.W., and Rogers, K.D. (2000) *Electrochim. Acta*, **45**, 3355.
147 Duffy, N.W., Peter, L.M., and Wang, R.L. (2002) *J. Electroanal. Chem.*, **532**, 207.
148 Gartner, W.W. (1959) *Phys. Rev.*, **116**, 84.
149 Tanaka, T., Nagatomo, T., Kawasaki, D., Nishio, M., Guo, Q.X., Wakahara, A., Yoshida, A., and Ogawa, H. (2005) *J. Phys. Chem. Solids*, **66**, 1978.
150 Kobayashi, T., Jimbo, K., Tsuchida, K., Shinoda, S., Oyanagi, T., and Katagiri, H. (2005) *Jpn. J. Appl. Phys.*, **44**, 783.
151 Aspnes, D.E. (1973) *Surf. Sci.*, **37**, 418.
152 Aspnes, D.E. (1980) *Handbook on Semiconductors*, vol. 2 (ed. M. Balkanski), North Holland, New York, p. 109.
153 Shirakata, S., Chichibu, S., Isomura, S., and Nakanishi, K. (1997) *Jpn. J. Appl. Phys. Part 2*, **36**, L543.

2
Tailoring of Interfaces for the Photoelectrochemical Conversion of Solar Energy

Hans Joachim Lewerenz

2.1
Introduction

In the search for terrestrially applicable carbon-neutral energy sources, photoelectrochemical systems exhibit several advantageous features: examples are the facilitated junction formation at the electrolyte interface, the scalability typically achieved by self-organized processes at the solid–liquid boundary [1, 2], low-temperature processing, and the *in situ* preparation and optimization of specific interface conditions [3, 4]. In addition, photoelectrochemical solar cells can operate in the photovoltaic [5–9] as well as in the photoelectrocatalytic [10–14] mode. The latter is of particular importance because global energy consumption is clearly dominated by the use of fuels compared to electricity [15]. Due to stability issues when operating a solar cell at a reactive phase boundary, such as a semiconductor–electrolyte contact, photovoltaic electrochemical solar cells have technologically not been realized despite the existence of systems that have demonstrated extended stability under operation [4, 7]. As limited *thermodynamic* stability is also a factor for solar fuel generating devices, a general strategy is needed for the future implementation of photoelectrochemical solar energy conversion systems.

At the present stage of the field, a multifaceted approach appears mandatory which focuses on fundamental research regarding charge and excitation energy transfer [16–18], materials development, particularly in conjunction with the developments in nanoscience [19–23], and control of interface behavior [24–26]. This latter aspect will be the focus of the present chapter.

Besides the investigation of fundamental properties of the solid–liquid interface, the situation with a rapidly deteriorating atmospheric situation [27] also demands the use of *rational screening* methods for the identification and examination of novel photoactive materials, material combinations, and composites [28–30]. It is obvious that a large-scale and international research and development effort is needed and first programs have been inaugurated[1),2)]. It is worth emphasizing that selection criteria regarding materials and systems should be considered carefully, in particular regarding cost arguments. As long as energy supply is subjected to the global market economy and rather deregulated financial instruments, systems could be realized that promise the largest monetary gain, being thus rather independent of the original costs of the respective devices. It is therefore important to explore the catalog of options and to *not* prematurely exclude materials, processes, and structures. In addition, the development of specific architectures, for instance in the light-harvesting process, might allow the use of scarce materials if only traces are needed for operation. A further aspect is to create mechanisms for the recycling of more expensive materials, as has been successful in photography where silver is routinely recovered.

2.2
Operation Principles of Photoelectrochemical Devices

2.2.1
Currents, Excess Carrier Profiles, and Quasi-Fermi Levels

2.2.1.1 Dark Current and Photocurrent

The semiconductor–redox electrolyte contact is characterized by the equilibration of the semiconductor Fermi level with the redox energy of the solution [31] and forms a rectifying contact [32]. The dark current–voltage behavior closely follows that of a solid-state diode (Schottky or p–n junction). In solid-state physics, the reverse saturation current j_0 differs between a Schottky diode where thermal emission of electrons is described by the effective Richardson constant [33] and the p–n junction where the well-known Shockley relation holds [34]. In (photo)electrochemistry, the current j_0 is determined by the Marcus–Gerischer distribution of the density of states of the oxidized and reduced species in solution [35, 36], the reorganization energy [37], the surface concentration of electrons (holes), as well as the concentration of oxidized (reduced) redox species and a factor that contains the transmission coefficient for outer sphere charge transfer [38]. Figure 2.1 shows an energy band schematic of a semiconductor–electrolyte junction in equilibrium and for a cathodic applied voltage. The resulting flattening of the conduction/valence band of the n-type semiconductor increases exponentially the electron concentration at the surface (within the limits of the Boltzmann approximation [39]) and the dark current–voltage behavior is expressed by

$$j_D = j_0(e^{eV/kT} - 1) \tag{2.1}$$

1) A large program: *Energy Frontier Research Centers* has been introduced in the USA with topical programs on solar fuels and photovoltaics.

2) A new funding agency, *Advanced Research Projects Agency-Energy* (ARPA-E) has been founded in the USA.

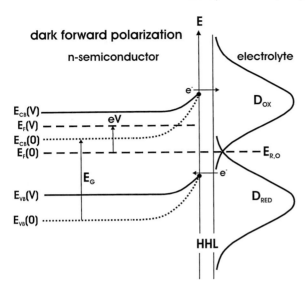

Figure 2.1 Energy band schematic for an n-type semiconductor–redox electrolyte junction. Dotted lines indicate conduction and valence band for the situation of equilibration of semiconductor Fermi level E_F and solution redox energy $E_{R,O}$; solid lines indicate the semiconductor bands when a cathodic potential V is applied, resulting in a flattening of the band bending and an exponentially increased surface concentration of electrons, n_s; $E_{CB,VB}$, conduction and valence band in contact; $E_{CB,VB}(V)$, bands for applied voltage V; D_{OX}, D_{RED}, density of states of the oxidized and reduced redox species; HHL, outer Helmholtz layer; E_G, semiconductor energy gap.

where V is the applied voltage and

$$j_0 = \text{const} \cdot f(\nu_{CB,VB}, T) n_s c_{R,O} \sqrt{\frac{1}{\lambda}} \exp\left[-\frac{(E_{CB,VB} - E_{R,O})^2}{4\lambda kT}\right] \tag{2.2}$$

Here λ is the reorganization energy, $c_{R,O}$ is the concentration of oxidized/reduced redox species, $n_s = n_s(V)$ is the surface concentration of electrons upon forward biasing of the semiconductor, ν is a frequency factor and $E_{CB,VB}$ and $E_{R,O}$ are the energies of the valence/conduction band and of the redox couple, respectively. In the case of illumination with light for which the photon energy exceeds the bandgap of the absorber ($h\nu > E_g$), a current of opposite sign with respect to the dark current flows as indicated in Figure 2.2. In the most simple approximation, the light-induced current j_L can be expressed by the number of absorbed photons multiplied by the elementary charge (internal quantum yield 1): $j_L = en_{Ph}(1 - R)$.

An expression where the current depends on materials parameters such as absorption, doping, and minority carrier diffusion length can be derived by considering the light-induced currents that originate from the semiconductor depletion region (drift current) and from the neutral region (diffusion current) [40]. An analytical expression is obtained by integrating the excess carrier generation rate

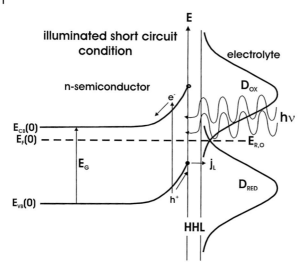

Figure 2.2 Energy band schematic for an illuminated n-type semiconductor under short circuit condition for example, operating at the maximum contact potential difference between semiconductor and redox electrolyte; j_L, light-induced current (see text).

in the space charge layer and by using the continuity equation with a diffusion term, carrier generation, and bulk recombination for the calculation of the diffusion current. With appropriate boundary conditions ($j = 0$ at $x = W$; see Figure 2.3), one obtains the total light-induced current with a superimposed thermal generation current (see Figure 2.3; Equation 2.5). The drift current is given by

$$j_{dr} = eI_0(1 - e^{-\alpha W}) \tag{2.3}$$

and the diffusion term yields

$$j_{diff} = -eI_0 \frac{\alpha L_p}{1 + \alpha L_p} e^{-\alpha W} \tag{2.4}$$

The total current, including the thermally generated part becomes

$$j_L = -eI_0 \left(1 - \frac{e^{-\alpha W}}{1 + \alpha L_p}\right) - \frac{ep_0 D_p}{L_p} \tag{2.5}$$

and a corresponding profile is shown also in Figure 2.3. As the model assumes instantaneous carrier transport for excitation in the depletion layer, the excess carrier profile is zero between the surface and $z = W$. The slope of $\Delta p(W)$ is a measure of the current from the diffusion region (see Equation 2.8). It should be noted that this model does not contain terms that describe surface recombination processes at the front and back contact or, for instance, carrier accumulation at the surface due to a slow charge transfer rate and simultaneously low surface recombination. Such features have been later included in attempts to describe spectral photocurrents at the semiconductor–electrolyte contact [41–44].

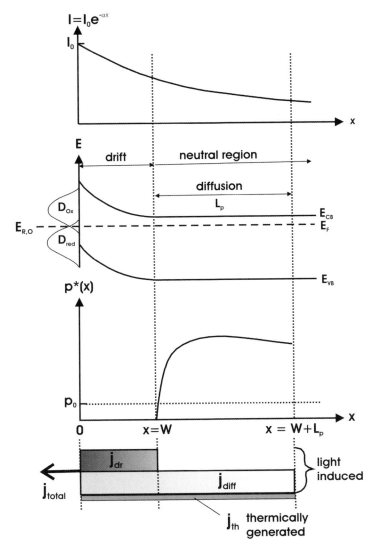

Figure 2.3 Absorption profile $I(x)$, band bending, excess carrier concentration profile, and current contributions according to the model of Gärtner for monochromatic light with absorption coefficient $\alpha(x)$; I_0, intensity of the impinging light; L_p, minority carrier diffusion length; W, depletion layer width; $p^*(x)$, minority carrier concentration in the illuminated semiconductor; p_0, dark concentration of minority carriers (holes) in the n-type semiconductor; D_{Ox}, D_{Red}, density of states for the oxidized and reduced redox species, respectively.

2.2.1.2 Excess Minority Carrier Profiles

In the following, an approach is used that has been developed for the quantitative description of combined stationary excess microwave conductivity experiments and photocurrent spectroscopy [45]. The model uses analytical expressions for the photocurrent and the stationary excess microwave conductivity which are obtained by integrating the transport equations for the diffusion region and the space charge layer. Here, the focus is on the photocurrent only.

In the *diffusion region*, the general statement for the continuity equation in the stationary case $dp^*(x)/dt = 0$ is

$$\alpha P_0 e^{-\alpha x} - \frac{D_h}{L_p^2}(p^*(x) - p_0) - \frac{q}{kT} D_p E_x \frac{dp^*(x)}{dx} + D_p \frac{d^2 p^*(x)}{dx^2} = 0 \tag{2.6}$$

where the terms describe carrier generation, recombination, drift due to the small electric field E_x during current flow, and diffusion, respectively. In Equation 2.6, the photon flux P_0 has been used instead of the light intensity I_0. They are related via $P_0 = I_0/qh\nu$. In the diffusion region, the solution for the excess carrier concentration $\Delta p(x) = p^*(x) - p_0$ becomes

$$\Delta p(x) = A^* \left[e^{-\alpha x} + e^{-x/L_p}\left(C + \frac{\Delta p(W)}{D} \right) + e^{x/L_p}\left(C' + \frac{\Delta p(W)}{D'} \right) \right] \tag{2.7}$$

In the expression for the excess carrier concentration at the boundary to the space charge layer, $\Delta p(W)$, the recombination velocity at the back contact, S_{rb}, is contained (see equations A8, A9 in [45].). The terms C, C', D, D' and the prefactor A^* contain expressions that are independent of the variable x. They are given in Appendix 2.A. Figure 2.4 shows examples of excess minority carrier profiles for a back contact recombination velocity S_{rb} of 10^7 cm s^{-1} which is in the range of the thermal velocity [46] for a set of minority carrier diffusion lengths that range from poor bulk mate-

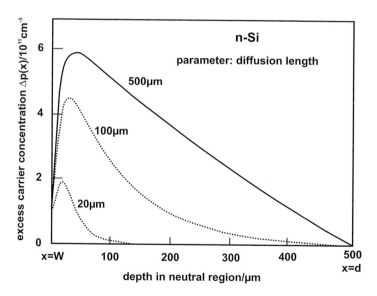

Figure 2.4 Light-induced excess minority carrier profiles for different diffusion lengths of n-Si; $\Delta p(W) = 10^{11}$ cm^{-3}; $P_0 = 6.25 \times 10^{15}$ cm^{-2} s^{-1}; $\alpha = 780$ cm^{-1}; $D_p = 11.65$ cm^2 s^{-1}; $d = 500\,\mu$m; $E_x = 0$; $S_{rb} = 10^7$ cm s^{-1}; solid line, $L_p = 500\,\mu$m; dashed line, $L_p = 100\,\mu$m; dotted line, $L_p = 20\,\mu$m.

rial properties ($L_p = 20\,\mu m$) to high-quality material where L_p reaches $500\,\mu m$. The excess carrier concentration at the boundary to the space charge layer, $x = W$, is typical for monochromatic irradiation [47]. As can be seen, for small diffusion length, for example reduced bulk lifetime, the profile exhibits a maximum near to the depletion layer boundary and $\Delta p(x)$ at this maximum is about twice as large as its value at $x = W$. The profile decreases rapidly over a short distance and excess carriers from regions beyond $100\,\mu m$ depth do not contribute to the current from the neutral region. For $L_p = 100\,\mu m$, the maximum of the distribution shifts to larger x values, peaking at about $30\,\mu m$ with subsequent exponential decrease. For the case where L_p equals the material thickness, bulk recombination does not contribute to the profile and $\Delta p(x)$ shows a linear decrease towards the back contact after having reached its maximum which is again shifted slightly towards higher x compared to the cases of $L_p = 20$ and $100\,\mu m$. The current in the space charge region is given by the superposition of the drift part and the diffusion component if current values similar to that of the drift component are reached:

$$j_L(x) = q\Delta p(x)\frac{q}{kT}D_p E_x - qD_p \frac{d\Delta p(x)}{dx} \tag{2.8}$$

Here, the Einstein relation between mobility and diffusion coefficient has been used. The increase of the slope of Δp at $x = W$ for larger L_p shows that a larger diffusion current enters the space charge region (for the calculation, $E_x = 0$ is assumed). It should be noted that the excess microwave reflectivity signal M_w is proportional to the spatial integral of the carrier profile [48]. The combination of photocurrent and M_w allows, for instance, the separate assessment of the charge transfer and surface recombination rate. This can be used to analyze the photoelectrochemical behavior of semiconductors with well-known solid-state properties, such as silicon [49], and will be presented in Section 2.2.4. In the situation shown in Figure 2.4 for $L_p = 20\,\mu m$, the small integral excess carrier profile and the small slope of $\Delta p(x)$ at $x = W$ (small current) show that, in this condition, the recombination processes dominate.

In Figure 2.5, the influence of the diffusion length on the charge collection efficiency, defined as current per number of photons $Q_c = j_L(W)/qP_0$, is plotted for three values of the excess minority concentration at $x = W$. For $\Delta p = 0$, which is the situation for the Gärtner model, Q_c reaches the largest values, and with increasing $\Delta p(W)$ the collection efficiency decreases. This is related to the reduced slope of $\Delta p(x)$ at the space charge layer boundary $x = W$ for larger values of $\Delta p(W)$. The decrease of Q_c with decreasing diffusion length shows the influence of bulk recombination.

In the *space charge region*, the excess carrier profile is determined from the transport equation (cf. Equation 2.8), the hole current entering the space charge region at $x = W$, $j_L(W)$, and the excess carrier generation rate integrated over the space charge region:

$$j_L(x) = q\Delta p(x)\frac{q}{kT}D_p E_x - qD_p \frac{d\Delta p(x)}{dx} = j_L(W) + qP_0(e^{-\alpha W} - e^{-\alpha x}) \tag{2.9}$$

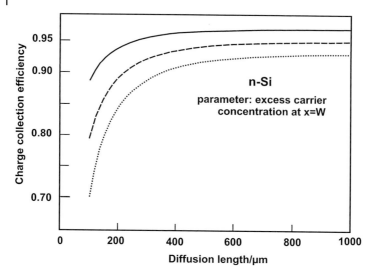

Figure 2.5 Charge collection efficiency Q_c as a function of the minority carrier diffusion length for three excess carrier concentrations at the space charge layer boundary; full line, $\Delta p(W) = 0$; dashed line, $\Delta p(W) = 5 \times 10^{11}\,\text{cm}^{-3}$; dotted line, $\Delta p(W) = 10^{12}\,\text{cm}^{-3}$; other parameters (except d) as for Figure 2.4.

Using the Poisson equation for expressing $E_x = f(x)$ in the depletion approximation gives

$$E_x = -\frac{kT}{qL_d^2}(W-x) \tag{2.10}$$

where L_d denotes the Debye length. Inserting Equation 2.10 into Equation 2.9 yields a differential equation for the excess carrier concentration $\Delta p(x)$ (e.g., see equation 42 in [45]). Assuming that (i) a notable space charge region exists with $W \geq 2L_d$ ($V \geq 0.05\,\text{V}$), (ii) significant excitation in the diffusion region occurs ($\alpha L_d \leq 1$), and (iii) a large diffusion length allows for current flow into the space charge region ($L_p \gg L_d$), a somewhat shortened expression for $\Delta p(x)$ is obtained that depends on the excess carrier concentration at the surface, $\Delta p(0)$:

$$\Delta p(x) = e^{s^2}[\Delta p(0)e^{-u^2} + ka(e^{-s^2}-1) + e^{-\beta'^2}(e^{-(s+\beta')^2}+1)] \tag{2.11}$$

The definitions for s, u, β', k, and a are given in Appendix 2.B. In the term $\Delta p(0)$, the surface recombination velocity S_r and the charge transfer rate k_r are contained (see Appendix 2.C), thus allowing one to calculate the excess carrier profile and the photocurrent as functions of the kinetic surface parameters S_r and k_r. An example is given in Figure 2.6 where the influence of the diffusion voltage in the space charge region on the spatial profile of the excess minority carrier concentration is shown. It has been assumed in the calculation that the sum of k_r and S_r is $10^2\,\text{cm}\,\text{s}^{-1}$. As surface recombination velocities and charge transfer rates can exceed values of $10^6\,\text{cm}\,\text{s}^{-1}$, this situation corresponds to a slow transfer rate with

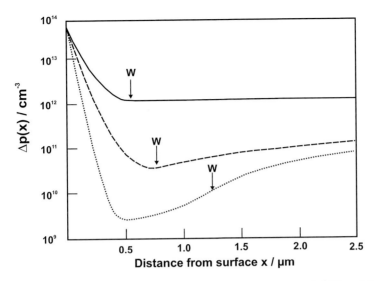

Figure 2.6 Spatial dependence of excess carrier concentration $\Delta p(x)$ for different diffusion voltages in the space charge region and for low excess carrier consumption at the surface ($x = 0$), expressed by $k_r + S_r = 100\,\text{cm s}^{-1}$; full line, 0.1 V; dashed line, 0.2 V; dotted line, 0.5 V; arrows indicate the extension of the space charge region; parameters as in Figures 2.4 and 2.5.

small surface recombination resulting in a considerable increase of the excess carrier concentration at the surface, as can be seen in Figure 2.6. In a linear plot of $\Delta p(x)$, not shown here, it can be seen that the slope of the carrier profile at $x = W$ becomes independent of the respective concentration $\Delta p(0)$ at the surface if the voltage drop in the space charge layer is large enough (here, larger than about 0.1 V).

2.2.1.3 Quasi-Fermi Levels

Fermi levels in solids and redox energies in electrolytes represent electrochemical potentials given by the chemical potential μ_i and a term that describes the Galvani potential φ related to the charge of the considered particles (electrons) [50, 51]:

$$\tilde{\mu}_i = \mu_i + e\varphi_i \tag{2.12}$$

with

$$\mu_i = kT \ln c_i^0 \tag{2.13}$$

In semiconductors, the Fermi level is given by the conduction or valence band edge, E_{CB} or E_{VB}, respectively, and by the effective density of states at the band edge, N_{CB} or N_{VB}, in relation to the carrier concentration in the bands, n and p:

$$E_F = E_{CB} - kT \ln \frac{N_{CB}}{n}; \quad E_F = E_{VB} - kT \ln \frac{N_{VB}}{p} \tag{2.14}$$

For stationary illumination and sufficiently long lifetimes (or diffusion lengths) of the photogenerated electrons and holes, excess minority and majority carriers exist that are stationary at the respective band edges in addition to the concentration values without illumination. This change in the total concentration of carriers alters the energy of the system (cf. Equation 2.13) and can be viewed as a new quasi-equilibrium situation. Attempts have been made to describe the situation in the terminology of the equilibrium between the carrier concentrations n and p in the dark and the Fermi level [52]:

$$n = N_{CB} \exp\left(-\frac{E_{CB} - E_F}{kT}\right) \tag{2.15}$$

$$p = N_{VB} \exp\left(-\frac{E_F - E_{VB}}{kT}\right) \tag{2.16}$$

The analogous expression for illuminated semiconductors is obtained by substituting $n \rightarrow n^*(x) = n + \Delta n(x)$, $p \rightarrow p^*(x) = p + \Delta p(x)$, and $E_F \rightarrow {}_nE_F^*(x), {}_pE_F^*(x)$, yielding expressions for the quasi-Fermi levels of electrons and holes in illuminated semiconductors:

$$n^*(x) = N_{CB} \exp\left(-\frac{E_{CB} - {}_nE_F^*(x)}{kT}\right) \tag{2.17}$$

$$p^*(x) = N_{VB} \exp\left(-\frac{{}_pE_F^*(x) - E_{VB}}{kT}\right) \tag{2.18}$$

The change of the energy of the system can be expressed by the change in the dark Fermi level:

$${}_nE_F^*(x) = E_F + kT \ln \frac{n^*(x)}{n} = E_F + kT \ln\left(1 + \frac{\Delta n(x)}{n}\right) \tag{2.19}$$

$${}_pE_F^*(x) = E_F - kT \ln \frac{p^*(x)}{p} = E_F - kT \ln\left(1 + \frac{\Delta p(x)}{p}\right) \tag{2.20}$$

For an n-type semiconductor, as considered above, the effect of illumination within the low injection limit [53] differs substantially between majority carriers and minority carriers. The mass action law in semiconductors, i.e., $n \times p = n_i^2$, where n_i is the intrinsic carrier concentration ($n_i = 1.45 \times 10^{10}\,cm^{-3}$ at 300 K for Si), yields a minority carrier concentration of $p = 2.1 \times 10^5\,cm^{-3}$ for a typical bulk doping level of $n = 10^{15}\,cm^{-3}$ for Si. With excess carrier concentrations in the range $\Delta n = \Delta p \approx 10^{13}$–$10^{14}\,cm^{-3}$ (cf. Figure 2.6), the relative change in concentration is large for the minority carriers, resulting in a pronounced deviation of the respective quasi-Fermi level from its energetic position without illumination. An example is given in Figure 2.7, where the spatial dependence of the quasi-Fermi levels for electrons and holes in an n-type semiconductor is shown. An increase of the minority carrier concentration from a bulk value of 10^5 to $10^{13}\,cm^{-3}$ under illumination results in a position of the quasi-Fermi level for holes 0.47 eV below the value in the dark, whereas the majority carrier quasi-Fermi level is shifted energetically

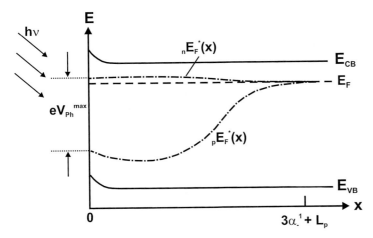

Figure 2.7 Schematic representation of the quasi Fermi levels of an illuminated n-type semiconductor with high absorption and low minority carrier diffusion length; α^{-1} denotes the absorption length; after $3\alpha^{-1}$, the light intensity at the surface, I_0, is reduced to 0.05I_0; the excess carrier profile $\Delta p(x)$ is shifted towards the interior of the semiconductor by the diffusion length and therefore differs from the generation profile ($G = \alpha I_0 \exp(-\alpha x)$).

upwards by 0.3 meV. Therefore, the change in $_nE_F^*(x)$ has been exaggerated for better visibility.

The spatial behavior shown in Figure 2.7 differs from that of the profile shown in Figure 2.6, for instance. In Figure 2.6, a minority carrier diffusion length has been assumed ($L_p = 500\,\mu m$) that equals the thickness of the absorber. Also, the absorption length is 13 μm. This allows the investigation of the photocurrents, microwave signals in high-quality Si samples [54]. In Figure 2.7, however, a situation is displayed where the sum of absorption length and minority diffusion length is smaller than the absorber thickness. Such a situation with low diffusion length and an absorption profile closer to the surface is typical for highly absorbing compound semiconductors such as the III–V compounds InP, GaAs, and some of their alloys, II–VI compounds (CdTe), and ternary and quaternary chalcopyrites such as $CuInSe_2$, $CuInS_2$, and $CuIn_xGa_{1-x}Se_2$. The upward curving of the hole quasi-Fermi level near the surface indicates a loss of stationary excess minority carriers due to surface recombination. Analogously, the quasi-Fermi level of the majority electrons is curved downwards. The attainable photovoltage is given by the difference in the quasi-Fermi levels at the surface.

2.2.2
Photovoltages and Stability Criteria

Based on (quasi-)thermodynamic considerations, it can be shown that the separation of the quasi-Fermi levels at the surface yields the maximum attainable photovoltage of the system [55] because it is related to the free energy difference ΔF with $F = U - TS$:

$$\left| {}_nE_F^*(0) - {}_pE_F^*(0) \right| = eV_{Ph}^{max} \tag{2.21}$$

Figure 2.7 shows the energy scheme for an n-type semiconductor close to the flatband situation, that is, at open circuit. The curvature of the quasi-Fermi levels towards the surface ($x = 0$) indicates surface recombination that results in a reduced excess carrier concentration. Accordingly, the attainable photovoltage is reduced. Due to the surface recombination, the energy bands show a remnant band curving because majority electrons can recombine with minority excess holes inhibiting complete flattening of the bands. For higher illumination levels, surface recombination channels may be filled and the residual band bending can then be smaller. Without surface recombination, the photovoltage is defined by the excess carrier lifetime, and the curvature towards the surface (energetically upwards for ${}_pE_F^*(0)$ and downwards for ${}_nE_F^*(0)$) is then absent. It should also be noted that the considerations so far have been made for monochromatic light. For solar irradiation, multichromic quasi-Fermi levels are obtained from the calculation of the spatial dependence of the total excess carrier concentration[3].

The reactive semiconductor–electrolyte interface makes stability a major issue in photoelectrochemical solar energy conversion devices, and aspects of thermodynamic and kinetic stability are briefly reviewed here. Thermodynamic stability considerations are based on so-called decomposition levels [56, 57] that are determined by combining the decomposition reaction with the redox reaction of the reversible hydrogen reference electrode. The anodic and cathodic decomposition reactions of a compound semiconductor MX can be written for aqueous solutions as

$$MX + zh^+ + aq \rightarrow M_{aq}^{z+} + X \tag{2.22}$$

$$MX + ze^- + aq \rightarrow M + X_{aq}^{z-} \tag{2.22a}$$

As an example, the light-induced anodic decomposition of CdS is discussed here which results in the formation of a passivating sulfur film according to $CdS + 2h^+(h\nu) \rightarrow Cd_{aq}^{2+} + S^0$. In the derivation of the decomposition potentials, the hydrogen electrode serves as source or sink of electrons according to

$$zH_{aq}^+ + ze^- \Leftrightarrow \frac{z}{2}H_2 \tag{2.23}$$

Combining Equations 2.22 and 2.22a with Equation 2.23 yields

$$MX + zH_{aq}^+ \Leftrightarrow M_{aq}^{z+} + X + \frac{z}{2}H_2 \tag{2.23a}$$

for the anodic decomposition and

$$MX + \frac{z}{2}H_2 + aq \Leftrightarrow M + X_{aq}^{z-} + H_{aq}^{z+} \tag{2.23b}$$

---

3) The calculation involves integration over the single wavelength profiles including diffusion and drift.

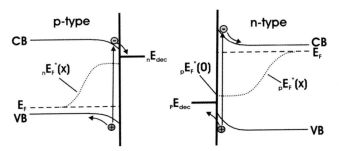

Figure 2.8 Energy band schematic for stability assessment of illuminated p-type (left) and n-type (right) semiconductors; the energetic positions of the minority carrier quasi-Fermi levels at the surface and that of the cathodic and anodic decomposition levels are shown (see text).

for the cathodic reaction. With the knowledge of the Gibbs reaction enthalpy $\Delta G°$, one can calculate the standard decomposition potentials in relation to the electrochemical potential scale:

$$\frac{\Delta G°}{zF} = {}_pE_{dec}; \quad -\frac{\Delta G°}{zF} = {}_nE_{dec} \quad (2.24)$$

where ${}_pE_{dec}$ and ${}_nE_{dec}$ are the anodic and cathodic decomposition energies, respectively. Thermodynamic stability against photocorrosion is described within this concept by the relative energetic positions of the quasi-Fermi levels at the surface and the respective decomposition levels as shown in Figure 2.8. For a p-type semiconductor, the minority carrier quasi-Fermi level has been chosen such that its value at the surface, ${}_nE_F^*(0)$, is energetically located below the cathodic decomposition energy ${}_nE_{dec}$. This indicates thermodynamic stability for the given illumination level (light intensity) and surface recombination. Accordingly, a stable situation has been depicted for the n-type semiconductor and the stability criteria are given by ${}_nE_F^*(0) < {}_nE_{dec}$ for cathodic stability and by ${}_pE_F^*(x=0) > {}_pE_{dec}$ for anodic stability. Two examples (for TiO$_2$ and GaAs) are shown in Figure 2.9 where the decomposition levels have been determined from the reactions

$$TiO_2 + 4HCl \rightarrow TiCl_4 + O_2 + 2H_2 \text{ (anodic)} \quad (2.25a)$$

$$TiO_2 + 2H_2 \rightarrow Ti + 2H_2O \text{ (cathodic)} \quad (2.25b)$$

and

$$GaAs + 6H_2O \rightarrow Ga(OH)_3 + H_3AsO_3 + 3H_2 \text{ (anodic)} \quad (2.25c)$$

$$GaAs + \frac{3}{2}H_2 \rightarrow Ga + AsH_3 \text{ (cathodic)} \quad (2.25d)$$

Also shown in Figure 2.9 are the thermodynamic potentials for water dissociation at a pH of 7. From the position of the conduction and valence band of titania relative to the cathodic and anodic decomposition of water, water splitting under

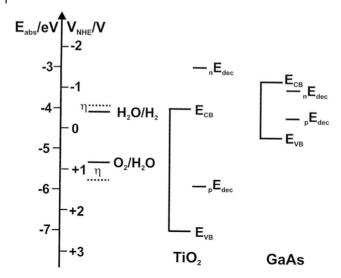

Figure 2.9 Examples for thermodynamic decomposition energies of TiO_2 and GaAs; note that both semiconductors are anodically unstable but TiO_2 is cathodically stable (see text).

illumination should be thermodynamically possible if the quasi-Fermi levels supersede the respective levels in solution. The overpotentials η, associated with the hydrogen and oxygen evolution reactions, result in an upward and downward shift of the solution levels, respectively, as indicated in the figure by the dotted levels.

The kinetics of the decomposition reactions such as Equations 2.25a–d involve individual reaction steps. For an anodic decomposition, for example, the individual decomposition levels are considered for the first two partial reactions:

$$MX + h^+(h\nu) + aq \rightarrow M^+_{ad,aq} + X_{ad} \tag{2.26}$$

$$M^+_{ad,aq} + h^+(h\nu) \rightarrow M^{2+}_{aq} \tag{2.26a}$$

$$X_{ad} + X_{ad} \rightarrow X_2 \tag{2.26b}$$

where M_{ad} and X_{ad} denote species that are adsorbed at the surface. The reaction scheme corresponds basically to the oxidation reactions of ZnO or TiO_2 where gases are evolved. Assuming that the first reaction is characterized by an activation barrier that brings its decomposition energy $_pE_{dec}(a)$ close to the valence band edge, the dissolution reaction cannot proceed for the quasi-Fermi level of holes as shown in Figure 2.10 because $_pE_F^*(0) > {_pE_{dec}}(a)$, although, when considering the overall reaction (see $_pE_{dec}$ in Figure 2.10), the semiconductor would be unstable with $_pE_F^*(0) < {_pE_{dec}}$. In this way, the rate-determining step of the dissolution reaction (2.26) can influence the overall stability.

Several approaches for the realization of stable photoelectrochemical systems have been introduced. A schematic overview is given in Figure 2.11. One distinguishes between chemical stabilization [58–62] (Figure 2.11a), the use of specific

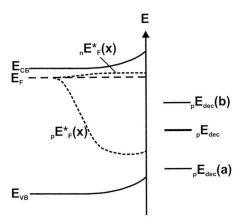

Figure 2.10 Energy scheme for individual steps, labeled (a) and (b), in the photocorrosion of an n-type compound semiconductor; the hole quasi-Fermi level $_pE_F^*(0)$ at the surface is chosen such that it is energetically located between the overall decomposition energy and that of the first reaction step (see Equation 2.26).

band structure properties [63–65] (Figure 2.11b), kinetic stabilization [66–68] (Figure 2.11c), and physical stabilization [30, 69–73] (Figure 2.11d). As Figure 2.11a shows, the principle of chemical stabilization relies on the formation of dangling bond-terminating surfaces; here an example is shown for the methylation of a Si(111) surface [59]. Other approaches include the H-termination of this surface [58, 60, 61] and the Se-termination of GaAs(100) [62]. Because even the best-prepared surfaces are characterized by an at least small number of atomic steps, edge atoms will exist that can be terminated differently. The hitherto best H-terminated Si(111) surfaces which exhibit an interface state density of $10^{10}\,\text{cm}^{-2}\,\text{eV}^{-1}$ [74] show a residual coverage of some Si surface atoms with OH and F [75]. Therefore, chemically passivated surfaces are susceptible to attack by water or other solution species, particularly under operation in solar devices when currents flow, which can result in oxide formation or semiconductor dissolution [76]. Figure 2.11b shows electron energy plotted versus density of states (DOS) for a group VIb transition metal dichalcogenide such as MoS_2 or WSe_2. The stability of these compounds has been attributed to the thermalization of photoexcited charge carriers to band edges that exhibit rather pure metal d-band character [63]. Therefore, the bonding between the metal and the chalcogenide is not affected and as a result these materials show surprising stability as photoelectrodes [64, 65]. The electrodes mostly used, however, are compounds such as MoS_2, $MoSe_2$, and WSe_2 [77, 78]. They are indirect-gap semiconductors with the top of the valence band at $\Gamma (k = 0)$ and the bottom of the conduction band approximately in the middle between Γ and K in the reduced zone scheme [79]. The predominant absorption features in these compounds are the A and B excitons, energetically located near the optical absorption edge. These excitons can be assigned to transitions at the K-point with K_1 and K_4 as initial states and K_5 as final state. These transitions are different from d–d transitions, allowed, but show a dependence on the polarization

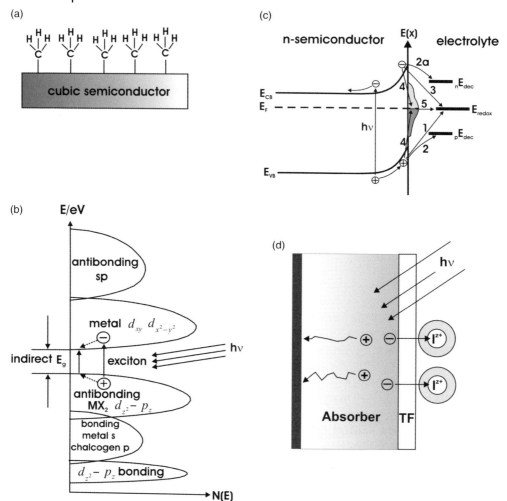

Figure 2.11 Strategies for stabilization of photoelectrodes against corrosion. (a) Chemical stabilization. (b) Stabilization by specific properties of the band structure as for group VIb transition metal dichalcogenides; the five metal d-band contributions, the indirect gap, and the direct excitonic gap are shown; also indicated is the thermalization process from the highly absorbing excitonic states to the indirect band edges. (c) Kinetic stabilization via reaction rate relations: (1) redox reaction of light-induced holes with the reduced part of the redox couple; (2) anodic and cathodic (2a) decomposition reaction; (3) back reaction of conduction band electrons with the redox couple; (4) surface recombination; (5) reaction with redox couple via surface states. (d) Physical passivation by an ultrathin film that can be tunneled through by the light-induced excess carriers; TF, tunneling film; I^{z+}, solvated ion in electrolyte (charge z+); an outer sphere charge transfer is indicated.

state of the light [80]. The A and B excitons correspond to the smallest direct gap in these materials, located in k-space at the K-point. As the indirect gap is smaller, the excess carriers will eventually relax to the top of the valence band and to the bottom of the conduction band. The Γ_4^- state at the top of the valence band, however, is rather an antibonding state between metal d_{z^2} and chalcogen p_z orbitals (with an approximately 40% contribution from chalcogen orbitals) instead of the earlier assumed nonbonding state. Due to the large covalent interaction between these orbitals, the corresponding bonding state is located at about 5 eV below the top of the valence band and cannot contribute in the usual photoelectrochemical processes. This explains the unusual stability of these materials as photoanodes in photoelectrochemical solar cells [81]. As the conduction band state K_5 has almost exclusive metal d-character, cathodic photocorrosion with p-type material, which is typically less pronounced under reducing conditions [82], is negligible.

Figure 2.11c shows an energy scheme of an illuminated semiconductor–electrolyte junction. Also shown are the decomposition levels and occupied (dark gray) and empty (light gray) surface states. If the reaction rates for the redox reactions via the valence band, k_1, or via occupied surface states [83], k_5, are fast compared to the decomposition reaction rate, k_2, the system would be stabilized. This is the case for a dye-sensitized solar cell [84, 85], where the charge injection rate from the excited dye into the TiO_2 conduction band and the reduction reaction of the excited highest occupied molecular orbital by the reduced species of the redox electrolyte are much faster than, for instance, the back reaction [86], thus stabilizing the dye molecule. If the rate of reaction (3) is not negligible, the current from the junction is reduced; similarly, surface recombination will affect the achievable photovoltage.

The concept of physical stabilization by interfacial films that allow efficient charge transfer but shield the semiconductor surface from the solution has been applied in a series of experiments [87–92]. The principle is depicted in Figure 2.11d where an interfacial film of tunneling thickness on top of a photoelectrode is shown. If the film thickness is around 1 nm, efficient carrier tunneling is possible without corrosive attack of the electrode. With such ultrathin films, their homogeneity and conformal attachment to the nanotopography of the absorber electrode are of crucial importance and several examples exist where such properties could be realized [93–97]. They will be described in detail in Sections 2.4 and 2.5. In the future, the use of atomic layer deposition [98] films might be promising for a more general approach to physical electrode stabilization. These films are conformal to nanostructures and can be grown in monolayers in a step-by-step process [99].

2.2.3
Photovoltaic and Photoelectrocatalytic Mode of Operation

2.2.3.1 Photovoltaic Photoelectrochemical Solar Cells
The principle of operation of a photoelectrochemical solar cell (PECS) is shown in Figure 2.12 where the absorber is a semiconducting material. The rectifying junction is formed between redox electrolyte and semiconductor, as also shown in

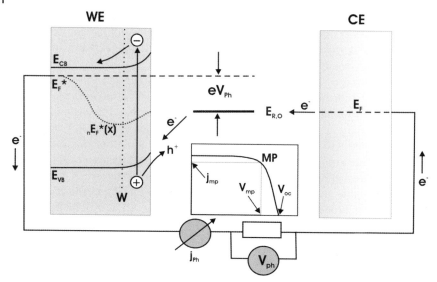

Figure 2.12 Operation principle of a photovoltaic PECS for an n-type semiconductor–electrolyte junction; $E_{R,O}$, redox level; WE, working electrode (absorber); CE, counter electrode; V_{Ph}, photovoltage; the electron quasi-Fermi level is located within the line width of the dark Fermi level and is not shown; MP, maximum power point of the solar cell (see inset); j_{mp}, V_{mp}, photocurrent density and photovoltage at the maximum power point, respectively.

Figures 2.1 and 2.2. The light-generated minority carriers diffuse and drift towards the electrolyte interface where charge transfer to the respective species (oxidized: electrons; reduced: holes) occurs. The majority carrier current results in injection of the opposite carrier (here electrons) at the counter electrode–electrolyte interface where the opposite redox reaction takes place. The semiconductor–electrolyte junction shown here is characterized by a photovoltage and a photocurrent, that is, the solar cell is operating at or near its maximum power which, in efficient devices, is rather close to the open circuit condition. This is indicated in the inset of Figure 2.12. Therefore, a residual band bending has been shown and the photovoltage under these conditions is given by the quasi-Fermi levels at the surface. Here, only the quasi-Fermi level for holes is shown because $_nE_F^*(x)$ only marginally differs from E_F, the Fermi level without illumination.

A dye-sensitized solar cell also operates in the photovoltaic mode. It is, however, based on a slightly different operation principle because the absorber is a dye instead of a semiconductor and the charge separation has been attributed to differences in the chemical potential on both sides of the dye [100]. It is therefore considered an excitonic solar cell [101].

2.2.3.2 Photoelectrocatalytic Systems

Photo(electro)catalysis encompasses a wide range of reactions [11, 102–105]. First, light-induced water dissociation is considered because of its relevance in solar fuel

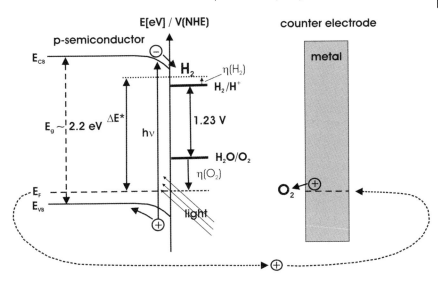

Figure 2.13 Energy/potential scheme for light-induced water dissociation for a p-type semiconductor as photocathode and a metallic anode; the left-hand side shows electron energies; on the right, at the electrolyte side, electrode potentials are given; η, overpotentials; ΔE^*, necessary photon energy for the dissociation reaction, corresponding to the splitting of the minority and majority quasi-Fermi level.

production. Subsequently, further important reactions, particularly at photoanodes, are briefly discussed. The determining parameter in water photoelectrolysis systems is the thermodynamic value for water splitting (1.23 V). Reaction kinetics result in overpotentials for both the hydrogen and the oxygen evolution reactions [106, 107]. In semiconductor-based systems, additional losses have to be considered. They comprise the series resistance of the electrolyte, the necessity of a residual band bending to drive the reactions, and energy losses due to the difference between band edge and Fermi level and the potential drop across the Helmholtz layer at working and counter electrode. Summation of the losses yields a selection criterion for the semiconductor band gap energy as visualized in Figure 2.13. For single-junction water photoelectrolysis, an energy gap of about 2.2 eV is necessary. Comparing this energy gap with the maximum theoretical efficiency for AM 1 (where AM represents air mass) insolation yields an efficiency of about 15% [108].

Therefore, two fundamental research strategies can be envisaged: (i) the development of robust devices made from abundant (nontoxic) materials that exhibit reduced efficiencies and (ii) the development of devices that use the photonic excess energy as, for instance, in tandem or other third-generation photoactive structures. Here, the theoretical efficiency can increase to values above 40% for a two-junction device, depending on the respective energy gaps. In the former approach, concepts and principles from photosynthesis have already been adapted. Examples include the preparation of macromolecules that contain centers with

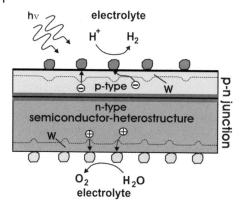

Figure 2.14 Schematic arrangement for a monolithic tandem device for light-induced water splitting; W, space charge region (extended under the catalyst particles that are assumed to form rectifying junctions); the dark gray areas represent the highly doped tunnel contacts on the back of each semiconductor.

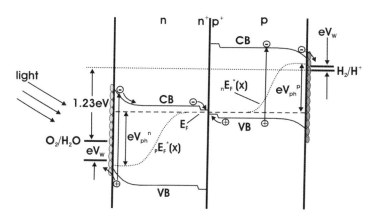

Figure 2.15 Energy scheme for a tandem structure with an n-type (left) and p-type (right) semiconductor for operation close to the maximum power point; eV_{ph}^n and eV_{ph}^p denote the photovoltages in the n- and p-type material, respectively; the light and dark gray ellipsoids symbolize catalytic particles for oxygen and hydrogen evolution, respectively; V_W denotes the working voltage for the respective current value which contains the reaction overpotential for O_2 and H_2 evolution.

catalytically active metals [109, 110], studies on excitation energy transfer [111, 112], and the use of chemically modified complete photosystems [113]. Following the second approach, a few tandem structures have been realized [12, 114] and the basic operation principle of monolithic tandem structures is shown in Figures 2.14 and 2.15, where the spatial arrangement and the energy band diagram, respectively, are shown.

In Figure 2.14, it has been assumed that the catalytically active particles form a rectifying junction with the absorber, forming partially overlapping semispherical space charge regions. In contact with a redox electrolyte such as H^+/H_2, the spatial depletion layer profile is amplified or weakened, depending on the relative energetic position of the redox level with respect to the work function of the catalyst, because the solid system assumes the electrochemical potential of the solution. At the back-to-back tandem structure shown, hydrogen evolution takes place at the p-type side and an oxidation reaction which could include Cl_2, Br_2, or O_2 evolution but also oxidation of organic compounds [115] can take place.

The energy band diagram in Figure 2.15 shows the structure operating near maximum power, that is, at rather large voltage and also large current flow (cf. Figure 2.12). The tunnel contact at the back of each semiconductor consists of the respective highly doped n^+ and p^+ regions where the majority carriers are transferred. The sum of the photovoltages at the maximum power point (MP), $eV_{ph}^{MP} = {}_nE_F^*(0) - {}_pE_F^*(0) = eV_{ph}^n + eV_{ph}^p$, has been chosen such that $eV_{ph}^{MP} > eV(H_2O) + \eta_{act} + \eta(j)$. The first term on the right-hand side of the inequality is the thermodynamic value for water splitting, the second term is the reaction overpotential and the third term is the overpotential due to the current passed. Then, $\eta_{act} + \eta(j) = eV_W$, the working potential at the given current density.

2.2.4
Separation of Charge Transfer and Surface Recombination Rate

The efficiency of photoelectrochemical devices is based on effective charge transfer while suppressing surface recombination and corrosion. While the photocurrent is a direct measure of the irreversibly transferred electrons, it is not trivial to obtain a measure for the losses at surfaces due to recombination. As will be shown in Section 2.3.1, stationary microwave reflectivity is a method that measures the integral of the excess minority carrier profile. Such profiles are shown in Figures 2.3–2.6. The simultaneous recording of photocurrent and excess microwave reflectivity in an electrochemical cell allows the assessment of the relative contributions of k_r and S_r for well-defined systems. These parameters are defined as follows:

$$k_r = \frac{j_{Ph}}{e\Delta p_s} \quad \text{and} \quad S_r = \frac{j_r}{e\Delta p_s} \quad (2.27)$$

where Δp_s is the excess minority carrier concentration at the surface, $\Delta p_s = \Delta p(0)$ (see also Equations 2.11 and 2.C.1). Expressing the time-dependent microwave signal $M_w(t)$ as dependent on the surface concentration of holes, Δp_s, can be achieved by assuming a linearity between the microwave signal with increasing surface concentration [116]:

$$M_w(t) = M_{w0} + b\Delta p_s(t) \quad (2.28)$$

where M_{w0} is the microwave signal for efficient charge transfer, that is, internal quantum yield about 1. Introducing the quantum yield

$$Q = \frac{hc}{e\lambda} \frac{j_{Ph}}{I_0} \quad (2.29)$$

where I_0 is the light intensity and the constants have their usual meaning. With Equation 2.27, the quantum yield can be expressed in terms of the carrier surface concentration and the charge transfer rate:

$$Q = \frac{hc}{e\lambda} \Delta p_s k_r \qquad (2.30)$$

Taking the value for Δp_s in Equation 2.28 and inserting it into Equation 2.30 yields an expression for the charge transfer rate dependent on the temporal profile of the measured quantum efficiency and microwave signals:

$$k_r(t) = B \frac{Q(t)}{M_w(t) - M_{w0}} \qquad (2.31)$$

and, accordingly, for the surface recombination rate, with $k_r/S_r = Q/(1-Q)$, one obtains

$$S_r(t) = B \frac{1-Q(t)}{M_w(t) - M_{w0}} \qquad (2.32)$$

Figure 2.16 shows chronoamperometric profiles of the photocurrent and the excess microwave signal obtained on an n-Si(111) sample in dilute NH$_4$F solution at anodic potential. The current profile is characterized by a first maximum that indicates the transition from the divalent dissolution regime where porous Si formation and photocurrent doubling persist [117, 118] to the tetravalent regime where oxide formation occurs. The current maximum (quantum yield >1.2) is followed by a plateau where the transition from photocurrent doubling due to electron injection from solution into the conduction band [119, 120] to a quantum yield slightly below 1 in the tetravalent regime with oxide formation occurs. The plateau is more extended for higher voltage, related to still efficient charge

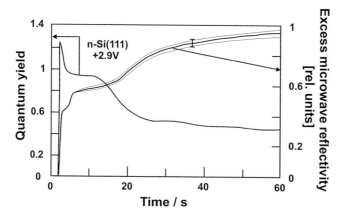

Figure 2.16 Temporal quantum yield and microwave reflectivity profiles of n-Si(111) in 0.2 M NH$_4$F solution, pH 4.6; light intensity, 1.3 mW cm^{-2}; potential, +2.9 V vs. SCE; the thin lines around the microwave reflectivity indicate the signal noise (error bar).

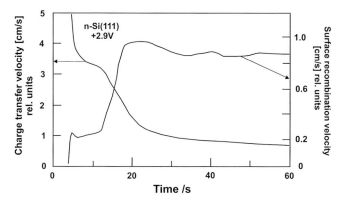

Figure 2.17 Temporal behavior of the charge transfer velocity and the surface recombination velocity as evaluated from the experimental data in Figure 2.16 (see text).

transport through thicker oxides at increased electrical field strength. For extended time, passivation via formation of oxide with increasing thickness sets in, resulting in a decrease of the photocurrent. The corresponding microwave signal is very small in the range of photocurrent doubling, indicating very low surface recombination and simultaneously high charge transfer rate, which then increases to a rather steady value at the current plateau and further increases upon passivation of the sample.

In Figure 2.17, the evaluation following the formalism outlined in Equations 2.31 and 2.32 is plotted. It can be seen that the region of photocurrent doubling is characterized by a high charge transfer rate which steeply drops in the temporal regime near the current plateau and is subsequently further reduced with the increasing passivation of the sample by oxide formation. The surface recombination velocity is very low at the onset of the experiment (2–3 s) and shortly reaches a stationary value in the region of the current plateau. Then, a steep increase of S_r is noted due to passivation-induced carrier recombination. Further passivation does not result in an additional increase of S_r, possibly related to the sustained interface condition (cf. Figure 2.16) that are attributed to the nanoporous structure of the oxide formed.

2.3
Surface and Interface Analysis Methods

Analysis methods, used for the investigation of modified surfaces and interfaces, are briefly reviewed. Emphasis is on the combination of chemical, structural/morphological, electronic, and optical characterization. Many techniques such as transmission electron microscopy (TEM), standard X-ray photoelectron spectroscopy (XPS) using Al or Mg K_α radiation, high-resolution scanning electron microscopy (HRSEM), and standard scanning probe microscopies (AFM in contact mode, STM) are well established and their principles can be found in the literature [121–125]. Therefore, features such as plasmon satellites, shake-up or shake-off lines, and the various approaches for background approximation [126–128] are only referred to; the same holds for the Doniach–Sunjic function [129] and a Cooper minimum in the excitation cross section [130]. The latter are of relevance for the interpretation of the results on Pt (and Ir) nanoisland deposits (see Figures 2.93 and 2.94).

Accordingly, methods are presented below that are either less well known or that are of specific use for the characterization of electrochemically and photoelectrochemically conditioned surfaces and interfaces.

2.3.1
In Situ Methods: I. Brewster Angle Analysis

Brewster angle spectroscopy (BAS), developed originally for the contactless characterization of electronic defects in semiconductors [131], was extended, shortly after the introduction of the method, to high-sensitivity surface optical characterization [132]. Recently, the method was used *in situ* to follow surface changes in real time upon conditioning of semiconductors [133, 134]. The latter technique is described here.

The Brewster angle is well known in the optics of dielectrics [135]. At the Brewster angle, the reflectivity for p-polarized light vanishes as the refracted beam is perpendicular to the reflected one. In a simple picture, the dipole radiation from the refracted light is zero in this direction. Originally, the idea was to consider a semiconductor as a dielectric for photon energies below its energy gap, which is the fundamental absorption edge. The absorption due to excitation of an electronic defect within the bandgap was expected to change the Brewster angle and the reflectivity at that angle [136], which was indeed observed for GaAs and $CuInS_2$ for instance [137], thus providing a contactless method for identification of electronic defects at room temperature. In a second application, it was shown that the high surface sensitivity of the method can be used for optical analyses [138, 139]. Because the change of the Brewster angle and the reflectivity at this angle are measured simultaneously, yielding two experimental parameters in one measurement, the method has a certain similarity to ellipsometry where the ellipsometer angles Δ and Ψ are recorded [140].

A block diagram of the experimental arrangement [141] including the extension of the original setup to *in situ* photoelectrochemical measurements [142] is shown in Figure 2.18. In the center of the apparatus, the (photo)electrochemical cell is mounted on the goniometer table. With the bias light (typically supplied by a W–I lamp), *in situ* and, depending on the rate of induced changes, also real-time photoelectrochemical experiments in a typical three-electrode potentiostatic arrangement can be carried out. Also, chemical etching can be followed. Such experiments will be presented in Section 2.4.1.

The optical analysis is based on p-polarized light that impinges on the sample ("S" in Figure 2.18) after having passed the monochromator, the apertures for

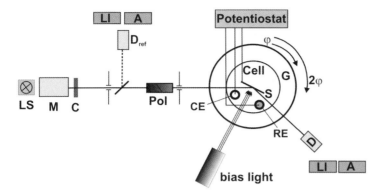

Figure 2.18 Principle of the Brewster angle spectroscopy technique; LS, light source; M, monochromator; C, chopper, D_{ref}, reference beam detector; LI, A, lock-in amplifier and current amplifier; Pol, Glan–Thomson polarization filter; G, goniometer table; S, sample in the electrochemical cell; RE, reference electrode; CE, counter electrode; φ, goniometer and sample turning angles (see text).

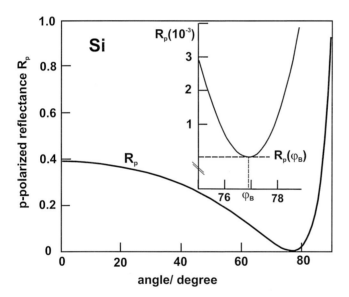

Figure 2.19 Angular dependence of the reflectance behavior of Si for p-polarized light for $\lambda = 500\,nm$; the inset shows R_p around the Brewster angle φ_B in units of 10^{-3}.

angle divergence reduction, and the polarizer. The large focal length of the Czerny–Turner monochromator used allows high angular precision measurements due to the small beam divergence. The step accuracy of the goniometer is 0.0004°. The polarization ratio of the Glan Thomson element is 10^6. Signal processing includes a lock-in technique, cooled photodetectors, and current amplifiers. The Brewster angle and the reflectivity at this angle are determined with high accuracy by a polynomial fit as shown in Figure 2.19 for Si [143]. In typical measurements, the

Brewster angle accuracy is of the order of 0.001° where the influence of the light beam divergence, the polarization state of the light, and the exact angle of incidence contribute to the overall resolution limit. Figure 2.19 shows the evaluation routine for a wavelength of 500 nm ($h\nu = 2.48$ eV) using literature data [144]. The inset shows the parabolic behavior of the p-polarized reflectance, R_p, in higher resolution around the Brewster angle φ_B. It is seen that R_p becomes extremely small at φ_B.

BAS measures the optical response of a sample using φ_B and $R_p(\varphi_B)$. The optical response is obtained as the wavelength (or frequency)-dependent complex dielectric function $\hat{\varepsilon} = \varepsilon_1 + i\varepsilon_2$ (where ε_1 and ε_2 are the real and imaginary parts of the dielectric function, respectively), and analytical expressions have been derived to determine $\hat{\varepsilon}(\omega)$ from φ_B and $R_p(\varphi_B)$. For a sample with purely dielectric properties, for example, $\hat{\varepsilon} = \varepsilon_1$, the Brewster angle is given by

$$\varphi_B = \arctan\sqrt{\varepsilon_1} \tag{2.33}$$

Typically, the analysis of samples involves inclusion of several components such as the outer roughness, interfacial films, and the bulk of the sample. Therefore, the overall optical response has to be deconvoluted into signal contributions from the respective parts of the sample. For a reasonable assessment in the calculation, additional knowledge is desirable. Roughness, for example, can be determined in AFM experiments and will be used in Section 2.4.1.

The real and imaginary parts of the dielectric function are obtained by solving a reduced fourth-order equation that connects $R_p(\varphi_B)$, φ_B, and the dielectric function [145]. The dielectric functions are expressed in terms of the Brewster angle, determined from $dR_p(\varphi)/d\varphi = 0$:

$$\varepsilon_1 = |\hat{\varepsilon}|^2 (1 + 2\cos^2\varphi_B) - |\hat{\varepsilon}|^2 \frac{\cot^2\varphi_B}{\sin^2\varphi_B} \tag{2.34}$$

$$\varepsilon_2 = \sqrt{|\hat{\varepsilon}|^2 - \varepsilon_1^2} \tag{2.34a}$$

The analysis of multiple-layer systems considers partially reflected and transmitted light beams at the interface of two or more phases (media). Figure 2.20 shows a simple three-phase vertical structure where the partial wave interference yields the reflectance R_p (and transmittance T). Mathematically, the expressions for R can be simplified by a matrix formalism [146]. In the three-layer system considered, consisting of ambient, intermediate film, and substrate, for nonmagnetic media a complex index of refraction can be defined for the ith layer:

$$n_i = \text{Re}\left(\sqrt{\varepsilon_i}\right) - \text{Im}\left(\sqrt{\varepsilon_i}\right) \tag{2.35}$$

The Fresnel equations for the reflectance coefficients r_{01} and r_{12} for the ambient–film and the film–substrate interface are given by

$$r_{i(i+1).p} = \frac{n_{i+1}\cos\varphi_i - n_i\cos\varphi_{i+1}}{n_{i+1}\cos\varphi_i + n_i\cos\varphi_{i+1}}; \quad i = 0,1 \tag{2.36}$$

The reflectance coefficient for the three-layer system is then calculated from the expression

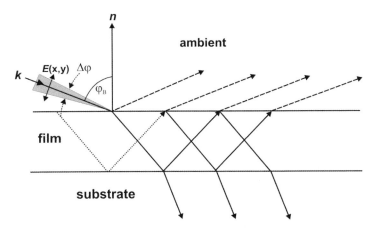

Figure 2.20 Brewster angle analysis of a multiple-layer structure, shown here for a three-phase system; the plane of incidence, defined by the surface normal **n** and the photon wave vector **k**, is the x–y plane; **E**(x,y) is the electric field vector for p-polarization; $\Delta\varphi$ denotes the angular measurement for determination of the Brewster angle (cf. Figure 2.19); the dashed lines indicate the low reflectance at that angle.

$$r_p(\varphi_B) = \frac{r_{01} + r_{12} e^{-i2\Phi}}{1 + r_{01} r_{12} e^{-i2\Phi}} \quad (2.37)$$

where Φ ($= (2\pi d/\lambda) n_1 \cos\varphi_1$) is the phase shift. The p-polarized reflectance $R_p(\varphi_0)$ is obtained by multiplication:

$$r_p(\varphi_B) \times r_p^*(\varphi_B) = |r_p(\varphi_B)|^2 = R_p(\varphi_B) \quad (2.38)$$

Hence, BAS uses the well-known formalism for multiple-layer analysis with the difference that the experimentally determined optical parameters are R_p and the Brewster angle of the layer structure. In Figure 2.20, the measurement technique to determine the Brewster angle by angle variation, as shown in Figure 2.19, is also depicted and, for intelligibility, the angle variation has been greatly enhanced ($\Delta\varphi$). Surface or interface roughnesses are modeled using effective medium theory in the formalism of Bruggeman [147] or Maxwell Garnett [148] where the roughened region is described as a layer with an effective dielectric function. The former is used for a more continuous distribution of roughness whereas the latter applies to large porosities, for instance.

2.3.2
In Situ Methods: II. Stationary Microwave Reflectivity

Figure 2.21 shows a schematic of the setup for simultaneous measurement of the stationary light-induced excess minority carrier microwave reflectivity and the photocurrent at the semiconductor–electrolyte contact. The sample is illuminated from the front side and photoelectrochemistry is performed using the standard

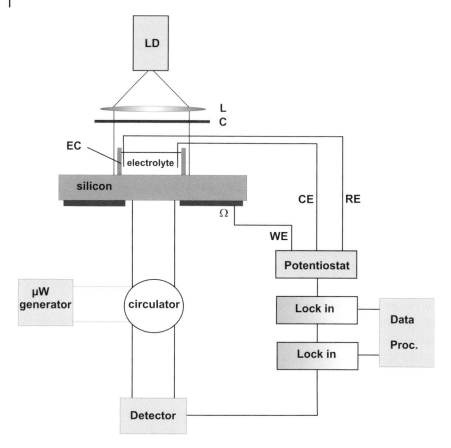

Figure 2.21 Experimental arrangement for simultaneous *in situ* stationary light-induced excess microwave reflectivity and photocurrent measurements; LD, laser diode ($\lambda = 830$ nm); L, collimating lens; C, chopper ($f = 80$ Hz); EC, electrochemical cell; Ω, circular ohmic back contact; CE, counter electrode; RE, reference electrode; WE, working electrode; data processing as indicated (see text).

three-electrode potentiostatic arrangement. The ohmic back contact of the sample has a circular hole for measurement of the semiconductor microwave response. For samples with low enough resistivity (in the Ω cm range), the applied potential variation across the sample surface can be neglected. The light from the laser diode is chopped and a lock-in technique is used for detection of both the photocurrent and the microwave reflectivity. The microwave signal is generated by a Gunn diode mounted in a resonance cavity [149] and the circulator separates the incident from the reflected microwaves. The X-band was used for analysis.

The excess conductivity of a semiconductor, $\Delta\sigma = e\mu_{e(h)}\Delta n(p)$, due to the induced change in carrier concentration $n(p)$ results in a corresponding change of the reflected microwave intensity ΔP_R. With the relative microwave reflectivity defined as

2.3 Surface and Interface Analysis Methods

$$R_{MW} = \frac{P_R + \Delta P_R}{P_{in}} \qquad (2.39)$$

the excess microwave reflectivity becomes

$$\Delta R_{MW} = \frac{\Delta P_R}{P_{in}} \qquad (2.39a)$$

In the low injection limit, ΔR_{MW} is given by a sensitivity factor S^*, the electrical field of the microwaves, the mobilities of holes and electrons, and the respective excess carrier concentrations. For the one-dimensional case (see Figures 2.3 and 2.4), the excess microwave signal is given by

$$\Delta R_{MW} = S^+ \int |E_x|^2 [\mu_h \Delta p(x) + \mu_e \Delta n(x)] dx \qquad (2.40)$$

For semiconductors with thickness $d < 1\,\mathrm{mm}$, the microwave field E_x can be considered as constant along the x-axis. Equation 2.40 then reduces to (with an according change of the sensitivity factor to S^*)

$$\Delta R_{MW} = S^* \int [\mu_h \Delta p(x) + \mu_e \Delta n(x)] dx \qquad (2.40a)$$

For further analysis, in particular regarding the signal from the majority carriers in n-Si where the electron mobility is about three times larger than that of the holes, the excess carrier profiles are considered separately in the space charge region ($0 < x < W$) and in the neutral region ($x > W$).

The charge distribution for the semiconductor–electrolyte system is given by the excess minority carrier profile $\Delta p(x)$ (2.40a) and the corresponding neutralizing majority carrier profile. The microwave signal due to the minority carriers, ΔR_{MW}^h, is given by the excess carrier distribution throughout the sample:

$$\Delta R_{MW}^h = S^* \left(\int_0^W \mu_h \Delta p(x) dx + \int_W^d \mu_h \Delta p(x) dx \right) \qquad (2.41)$$

In the semiconductor's neutral region, the majority carrier profile will follow that of the minority carriers in the bulk. In the space charge layer, the neutralizing charges are majority carriers, confined to the end of the space charge layer towards the bulk and counter ionic charges in the electrolyte Helmholtz layers. Only majority carriers within the semiconductor contribute to the microwave signal. The microwave response of the electrons from the space charge regions can be obtained by using the relation between current (here, photocurrent) and carrier concentration $\Delta n \propto j_{Ph} R_S C_{SC}/q$, where the R_S term describes the series resistance including electrolyte, back contact, and bulk semiconductor losses and C_{SC} is the space charge layer capacitance. With C_{SC} in the nF range for not too highly conductive semiconductors and R_S in the range of a few $\Omega\,\mathrm{cm}^{-2}$, the according time constant is in the ns range. For minority carrier lifetimes as typical for Si, this contribution can therefore be neglected and the expression for the microwave response then reads (with $S = S^*(\mu_h + \mu_e)$)

$$\Delta R_{MW} = S \left(\frac{\mu_h}{\mu_e + \mu_h} \int_0^W \Delta p(x) dx + \int_W^d \Delta p(x) dx \right) \qquad (2.42)$$

Thus, the recorded signal corresponds to the integral of the excess minority carrier profile throughout the sample modulated by the relative carrier mobilities. This finding was the basis for the analysis of the charge transfer and surface recombination velocity profiles above (cf. Figures 2.16 and 2.17).

2.3.3
X-ray Emission and (Photo)Electron Spectroscopies

In this section, spectroscopies that use higher energy electrons and photons than in standard optics are reviewed. Techniques that are directly related to the results presented below in Sections 2.4–2.6 are described, and also a short overview is given of X-ray analyses that have been selected because of their application in the analysis of biomolecules including the energetics of the metallic centers, buried interfaces in solar cells, and femtosecond time resolution.

Among the related methods, specific experimental designs for applications are emphasized. As in-system synchrotron radiation photoelectron spectroscopy (SRPES) will be applied below for chemical analysis of electrochemically conditioned surfaces, this method will be presented first, followed by high-resolution electron energy loss spectroscopy (HREELS), photoelectron emission microscopy (PEEM), and X-ray emission spectroscopy (XES). The latter three methods are rather briefly presented due to the more singular results, discussed in Sections 2.4–2.6, that have been obtained with them. Although ultraviolet photoelectron spectroscopy (UPS) is an important method to determine band bendings and surface dipoles of semiconductors, the reader is referred to a rather recent article where all basic features of the method have been elaborated for the analysis of semiconductors [150].

2.3.3.1 Selected X-ray Surface/Interface Analysis Methods
Figure 2.22 shows a schematic of the basic second-order processes that occur upon X-ray photon absorption by an atom or a semiconductor, for instance. The excitation of a K (1s) core electron into continuum states above the vacuum level causes a relaxation of an outer electron which excites either a second electron that leaves the system (Auger process) or produces lower energy X-rays (compared to that of the incident photons). The latter is termed X-ray emission and the related spectroscopy is XES [151]. In Figure 2.22b, the processes occurring with semiconductors are visualized, where the core level region and the region of extended valence and conduction band states are distinguished. The energetic region around the energy gap E_g has been magnified, and therefore the lengths of the arrows in the Auger process on the right-hand side of Figure 2.22b are not equal although energy conservation holds. The method is also applicable for the investigation of buried or hidden interfaces and materials components due to the larger escape depth of the fluorescent X-rays. This will be used in Section 2.4.3 in the analysis of Cu–S remnant phases in $CuInS_2$ solar cell absorbers where the sulfur signal ($Z = 16$) is measured.

The relative yield per K-shell vacancy is considerably lower for X-ray emission from lighter elements (see Figure 2.23) and spectroscopic experiments have

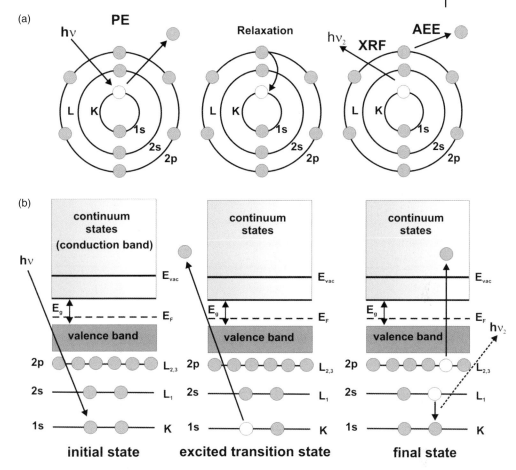

Figure 2.22 Schematic of the second-order processes in Auger electron emission (AEE) and X-ray fluorescence (XRF) upon photonic excitation with X-rays. (a) Processes in an atom; PE, photoemission process. (b) Processes in a semiconductor; note that the energy is not drawn to scale in order to emphasize the region around and above the semiconductor band gap E_g; incoming photon energy, $h\nu$; X-ray fluorescence energy, $h\nu_2$; the excited transition state corresponds to the photoemission process; the final state is characterized by either an Auger process or X-ray emission.

become possible only with the increased photon flux at third-generation synchrotrons [152]. XES has been extended to resonant elastic and resonant inelastic X-ray spectroscopy, labeled REXS and RIXS, respectively. A corresponding energy term scheme is shown in Figure 2.24. In XES, the final state is a core hole of this so-called spectator decay [153]. In the two resonant techniques, the photon energy of the incoming X-rays is tuned such that excitation occurs into binding but unoccupied states. In REXS, the final state is the ground state due to participator decay by recombination of the excited electron with the core hole; in RIXS, the final state is given by one electron and one hole.

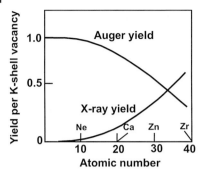

Figure 2.23 Relative probabilities for X-ray emission versus Auger electron emission the decay of K (1s) holes as a function of the element atomic number.

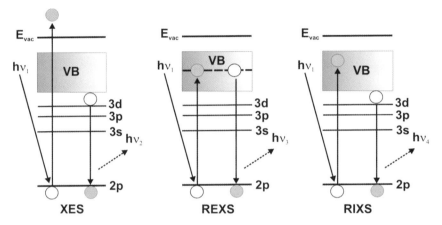

Figure 2.24 Comparison of X-ray emission processes: XES, standard X-ray emission process; REXS, resonant emission where excitation and decay involve the same energy level; RIXS, resonant inelastic X-ray scattering where the decay occurs from d to p states (dipole allowed) (see text).

In photosynthesis, the catalytic enzymes are metalloproteins. By a change of valency, spin, and hybridization with ligands, these metal centers determine biological functions and their analysis is particularly important for the design of stable mutants and derived macromolecular entities. Obtaining direct information has been difficult due to forbidden d–d transitions, but with XES techniques (here, RIXS) it is possible to excite a p-level and observe the d → p decay which is dipole allowed (e.g., the right-hand side of Figure 2.24). A problem is radiation damage of biological specimens, and as a beginning, experiments were made on rather stable molecules such as myoglobin, at 77 K. Shifting the sample along the beam allows one to investigate undamaged molecules. The high- and low-spin configuration of the center iron of the heme site of myoglobin could be clearly distinguished [154].

Due to the resonant optical behavior at the core shell absorption edges of elements, it has been shown that Cherenkov radiation can be generated in comparably narrow spectral segments in the soft X-ray region [155]. Cherenkov radiation occurs if relativistic electrons penetrate a sample if the velocity of the particle, v, is larger than the phase velocity of the electromagnetic radiation in the material, that is, $v > c/\sqrt{\varepsilon(\omega)}$. Of particular interest is the spectral region of the so-called water window where water is transparent and absorption from the carbon K edge (4.37 and 2.33 nm for oxygen) makes *in situ* investigations of biological samples possible [156].

High time resolution can be obtained by femtosecond laser slicing, a technique that has been introduced recently [157] and that uses the electron bunch–laser interaction in a wiggler. The method is based on the violation of the Lawson–Woodward theorem [158] that neglected ponderomotive forces that occur, for instance, in the energy transfer between ultrarelativistic electrons and the high electric field of a femtosecond laser (Ti:sapphire) [159]. The laser pulse induces an energy modulation of electrons that traverse the wiggler, induced by higher order laser field inhomogeneities. The result is a net force acting on the center of oscillation of the particles. This force, the ponderomotive force, Φ_{POND}, is proportional to the gradient of the intensity of the wave field $E_0(r)$:

$$\Phi_{POND} = \frac{e}{4m\omega^2} \nabla |E_0(r)|^2 \qquad (2.43)$$

The force is consistent with the concept of radiation pressure. The electrons will experience a deflection towards positions of least electrical field strength. The deflection of the electrons from the bunch occurs on the time scale of the femtosecond laser pulse and, using a bending magnet, these corresponding photons emitted by those electrons that interacted with the laser are emitted and can be used for measurements as shown in Figure 2.25. The resulting electron cavity emits femtosecond terahertz radiation [160]. Applications encompass water chemistry, protein dynamics (with vibrational periods in the range of 100 fs), time-resolved X-ray diffraction, as well as analysis of phase transitions [161, 162].

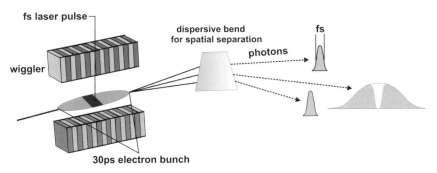

Figure 2.25 Schematic for femtosecond laser slicing; the fs X-ray photon signals and the missing signal in the unaltered signal are shown.

2.3.3.2 In-System Synchrotron Radiation Photoelectron Spectroscopy

Core-level analysis was made using the Solid-Liquid Analysis System (SoLiAS) of the collaborating research group at undulator 49/2 at BESSY. The basic setup of SoLiAS is shown in Figure 2.26 which is self-explanatory together with the figure caption. Figure 2.27 shows the design of a low-contamination apparatus for (photo) electrochemical conditioning with successive transfer to an outgassing chamber and to the analysis chamber [163]. (Photo)electrochemistry is performed on a droplet that is brought onto the sample through the light guide above it which also holds the cylindrical Pt counter electrode. The ultrapure solutions are all purged with N_2. The sample, mounted on a vacuum stub, is brought from the machine side into the cell-containing chamber that has been previously attached and purged with dry N_2 for about a day. The electrolyte droplet actually used is obtained after letting a series of drops fall beneath the sample, thus avoiding possible contamination from the hydrocarbons that can reside on the surface of water. Photoelectrochemistry is performed using a three-electrode potentiostatic arrangement and the experiment is terminated by jet-blowing the electrolyte drop off the surface with N_2. Subsequently, the sample is turned on its manipulator to be rinsed with

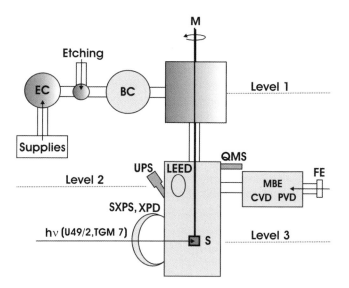

Figure 2.26 Schematic of the SoLiAS used for SRPES at BESSY; level 1 is the conditioning level with the electrochemical vessel (EC; see Figure 2.27) attached, facilities for etching, and a buffer chamber (BC) for outgassing of samples that underwent wet treatments; after transfer to the manipulator M, measurement levels 2 and 3 can be assessed; at level 2, UPS LEED and quadrupole mass spectroscopy (QMS) are available and at level 3 SRPES measurements are done. The sample enters via the fast entry lock at level 3 (where also physical and chemical vapor deposition, PVD and CVD, respectively, and molecular beam epitaxy, MBE, are provided) and is introduced into EC from the machine side thus ensuring that the electrochemical cell is kept under constant nitrogen atmosphere.

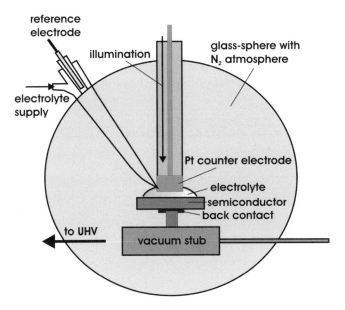

Figure 2.27 Electrochemical vessel for in-system surface analyses (see text).

ultrapure deoxygenated water, then turned back to its original position and dried in an N_2 stream. This procedure can result in the loss of information on surface chemistry because less strongly bound reaction intermediates could be washed off. Subsequently, the sample is transferred to the buffer chamber (see Figure 2.26) and parked until the pressure is low enough to allow transfer into the analysis chamber of the SoLiAS where the base pressure is in the region of 10^{-10} hp. The remnant contamination by hydrocarbons, hydroxyl groups, oxygen, and sulfur is in the region of 0.1 ML on samples where electrochemical currents are flowing.

For assessment of the surface chemistry and the changes induced by (photo) electrochemical treatment, mainly the chemical shift and the onset of the valence band emission are analyzed. The former gives information on the partial charge of respective elements and the latter indicates whether a band bending exists. For surface chemical changes, the core-level binding energy position is analyzed. The experimentally measured binding energy, E_B^{exp}, is given by the terms

$$E_B^{exp} = E_B + \Delta E_{CH} + \Delta E_{BB} + \Delta E_M + \Delta E_R \tag{2.44}$$

In this equation, the recorded binding energy is given by the binding energy of the element, E_B, the shift of the core-level position due to ionization and chemical interaction with the environment, ΔE_{CH}, a possible band bending in semiconductors, ΔE_{BB}, the Madelung term, ΔE_M, and the relaxation energy, ΔE_R.

The second term on the right-hand side of Equation 2.44, the chemical shift, is given by the effective charge of the considered atom and by the influence of the neighboring atoms. In a simple spherical charge model, this energy change can be quantified. The energy of a spherical charge is given by $E = z_V e^2/r_V$, where z_V is the number of valence electrons and r_V the average radius of the valence electrons. A change of the valency due to an altered valence electron distribution (chemical interaction) yields $\Delta E = \Delta z_V e^2/r_V$. The term Δz_V contains partial charges, and, for simplicity, r_V has been assumed to remain unchanged. Including an interatomic potential change, ΔV_{ij} (where V_{ij} is the potential of atom i on atom j), the binding energy shift reads

$$\Delta E_B = \frac{\Delta z_V e^2}{r_V} - \Delta V_{ij} \qquad (2.45)$$

Band bending occurs if semiconductors equilibrate with their surface states which shift the Fermi level into a new equilibrium position by ΔE_{BB} that differs from that given by the doping level. In most cases, depletion layers are formed. The onset of photoemission from the valence band with respect to the Fermi level, E_{VB}, is then shifted and accordingly the overall energy distribution curve is shifted by typically a few tenths of an electron volt (cf. Figure 2.28). For p-type semiconduc-

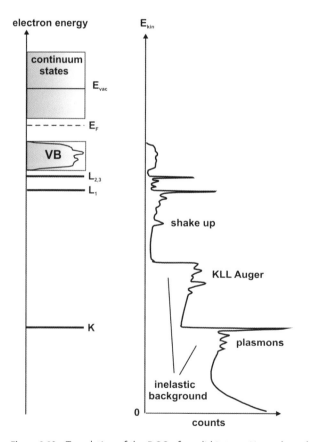

Figure 2.28 Translation of the DOS of a solid into an X-ray photoelectron spectrum (see text).

tors, shifts due to depletion layer formation increase the E_F to E_{VB} distance, and a decrease will occur for n-type semiconductors.

The Madelung term, an extra-atomic contribution, results from the additional electrical potential of the material's atoms, for instance outside the nearest-neighbor tetrahedron atoms in SiO_2 [164], acting on the excited atom with $\Delta E_M = \sum_i \rho_i / R_i$, where ρ_i is the charge on neighbor atoms (ions in ionic crystals) at positions R_i. In SiO_2, the ionicity is 51%. In addition, the Madelung contribution is not independent of the partial charge of the excited atom because the charge correction also depends on the charge of the central atom considered and the strict separation of the terms in Equation 2.44 is a simplification. Because the interatomic distances are also influenced by the charge of the probed atom, it is difficult to assess correctly the binding energy change with atomic charge.

The relaxation effects, ΔE_R, include the dynamic reaction of the system's electrons, resulting from an electron density shift of the neighboring atoms towards the oxidized atom, induced by the core hole formation in the photoexcitation process. This excitation results in an unbalanced nuclear charge. The increase in positive charge pulls the electron levels of the probe atom to higher binding energies and the valence electronic charge can react, depending on its polarizability, to the formed local potential well. The extra-atomic relaxation energy can be approximated by $\Delta E_R = q^2 / 2r_{SCR}(1 - 1/\varepsilon_M)$, with r_{SCR} being the minimum electronic screening distance and ε_M the dielectric constant of the surroundings. As Auger parameter shifts, $\Delta \alpha$, are directly correlated to the polarizability of a chemical state, it has been shown that $\Delta \alpha = 2\Delta E_R$ [165]. A comparably slow relaxation of the electrons in outer shells can result in insufficient screening of the core hole and higher measured binding energy. A related effect will be discussed below in the analysis of SRPES skew lines observed for Pt nanoislands that have been electrodeposited onto Si surfaces.

Figure 2.28 shows the translation of an original DOS distribution which includes core levels and a valence band structure of a semiconductor to an energy distribution curve (EDC) upon excitation with high-energy X-rays allowing K-shell excitation. The EDC is characterized by the original band structure, revealing the K, L levels and the valence band. In addition, typical features such as the inelastic background, shake-up or shake-off satellites, and plasmon excitations are seen. Also, the KLL Auger decay, located energetically between the K and the L core levels (see also Figure 2.22), is shown. The binding energy is typically measured with respect to the Fermi level: $E_B = 0 = E_F$. An existing band bending shifts the onset of the valence band emission and thus the whole EDC, including the core-level positions. As this is of importance in the SRPES analysis of step-bunched Si, the effect is shown in Figure 2.29. For n-type semiconductors the measured binding energy, E_B^{bb}, is lower due to depletion layer formation; for p-type semiconductors, E_B increases. Also shown is the photoeffect due to the photons from the probing light which results in a partial reversal of the energy band bending, indicated by eV_{Ph} in the figure.

SRPES allows one to tune the photon energy to the desired surface sensitivity within the limits of XPS. As the mean inelastic scattering length for photoelectrons

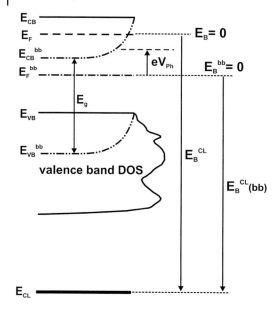

Figure 2.29 Influence of semiconductor band bending (due to equilibration with intrinsic or extrinsic surface states) on the energetic position of the EDC and of the measured core-level binding energy, E_B^{CL}, with respect to E_F (defining $E_B = 0$); the superscript "bb" indicates the situation for a depletion layer with a band bending given by $eV_{bb} = E_{CB} - E_{CB}^{bb}$; eV_{Ph} symbolizes (partial) reversion of the band bending due to a photovoltage induced by the synchrotron light.

in solids, λ_{esc}, has its minimum around 50 eV [166], an elastic escape depth of about 0.3 nm can be reached if the photon energy is adjusted to the respective core-level binding energy according to $h\nu \cong 50\,\text{eV} + \Phi + E_B$. For semiconductors, the work function Φ is given by its relation to the electron affinity χ, the energy gap, and the energetic distance between Fermi level and band edge: $\Phi = \chi + E_g - (E_F - E_{VB})$. In addition, depth profiling of the near-surface region is possible without the destruction usually accompanying sputter processes, even if they are as soft as low-energy ion scattering [167]. In this case, the photon energy is tuned to energies where λ_{esc} reaches values of 2–3 nm. The surface sensitivity and the high spectral resolution can be seen (Figure 2.30) in the XP spectrum of the Si 2p core level, measured for an H-terminated and an oxidized sample [168]. The oxide thickness, determined from $d_{ox} = \lambda_f \ln(I_f I_S^\infty / I_S I_f^\infty + 1)$ is about 0.5 nm. The parameter λ_f is the mean inelastic scattering length of the oxide film, I_S and I_f are the measured intensities of the Si substrate and the oxide film, and I_S^∞ and I_f^∞ are the photon energy-dependent sensitivity factors with $I_S^\infty / I_f^\infty \approx 0.8$ at $h\nu = 170$ eV. The FWHM of the Si $2p_{3/2}$ line is 0.45 eV for the H-terminated sample that shows virtually no oxide. Because the Si core-level shift due to the shift of partial charges from Si to H by about $\Delta\rho = 0.1$, resulting in $\Delta E_B \approx 0.25$ eV [169], is contained in the signal, the actual resolution is even higher.

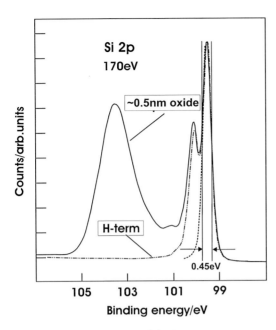

Figure 2.30 SRPE spectrum of the Si 2p core level for a chemically H-terminated sample and a sample where silicon oxide has been formed due to hole injection into the Si valence band and/or surface states; the pronounced peak at $E_B = 103.5\,\text{eV}$, corresponding to a thickness in the range of an oxide monolayer, shows that the surface sensitivity reaches far below the monolayer range; the FWHM of 0.45 eV shows the high spectral resolution.

2.3.3.3 High-Resolution Electron Energy Loss Spectroscopy

HREELS is a technique that enables the vibrational characterization of surfaces in UHV environments. It uses low-energy electrons as probe and monitors the energy loss of the electrons due to interactions with a surface [170, 171]. The principal electron scattering events can be divided into three types: (i) impact scattering, (ii) dipole scattering, and (iii) negative ion resonances where the impinging low-energy (typically a few eV) electron becomes trapped in an unoccupied orbital of a surface species forming a short-lived intermediate negatively charged ion state [172]. Whereas the dipole interaction is long range, the other two scattering events are short range. Impact scattering can be observed at off-specular angles, and dipole losses are obtained in the specular direction. A schematic of the experimental arrangement is shown in Figure 2.31. In the experiments described below, the energy loss has been measured in the specular direction.

2.3.4
Tapping-Mode AFM and Scanning Tunneling Spectroscopy

For imaging of metal nanoislands (photo)electrodeposited on semiconductor surfaces, tapping-mode AFM (TM-AFM) was used due to the reduced adherence of

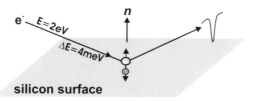

Figure 2.31 Principle of an HREELS experiment; incident and reflected beam angle, 55°; beam energy, 2 eV; energy resolution, 4 meV as indicated (see text).

the material to the surface. Highly local electronic properties were analyzed by scanning tunneling spectroscopy (STS) performed on top of nanoparticles and on the areas beside them. Therefore, selected aspects of these methods are briefly described.

2.3.4.1 Tapping-Mode AFM

TM-AFM is also known as *intermittent contact* AFM because, in this mode, the tip oscillates and touches the sample only at the lower inflection point of its periodic movement thus reducing the influence of the scanning movement of the sample. This dynamic force microscopy method [173] is preferred for imaging of soft objects which include anodic oxides and biological samples. In addition, it is applied when the adhesion of objects on surfaces is weak. This is the case for metals electrodeposited on Si and InP (Sections 2.5 and 2.6). The regime along the force–distance curve where TM-AFM operates is shown in Figure 2.32. It is characterized by the oscillating distance of the tip between the contact and the noncontact region.

The total force that governs the motion of the tip is given by the balance

$$F(x) = -k_{CL}x - \frac{m\omega_0}{Q}\frac{dx}{dt} + F_{T\text{-}S} + F_0 \cos\omega t \tag{2.46}$$

where k_{CL}, Q, and ω_0 are the cantilever spring constant, the quality factor, and the (angular) resonance frequency of the free cantilever, respectively. F_0 and ω are the amplitude and the frequency of the driving force of the tip oscillation. In Equation 2.46, the total force is given by the elastic response of the cantilever, hydrodynamic damping due to the medium, the tip–sample interaction force $F_{T\text{-}S}$, and the periodic driving force [174]. In fluid TM-AFM, the tip–sample force can be described using the Derjaguin–Muller–Toporov model [175]. The tip–sample interaction is then given by

$$F_{T\text{-}S} = -\frac{Hr_{tip}}{6a_0^2} + \tilde{E}\sqrt{r_{tip}}(a_0 - d)^{3/2} \tag{2.47}$$

where H is the Hamaker constant, a_0 an intermolecular distance, $\tilde{E}$ the reduced elastic modulus of tip and sample, and d the tip–sample separation. In the steady state, the periodic motion of the tip can be described by a sinusoidal oscillation that is given by the (mean) deflection x_0, the phase shift φ and the amplitude A_0:

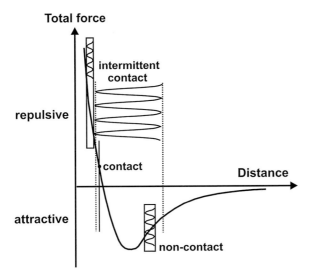

Figure 2.32 Overview of some operational modes of AFM; TM-AFM operates in the distance versus force regime denoted by "intermittent contact"; contact and noncontact modes and their positions on the force–distance curve, given by the Lenard-Jones potential $V(r) = A/r^{12} - B/r^6$ (where B contains the Hamaker constant), are indicated; the onset of the contact regime is indicated by a perpendicular bar; the wavy lines in contact and noncontact mode represent the much smaller vertical movement of the tip compared to the tapping mode.

$$x(x_{TS}, t) = x_0(x_{TS}) + A_0(x_{TS})\cos[\omega t - \varphi(x_{TS})] \quad (2.48)$$

with x_{TS} being the tip–sample separation in the absence of interactions. The change in the amplitude of the cantilever due to the interactions can be approximated for small contact times of the tip with the sample by [176]

$$A_0 = A\sqrt{1 - 4\left(\frac{\langle F_{T\cdot S} \rangle}{F_0}\right)^2} \quad (2.49)$$

The expression $\langle F_{T\cdot S} \rangle$ represents the virial, that is, the time average of the tip–sample interaction force, and A is the free oscillation amplitude given by QF_0/k_{CL}. Experimentally, the cantilever–tip ensemble oscillates at a frequency close to its resonance and the imaging of the sample takes place with the feedback adjusting the tip–sample separation such that the oscillation amplitude remains at its fixed value. This high-amplitude dynamic force microscopy allows high-resolution imaging of nanoscopic entities on surfaces and soft matter such as biological molecules [177].

2.3.4.2 Scanning Tunneling Spectroscopy

The parameters controlled in an STM experiment are current, I_t, the energy of the tunneling electrons, and the width of the barrier, that is, the height x of the tip over the sample. In STS, the behavior of a parameter pair is analyzed while the third parameter is kept constant. This results in the operation modes I_t versus x at constant V, also known as distance tunneling spectroscopy (DTS) [178], I_t versus V at constant x which will be employed in Section 2.5, and, finally, x versus V at constant I_t [179]. The tunneling current is expressed by the convolution of the DOS of the tip and the sample, D_T and D_S, respectively, the respective Fermi distributions $f(E)$, and the tunneling matrix element M_{TS} [180]:

$$I_t = \frac{4\pi e}{\hbar} \int_{-\infty}^{\infty} f_S(E) f_T(E) D_S(E_F - eV + E) D_T(E_F + E) |M_{TS}|^2 \, dE \tag{2.50}$$

The matrix element for the tunneling process is described by perturbation theory by an integral over an energy surface composed of the tip and sample wave functions Ψ and their spatial derivatives:

$$M_{TS} \propto \int_{SE} \left(\Psi_S^* \frac{\partial \Psi_T}{\partial x} - \Psi_T \frac{\partial \Psi_S^*}{\partial x} \right) dS \tag{2.51}$$

For a low temperature, not too high a voltage, and a constant matrix element, Equation 2.51 reduces to

$$I_t \propto \int_0^{eV} D_S(E_F - eV + E) D_T(E_F + E) \, dE \tag{2.52}$$

Finally, for a constant DOS of the tip in the considered energy range, the tunneling conductance measures the DOS of the sample:

$$\frac{dI_t}{dV} \propto D_S(E_F - eV + E) \tag{2.53}$$

In most cases where Pt or Pt–Ir tips are used, this condition is difficult to verify because of d-band contributions, but it might be realized for a blunt and disordered tip. Figure 2.33 shows the influence of the applied voltage on the measurement of an n-type semiconductor DOS under these approximations. At low temperature, E_F is located between the donor levels and the conduction band because the electrons on the donors are not thermally excited. For negative voltage (increase in energy), applied to the sample, the density of the occupied surface states and the valence band region is examined; for positive voltage (Figure 2.33b), the DOS of unoccupied states is measured. In our ambient atmosphere STS experiments, the thermal generation of carriers results in empty donor states, a substantial intrinsic carrier concentration, and equilibrium formation between the semiconductor and its surface states, thus changing the relations in Figure 2.33 substantially. Assuming thermal equilibrium of the semiconductor with its surface states and partial Fermi level pinning [181], a band bending occurs, as seen in Figure 2.34a. Negative biasing then results in a partial reduction of the band bending and an upward energy shift of the band edges, which, at higher bias, leads to the formation of an accumulation layer where the Fermi level at the surface is closer to the conduction band edge than by doping in the bulk. Further application

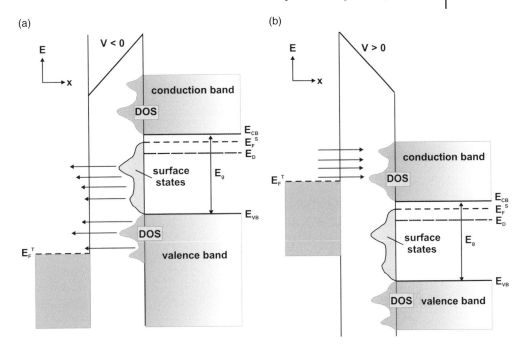

Figure 2.33 Energy relation for scanning tunneling experiments with an n-type semiconductor at low temperature and applied voltage, constant tunneling matrix element, and constant tip density of states; E_F^S, Fermi level of the semiconductor; E_F^T, Fermi level of the tip; E_D, donor levels: (a) negative voltage applied to sample, $V_a < 0$ (tip grounded); (b) $V_a > 0$.

of a negative voltage induces band edge shifts and the valence band DOS is measured. Upon positive biasing, the already existing band bending is increased until the condition of strong inversion is reached where the energetic distance between the Fermi level and the valence band maximum, E_F–E_{CB}, reaches that of E_{CB}–E_F, given by the doping level. In order to reach accumulation ($V < 0$) or strong inversion ($V > 0$), larger voltages than in the ideal case have to be applied due to the charging and discharging of surface states. Therefore, the deduced magnitude of the energy gap can be larger than the actual value, depending on the strength of the Fermi level pinning. The reason is the upward and downward band edge shift that occurs upon biasing the sample towards accumulation and strong inversion, respectively, as shown in Figure 2.34.

In the experiments on nanoscale Pt islands on Si (Section 2.5), the situation is even more complex because, besides the tunnel gap between Pt and the tip, an interfacial oxide film also exists between Si and Pt. At this interface, the photocurrent–voltage characteristics indicate the presence of Si surface states; details are given together with the experimental data in the appropriate section 2.5.3.2.

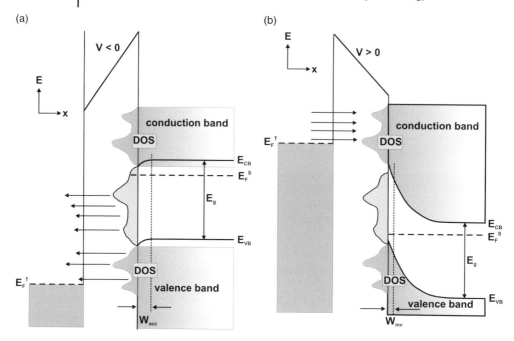

Figure 2.34 Energy relation for STS experiments (cf. Figure 2.33) at ambient temperature: (a) negative bias, forming an accumulation layer of width W_{acc}; (b) positive bias, forming a strong inversion layer of width W_{inv}. Note that the applied bias can be larger than the energy gap E_g due to charge exchange with surface states.

2.4
Case Studies: Interface Conditioning

(Photo)electrochemical processing procedures and their results are shown in Figures 2.35 and 2.36. In Figure 2.35, those crystal facets of Si are shown that have been electrochemically H-terminated in a (1×1) manner. Since the H-terminated surfaces are characterized by the lowest surface energy, they are the most stable crystal faces. The electrochemical H-termination of Si has been successful for the (111), the (113) [163], and the (110) [182] surfaces. The processing that led to the preparation of Si (111):(1×1)-H and Si (113):(1×1)-H surfaces is a multistep procedure and is therefore presented here as an example of the combinatorial possibilities of (photo)electrochemical conditioning in Figure 2.36. The conditioning uses anodic oxide formation with white light in a stepwise procedure that is assumed to prepare Si–oxide interfaces with high structural quality. The subsequent chemical/electrochemical etching includes mostly chemical oxide removal, then electron injection from a solution species into the Si conduction band (dark current transient), and subsequent chemical etching. Vibrational analysis using HREELS (Figure 2.37) shows predominantly the Si–H bending mode at 81 meV, the corresponding stretching mode (262 meV), and the scissor mode, typical for

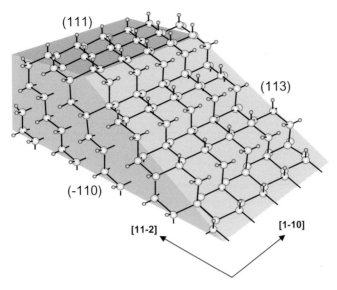

Figure 2.35 Crystal faces of Si that have been electrochemically terminated with monohydride bonds (see text).

the H–Si–H vibration, at 115 meV. The scissor mode is active because on the (113) surface, Si atoms with two dangling bonds (cf. Figure 2.35) coexist with single H-terminated surface atoms.

The mesa-type appearance of remnants on Si surfaces after photocorrosion in fluoride-containing solution in the divalent dissolution region (see Section 2.4.1), measured by contact-mode AFM (CM-AFM) (Figure 2.38), can be reproduced by combinations of the three low-index (1×1) H-terminated surfaces (111), (113), and (110) as shown on the left-hand side of Figure 2.38.

More generally, procedures include cyclic voltammetry, chronoamperometry/coulometry, their combination, and, in many cases, photoinduced modifications; pulse methods have also been applied. The abundance of possible treatments is reflected by the parameter space that includes pre-treatments, solution composition and pH, potential ranges, scan velocity, light intensity, and temporal regimes, for example. The surface (electro)chemistry, unique for each semiconductor, the energetic position of the band edges with regard to solution potentials, and the bulk semiconductor properties (doping, conductivity, band structure) demand a combination of rational and intuitive approaches for optimization of the energy conversion properties of semiconducting electrodes.

As interfacial properties control charge transfer, surface recombination, and stability, their optimization is a key to efficient and stable operation in solar energy-converting structures and devices. The examples considered here are related to the development of either photovoltaic or photoelectrocatalytic efficient systems that will be described in Sections 2.5 and 2.6. The case studies begin with the more detailed continuation of the above described formation of nanotopographies on Si

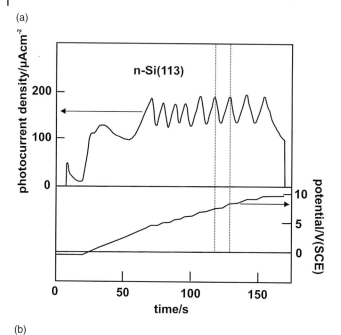
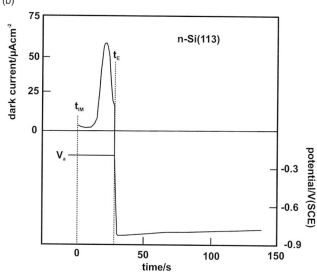

Figure 2.36 Example of photoelectrochemical conditioning for H-termination of crystalline Si, n-Si(113). (a) Photoanodization in potassium hydrogenphthalate buffer; for growth of thicker oxides, the potential was increased stepwise when the anodic current reached $100\,\mu A\,cm^{-2}$; initial potential, $-0.4\,V$; end potential, $9.5\,V$. (b) Oxide removal in two steps: (1) immersion of the electrode into $0.2\,M\ NH_4F$ (pH 4) at $t = t_{IM}$ with subsequent chemical etching and dark current transient; interruption at $t = t_E$ when dark current is 0.25 of its maximum value; (2) etching of the sample without applied potential (V_a) where V reaches the flatband potential of H-terminated Si.

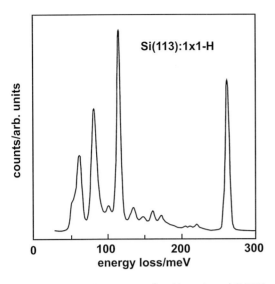

Figure 2.37 HREEL spectrum of an H-terminated Si(113) surface, conditioned as shown in Figure 2.36; $T = 200$ K; the energy loss peaks are discussed in the text.

electrodes. Despite the rather fundamental character of these studies, they have eventually led to the development of so-called nanoemitter solar cells (see Section 2.5.3) and also resulted in the formation of colored Si [183].

The *in situ* interface conditioning of p-InP by photoelectrochemical processes, described in Section 2.4.2, is a key procedure for the preparation of efficient and stable photovoltaic and photoelectrocatalytic solar cells and surface analyses will be presented that describe the induced chemical and electronic changes. The ternary chalcopyrites $CuInSe_2$ and $CuInS_2$ have meanwhile been developed for use in commercially available solid-state solar cells. For the sulfide-based cell, the use of a toxic KCN etch step of Cu-rich $CuInS_2$ to remove Cu–S surface phases is considered as deleterious for wide-scale application and an electrochemical method will be presented in Section 2.4.3 that replaces the chemical etching procedure.

2.4.1
Silicon Nanotopographies

2.4.1.1 Nanostructures by Divalent Dissolution

Silicon nanostructures can be obtained in acidic fluoride-containing media or in alkaline solution at potentials negative from open circuit potential. The (photo) current–voltage characteristic of Si in fluoride-containing electrolytes reveals a series of phenomena which have attracted considerable attention in chemistry, physics, and solar energy conversion. Figure 2.39 provides an overview of these processes for both n- and p-type Si. The *I–V* curve is characterized by two maxima and periodic variations at higher applied anodic potential. These photocurrent

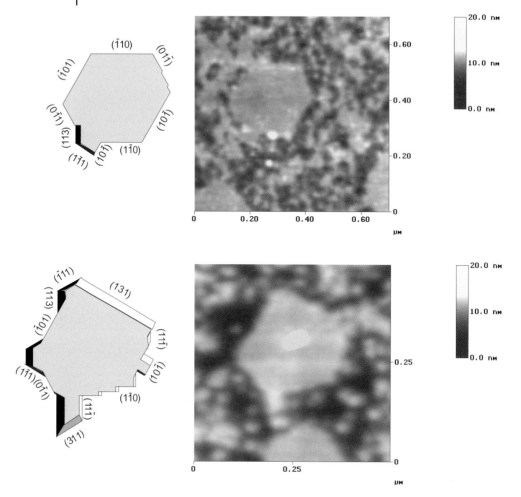

Figure 2.38 CM-AFM images of mesa-type structures obtained during divalent dissolution of Si (region I in Figure 2.39) together with a crystallographic reconstruction of the structural features using the (1×1) H-terminated low-index faces (111), (113), and (110) (cf. Figure 2.35).

oscillations will be treated separately because of the very specific nature of the process and its interrelation with silicon oxide nanotopography.

In the potential range below the first maximum, Si dissolves by two charges only, according to

$$Si_S + h_{VB}^+ + 6HF \rightarrow SiF_6^{2-} + 4H_{aq}^+ + H_2 + e_{CB}^- \qquad (2.54)$$

A Si surface atom, preferably at a kink site, is oxidized by a hole from the valence band that is either potential induced (p-Si) or light generated (n-Si). The second charge is transferred by electron injection from a solution species into the Si conduction band. The region is characterized by porous Si formation [184–187]

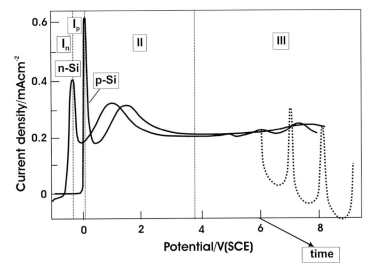

Figure 2.39 Photocurrent–voltage and dark current–voltage characteristics of n- and p-type Si, respectively, in dilute NH_4F solution at pH 4 including a chronoamperometric profile at $V = +6\,V$ (SCE) (see text).

and, in dilute NH_4F, the initial stages of the dissolution of the H-terminated Si surface can be examined. In CM-AFM images, recorded for increased anodic dissolution charge, the successive removal of the (111) H-terminated terraces is observed, eventually leading to the formation of so-called mesas whose top surfaces are still (1×1) H-terminated Si(111) faces [188, 189] (Figure 2.38). In this dissolution process, the original zigzag appearance of the (111) terraces with step edges in the [−1 2 −1] and [2 −1 −1] directions is successively lost by corrosion. The chemistry associated with this dissolution has been analyzed by SRPES. The SRPE spectrum at open circuit potential is virtually identical to the one displayed in Figure 2.30. Here, two spectra are shown recorded after light-induced anodization in the electrochemical vessel (Figure 2.27). For sample emersion near the inflection point of the I–V curve, Figure 2.40a shows a series of components in the deconvolution: (i) a surface core-level shift, due to the partial charge redistribution between Si and H on (1×1) H-terminated surfaces which amounts to $\Delta E_b \approx 0.25\,eV$ and (ii) at higher binding energy, species shifted by 0.8, 1, 1.3, 2.5, and 3.8 eV are found. In the deconvolution, the contributions α and β from stress between Si and its oxide [190] are neglected, because of their very small influence around the main line. The intensity from the surface core-level signal (dotted line) is larger than that in Figure 2.40b and indicates that more single H-terminated (111) surface areas exist than after conditioning at more anodic potentials. The influence of increased anodic potential (Figure 2.40b) is revealed in the more pronounced signals from higher oxidized Si species, particularly at $\Delta E_b = 4\,eV$, where the line of silicon dioxide is expected. With oxide formation, the signals for the reaction intermediates upon dissolution decrease as can be seen by comparing

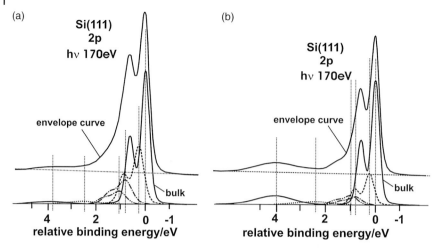

Figure 2.40 SRPES data for anodic photocorrosion of n-Si(111) in dilute NH₄F solution in the divalent dissolution regime (cf. Figure 2.39, region I): (a) sample emersion slightly anodic from open circuit potential; (b) sample emersion near first photocurrent maximum.

the corresponding lines in Figures 2.40a and 2.40b. This can be attributed to different routes in the dissolution where the mechanism of Equation 2.54 is partly substituted by the tetravalent oxidation

$$Si_S + 4h^+_{VB}(hv) + 2H_2O \rightarrow SiO_2 + 4H^+_{aq} \quad (2.55)$$

For the divalent dissolution route, the reaction starts at a kink site, for example, a surface H atom that has only two backbonds with the Si lattice. The capture of a light-induced hole into this presumed surface state generates a surface radical that is energetic enough to split water and in the process an electron is injected from solution into the Si conduction band. This has been observed as photocurrent doubling [118, 191]. The next step involves the exchange of OH and F and then a two-step solvolytic splitting of the remaining backbonds by water. In solution, OH ↔ F exchange occurs, leading to the formation of HSiF₃ which further reacts with water to SiOHF₃ + H₂. Finally, the reaction with HF results in the formation of $SiF_6^{2-} + H_2 + 2H^+_{aq}$. A corresponding scheme is presented in Figure 2.41, which is a slightly modified version of the original Gerischer model.

In connection with density functional theory (DFT) [192] on the partial charge of postulated reaction intermediates [193, 194], a reaction sequence has been postulated by comparing the binding energy shifts of the intermediates with their partial charge. Interestingly, the basic statements of a model, proposed by Gerischer et al. [193], have been largely confirmed by this combined SRPES and DFT study and a schematic of the reaction sequence is presented in Figure 2.41a. The partial charge on the reaction intermediates =Si(R) (the radical state on the right-hand side of the upper reaction in Figure 2.41), =Si–H–OH, =Si–H–F, and Si(OH)F₃ and that of the H-terminated Si surface, ≡Si–H, have been calculated by DFT [195] or, for the latter, by an electronegativity consideration that gives the

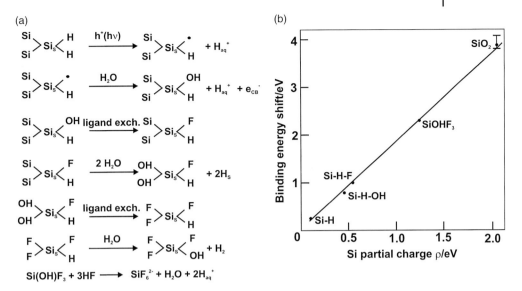

Figure 2.41 (a) Reaction schematic of the divalent dissolution of Si, beginning at a kink site atom Si_S; h^+, light-induced defect electron; CB, VB, conduction and valence band, respectively; H_S, hydrogen atom that bonds to the underlying Si atom that has become a surface atom (see text). (b) Plot of the binding energy shift of the Si 2p core level, measured by SRPES, versus the calculated partial charge of reaction intermediates.

surface core-level shift based on a microcapacitor model [196]. The corresponding partial charges are $\delta = 0.24$, 0.45, 0.54, 1.29, and 0.1, respectively. The partial charge on SiO_2 has been calculated as 2.05 [197]. The relation between partial charges and core-level shifts measured by SRPES yields a linear behavior (Figure 2.41b). Because the reaction intermediates, present in submonolayer amounts, experience predominantly the environment of the Si crystal, the screening should resemble that of bulk Si, except for special situations [198]. The result confirms substantial parts of the dissolution model. The processing in the electrochemical vessel including rinsing and drying procedures will only maintain species at the surface that are strongly bound, not too short-lived, or, for instance, held by capillary forces on rougher surfaces. Because the intermediates =Si–H–OH and =Si–H–F are observed, they could be associated with the rate-determining step which would be electron injection into the conduction band, known to be slow [199] and, also, the ligand exchange appears to be slow as both species are found on the surface. The higher oxidized $Si(OH)F_3$ species could be a precipitate or held by the aforementioned forces.

2.4.1.2 Step Bunched Surfaces

Silicon etching in alkaline solutions has a long history [200] and has found numerous applications [201–203], and, accordingly, models of Si dissolution have been developed [204–206]. Dissolution is typically assumed to originate at an

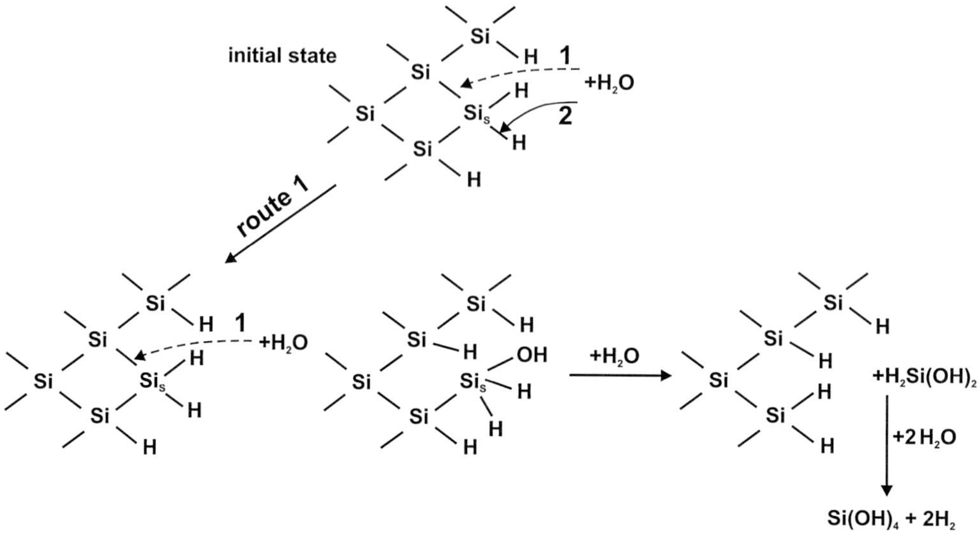

Figure 2.42 Schematic of etching sequence for Si(111) in alkaline solutions, indicating two possible routes labeled 1 and 2; a kink site surface atom with two backbonds is shown; route 1 [204] is presented here (see text).

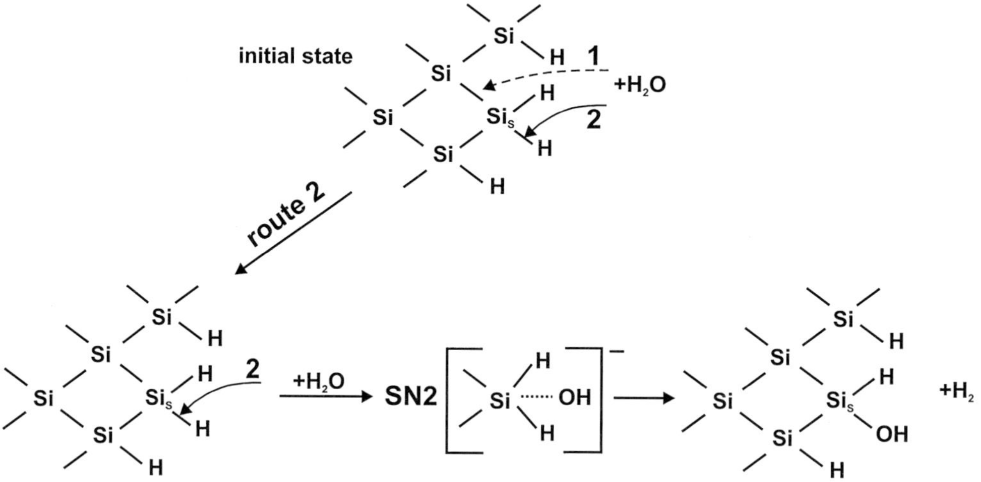

Figure 2.43 Reaction sequence for route 2 in Figure 2.42; chemical attack occurs at the Si–H bond of the kink site atom; subsequent OH formation takes place via an $S_{N}2$ reaction (after [206]; see text).

H-terminated kink site atom on a (111) surface that reacts with water (see Figure 2.42). Models vary depending on whether a Si–H bond on the kink site atom is first attacked or whether solvolytic splitting of a backbonds initiates the dissolution reaction. Figures 2.42 and 2.43 show this situation including the initial attack for

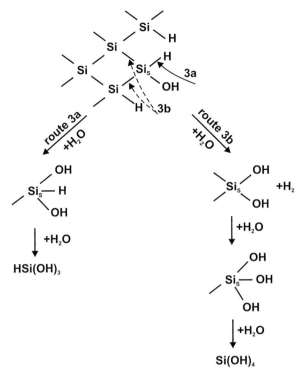

Figure 2.44 Consecutive steps in Si dissolution depending on either attack of the Si–H bond (route 3a) or on the Si backbond (route 3b); further dissolution steps for each route are shown (see text).

the two presumed reaction routes, labeled 1 and 2. In the dissolution sequence of route 1 [204] (Figure 2.42), formation of a –Si–H$_2$–OH intermediate occurs. Further attack by a water molecule yields an H-terminated surface and the species H$_2$Si(OH)$_2$ which further reacts to Si(OH)$_4$ under simultaneous hydrogen evolution.

Route 2 [206] (Figure 2.43) involves the intermediate formation of pentavalent Si in an S$_N$2 reaction, resulting in a =Si–H–OH surface species. Then, two scenarios are proposed (see Figure 2.44): (i) solvolytic attack of the remaining Si–H bond or (ii) backbond splitting. In the former process, successive oxidation results in formation of Si(OH)$_4$ and the latter sequence produces an HSi(OH)$_3$ molecule. In process (ii), the intermediate =Si–(OH)$_2$ is formed that reacts with water to Si(OH)$_4$.

As Figure 2.45 shows, the Si etching rate in alkaline solution depends strongly on the applied potential with a maximum at open circuit potential [204, 207]. The etch rate decrease in the cathodic region, for example, is related to the kinetic stabilization of the surface kink site atoms by electrons from the conduction band of n-Si under accumulation conditions. The corresponding reaction sequence is shown in Figure 2.46. The scavenging of a conduction band electron results in

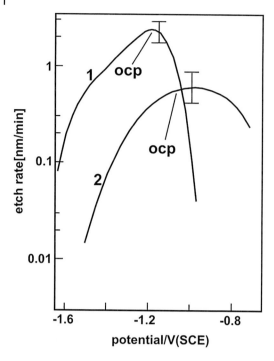

Figure 2.45 Influence of electrode potential on the etch rate of Si in alkaline solution: (1) after Glembocki et al. [207]; (2) Allongue et al. [204] (see text); the open circuit potential (ocp) is indicated in both cases.

formation of a surface radical. Capture of a second electron produces a negatively charged species that interacts with a water molecule to restore the original configuration. In this process, two OH^- ions and an H_2 molecule are formed and a local increase in the pH occurs. The reaction of the radical with water to form =Si–OH–H, associated with electron injection into the conduction band (cf. Figure 2.41b), is also possible but the rate of electron injection from the conduction band of a semiconductor in accumulation condition will be much larger compared to that of the electron injection step from solution [208].

A conditioning protocol is shown in Figure 2.47: beginning at open circuit potential at −1.2 V, the potential is linearly increased to −1.57 V with a scan rate of 20 mV s^{-1}. Then, at a current density of −110 µA cm^{-2}, the potential is held for 40 s. Subsequently, the sample is rinsed and transferred to the UHV system for analysis and Figure 2.48a shows the envelope curve for the Si 2p core level. Also shown, indicated by the arrows, are species that have been identified by deconvolution, and their respective contribution, normalized to the bulk Si line, and the chemical identification are given in Table 2.1. In Figure 2.48b, the O 1s line is depicted, showing a contribution from hydroxyl species at 532.1 eV and from molecular water at 532.8 eV. The signal shifted by 0.2 eV is attributed to Si–H bonds, the shift $\Delta E_b = 0.5$ eV corresponds very likely to =Si–H$_2$ bonds, and the remaining two

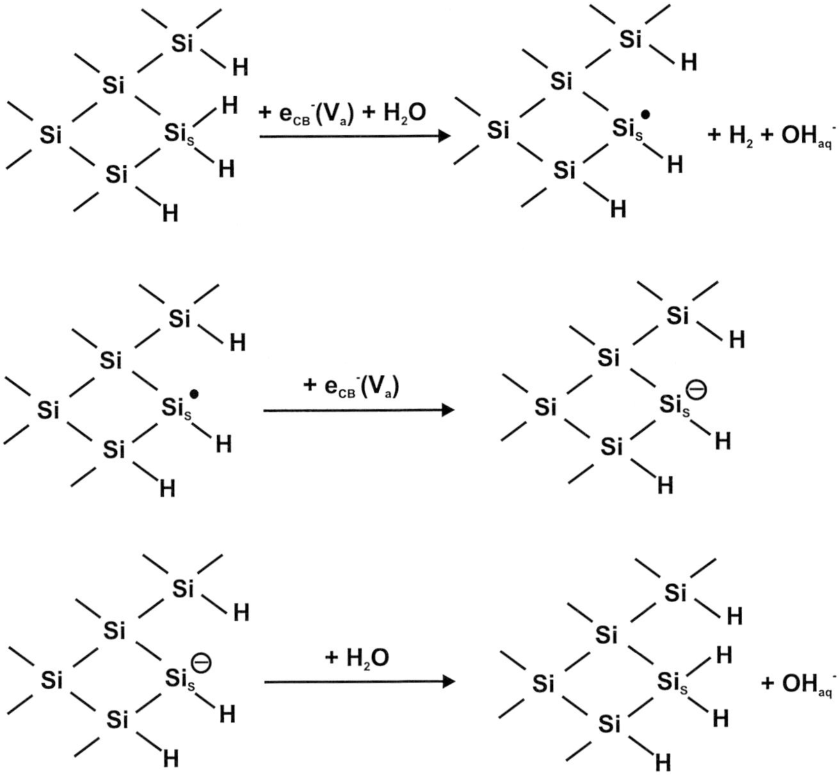

Figure 2.46 Reaction sequence for n-Si(111) under accumulation conditions (potential cathodic from flatband) in alkaline solution; note the restoring of the initial condition and the occurrence of a negatively charged Si surface atom.

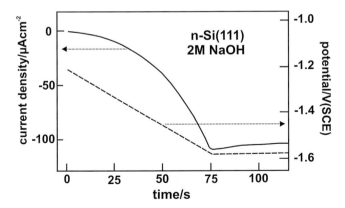

Figure 2.47 Conditioning protocol for mixed chemical/electrochemical preparation of step bunched Si.

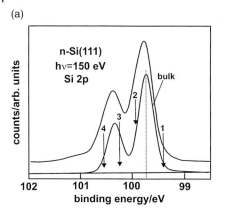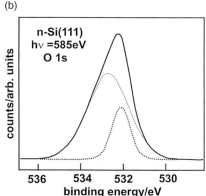

Figure 2.48 SRPES data for (a) the Si 2p and (b) the O 1s core level at high surface sensitivity, tuned by the photon energy; the arrows in (a) indicate positions of reaction intermediates obtained from deconvolution. The distance of the respective arrows to the base line for deconvolution gives the maximum value of the $2p_{3/2}$ signal relative to the other deconvoluted signals (see also Table 2.1).

Table 2.1 Reaction intermediates identified by SRPES Si 2p spectrum deconvolution after treatment in 2 M NaOH according to the conditioning procedure shown in Figure 2.47 (for species identification, see Figure 2.46).

Binding energy (ΔeV)	Surface chemical species	Relative contribution
−0.33	=Si–H–e$^-$	0.08
+0.20	=Si–H	0.68
+0.50	=Si–H$_2$	0.12
+0.80	=Si–H–OH	0.12

signals ($\Delta E_b = 0.8$ and -0.3 eV) are attributed to =Si–H–OH and the negatively charged surface species (second row down in Figure 2.46), respectively.

The surface is predominantly H-terminated and, if one includes the intermediate with a negative charge on a dangling bond, =Si–H–e$^-$, which belongs to the reaction circle that reinstalls H-termination (Figure 2.46), the H-termination reaches almost 80%. The finding of the =Si–H–OH species points to the reaction mechanism characterized by route 2 in Figure 2.43. On the basis of the present results, it cannot be decided how the reaction proceeds further, that is, whether route 3a or 3b is followed. The associated higher oxidized Si surface atoms are obviously short lived and weaker bound intermediates which do not withstand the rinsing, drying, and outgassing procedure in the sample transfer to the UHV analysis chamber.

The surface topography that arises from the treatment dipicted in Figure 2.47 is shown in Figure 2.49. The CM-AFM image shows the occurrence of steps that are several atomic bilayers high. Steps are about 3–5 nm and the atomic bilayer

(a)

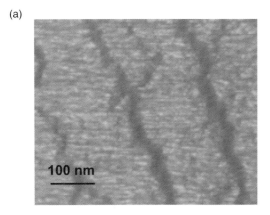

(b)

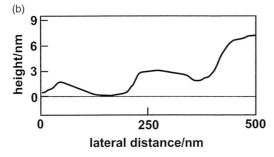

Figure 2.49 CM-AFM measurements on n-Si, conditioned according to Figure 2.47: (a) top-view scanning image; (b) cross-sectional height profile.

height is 0.314 nm for the (111) face. The steps are thus roughly 10–15 times larger than the atomic step height. Due to the curvature of the probing AFM tip (10–30 nm tip radius), the features observed in Figure 2.49 appear somewhat blurred. The phenomenon of a step height increase is known as step bunching [209, 210]. Although known for some time [211], only recently has a plausible model been suggested for the explanation of the basic mechanism acting to produce larger steps [212] on vicinal Si surfaces (miscut of 3.5° towards the <11-2> direction) exposed to aqueous electrolytes. This model, however, describes the chemical dissolution only and does not consider electrochemical processes. Basically, it considers an increased viscosity due to formation of polymerized dissolution products of the type $Si(OH)_4$ where –Si–OH and HO–Si– form oxobridges –Si–O–Si– by release of water molecules. The accordingly increased viscosity is thought to inhibit product dissolution and results in a decrease of the etching rate where the original atomic step density is high because, at such regions, more kink sites will be available for etching. Therefore, the condition for step bunching, $-\partial k_e/\partial \rho_{step} \neq 0$ [211, 212], is fulfilled here (k_e is the etch rate and ρ_{step} the step density). In the electrochemical preparation of step bunched Si surfaces, the etch rate changes with applied cathodic potential and becomes considerably smaller near

−1.6 V (cf. Figure 2.45 and the conditioning protocol) than at open circuit potential. Obviously, the product dissolution is still strongly enough inhibited to result in step bunching although the surface considered here had a nominal miscut of only 0.5° which yields a considerably lower initial step density than on a vicinal surface.

Surface electronic properties of step bunched Si have been analyzed by SRPES, measuring the valence band onset of photoemission and the shift of the Si 2p core level with changing photon energy. The latter results in a corresponding change of the mean inelastic scattering length of photoelectrons and thus provides depth information. This dependence is shown in Figure 2.50. One observes that the position of the 2p line changes from $E_b = 99.72$ eV for high surface sensitivity to

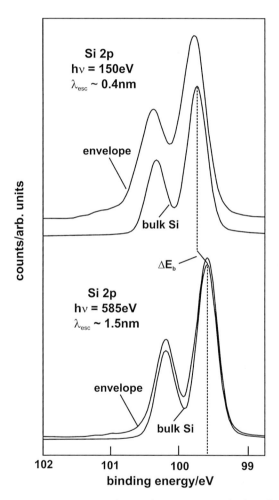

Figure 2.50 SRPE envelope and deconvoluted bulk Si spectra for two different photon energies corresponding to different escape depths of photoelectrons, λ_{esc}. Top: high surface sensitivity. Bottom: escape depth about 1.5 nm. The binding energy shift of about 0.15 eV is indicated (see text).

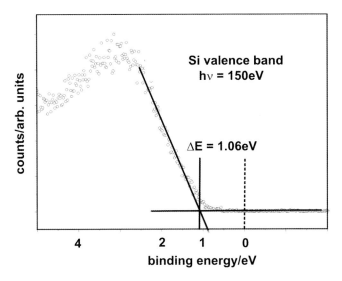

Figure 2.51 X-ray photoelectron valence band spectrum at high surface sensitivity ($h\nu = 150\,eV$) showing the energetic difference between Fermi level and valence band photoemission onset to be 1.06 eV.

99.65 eV for larger escape depth. Because the Si surface species are much more prominent for $h\nu = 150\,eV$, the deconvoluted bulk Si line has been used for comparison. In Figure 2.51, the photoemission onset at the Si valence band is extrapolated from an XPS valence band spectrum. The energetic difference between E_F and E_{VB} is 1.06 eV which is larger than expected from the doping of the n-type sample where the difference $E_{CB} - E_F = 0.26\,eV$. Hence, $E_F - E_{VB} = E_g - 0.26\,eV = 0.86\,eV$. This value is increased by 0.2 eV as also observed for the Si 2p core level where the shift was about 0.15 eV. This difference can be related to the spatial profile of the electrostatic (Galvani) potential in the accumulation layer that does not reach its bulk value at a depth of 1.5 nm. The sample is in an accumulation condition with a high electron concentration near the surface as depicted by the energy band schematic in Figure 2.52. The extension of the accumulation layer can be described using the Thomas–Fermi screening approximation [213] for different majority carrier concentrations n_e. In the free electron approximation, for $n_e = 10^{18}\,cm^{-3}$, the screening length is given by $k_{TF} = \sqrt{4(3n_e/\pi)^{1/3}/a_0}$, with $L_{TF} = 1/k_{TF}$, yielding a length of about 2 nm. The electrostatic potential changes according to $\varphi(x) = (q/x)\exp(-k_{TF}x)$ and the accumulation layer band bending extends about 2 nm below the outermost surface. Instead of a linear–parabolic dependence that characterizes the energy band change for nondegenerate semiconductors, the potential dependence is here exponential, as is indicated in the figure.

The lateral potential distribution across the surface has been measured by Kelvin probe AFM [214], which monitors contact potential differences across the surface and has been successfully applied, for instance, in the electrostatic

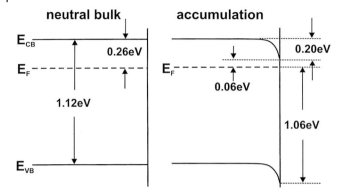

Figure 2.52 Energy versus space diagram for n-type Si in flatband situation (left) and in the accumulation situation as determined by photoelectron spectroscopy (right).

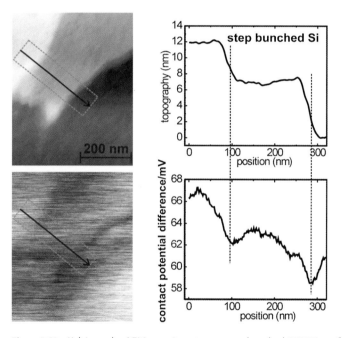

Figure 2.53 Kelvin probe AFM experiment on a step bunched Si(111) surface. Top: topographic features of the investigated step bunched sample; bottom: variation of the electrostatic potential.

characterization of small-grain polycrystalline solar cells [215]. A typical measurement on step bunched Si is shown in Figure 2.53. It can be seen that the surface nanotopography and the contact potential difference are related: the electrostatic potential drops mostly near step edges and reaches its smallest value at the inflection point of the step edges. This indicates a spatially inhomogeneous distribution of the surface charge in the accumulation layer having its maximum values around step edges. This property of the surface has been used for site-specific adsorption of heterodimeric enzymes in a precursor experiment for biologically inspired solar fuel generation systems [216].

2.4.1.3 Oxide-Related Nanotopographies

The current–voltage characteristic in Figure 2.39 shows that at higher anodic potential, current oscillations occur for both n- and p-type Si. Whereas the former is induced by light-generated holes, the latter results from hole flux due to the accumulation layer that is formed upon anodic polarization. The origin of these oscillations has been described by several authors and an unambiguous interpretation has not yet been achieved [217–221]. In addition, a macroscopic model has been introduced that describes oscillatory phenomena on a larger spatial scale [222, 223] than the more nanoscopic theories.

Basically, in the nanoscopic models, the assumed mechanisms encompass self-oscillating nanoscopic domains [217], the assumption of a so-called current burst [218], and the stress-induced formation and temporal evolution of nanopores in silicon oxide that forms under anodic polarization conditions [224]. Surface analysis by XPS and Fourier transform infrared (FTIR) spectroscopy shows that for illuminated n-Si, at +6 V (SCE), an anodic oxide exists that has an integral average thickness of about 10 nm [224]. In Figure 2.54, the FTIR signal has been evaluated with regard to the oxide thickness [225] and is contrasted with the oscillatory current behavior. One sees that a distinct phase shift exists: for the increasing branch of the photocurrent, the oxide has its minimum thickness of about 17 ML, at the current maximum and minimum, the thickness of 22 ML is virtually equal, and the largest thickness is found for the decreasing branch of the current after

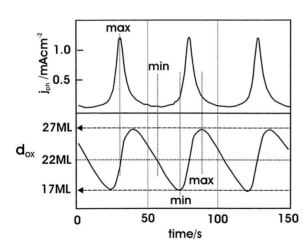

Figure 2.54 Photocurrent oscillations for n-Si(111) at $V = +6$ V in 0.1 M NH_4F, pH 4 (top) and oxide thickness variation determined by *in situ* Fourier transform infrared spectroscopy experiment measuring the asymmetric Si–O stretching mode; note the phase shift between current and thickness.

the maximum. Taking 0.45 nm as the thickness for an oxide monolayer [226] yields an average thickness of about 10 nm with a variation of ±2.3 nm. The occurrence of sustained oscillations at minimum oxide thickness values around 7.7 nm is difficult to envisage; therefore, the existence of pores within the oxide that allow electrolytic conductivity was postulated rather early [227] and, later, a theory was developed based on the temporal evolution of oxide nanopores [220, 221, 228]. Indeed, nanopores within anodic oxides formed under dynamic competitive conditions (i.e., anodic oxide formation and oxide etching in dilute ammonium fluoride solution) have meanwhile been seen under various experimental conditions [229, 230]. Typical images are shown in Figure 2.55 where nanopore fields are displayed for emersion of the Si sample at different phases of the oscillating current as indicated in the figure. It can be seen that the pore contrast and size are largest for the smaller oxide thickness. Such a nanoporous oxide will be discussed further below as a template for the preparation of nanoemitter solar cells.

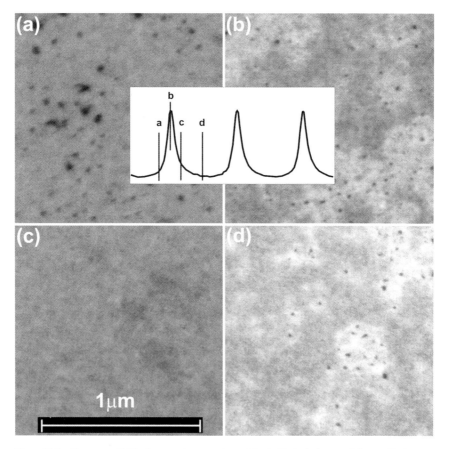

Figure 2.55 Nanopore fields for sample emersion at the indicated phases of the oscillating current; experimental conditions as in Figure 2.54.

The underlying concepts of the model, based on interfacial stress between Si and its anodic oxide, are briefly reviewed: at its core is the dynamic competition between oxide formation (cf. Equation 2.52) and dissolution in dilute (0.1–0.2 M) NH_4F solutions with pH of about 4. The processes of light- or potential-induced oxidation and etching in HF and HF_2^- occur simultaneously and the overall reactions are as follows:

$$\text{n-Si} + 4h_{VB}^+(h\nu) + 2H_2O \rightarrow SiO_2 + 4H_{aq}^+ \tag{2.56}$$

$$\text{p-Si} + 4h_{VB}^+(V_a) + 2H_2O \rightarrow SiO_2 + 4H_{aq}^+ \tag{2.56a}$$

$$SiO_2 + 6HF \rightarrow SiF_6^{2-} + 2H_{aq}^+ + 2H_2O \tag{2.57}$$

$$SiO_2 + 3HF_2^- \rightarrow SiF_6^{2-} + H_2O + OH_{aq}^+ \tag{2.57a}$$

Equations 2.56 and 2.56a describe oxidation via holes from the valence band that originate either from photon excitation of minority carriers in n-Si or from the formation of an accumulation layer by anodic polarization (V_a) of p-Si, respectively. Equation 2.57 is the etching reaction for HF that generates hydronium ions (for simplicity, only the solvated protons H_{aq}^+ are written) and Equation 2.57a describes the etching by the hydrogen bridge complex HF_2^- which only takes place at moderately acidic pH.

The competitive oxidation and etching processes are analyzed using the concept of highly local thickness oscillators; they represent the thickness behavior of the oxide at a given position (coordinates x, y on the Si surface plane). The upper inflection point of such thickness oscillators is given by the maximum oxide thickness that can be reached under the anodization conditions (light intensity, potential), reduced by the etching occurring already during oxide formation (etch rate depending on molarity and pH). Figure 2.56 shows that the oxidation process is fast compared to typical etching times in dilute fluoride-containing solutions. Therefore, the periodic behavior of the thickness oscillators is dominated by the oxide etching process. The cycle times of the thickness oscillators determine the temporal evolution of the system. The oscillating current, which is a photocurrent for n-Si and a dark current for p-Si, is derived from an initial current peak (due to oxidation of the bare H-terminated Si surface) whose temporal behavior is deduced from the current profile in Figure 2.56 and the temporal evolution of the thickness oscillators. In this model, the oscillating current $j(t)$ is proportional to a synchronization function $p(t) = \sum_i p_i(t)$; $p_i(t)$ define the synchronization of the thickness oscillators in the ith cycle of the oscillation and they are derived from the preceding synchronization function $p_{i-1}(t)$ via a Markov chain [220, 221, 231]:

$$p_i(t) = \int_{\Delta t_i} p_{i-1}(\tau) q_{i,\tau}(t-\tau) \, d\tau \tag{2.58}$$

The integrand $q_{i,\tau}$ represents the probability distribution of the cycle time of the thickness oscillators. In this formalism, the temporal development of the system is described in some mathematical analogy of the Feynman path integrals that also use a recursive description and probability theory [232, 233] for particle propagation in quantum electrodynamics.

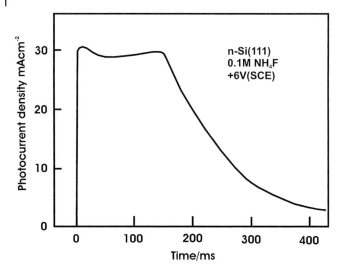

Figure 2.56 Photocurrent decay (oscilloscope measurement) during the first initial oxidation phase of n-Si(111) in dilute NH₄F, pH 4; W–I lamp, ms shutter; in this first oxidation phase, about 10 monolayers of oxide are formed (see text).

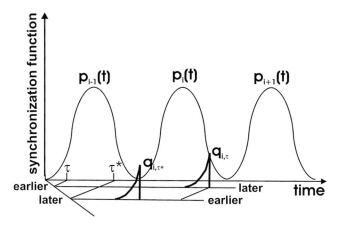

Figure 2.57 Plot of the temporal behavior of the synchronization functions $p_i(t)$ and of the probability distribution of the periods of the oxide thickness oscillators $q_{i,\tau}$; a plausible but arbitrary shape of $q_{i,\tau}$ has been assumed; initial and final conditions for two different electrode areas where oxidation sets in at different times are shown. Also shown is the contraction of the time interval for the later started oxidation process (see text).

Synchronization depends crucially on the maintenance of the temporal shape of the functions $p_i(t)$. Therefore, the temporal behavior of the thickness oscillators, described by $q_{i,\tau}$, must contain a synchronizing feature. Mathematically this can be achieved by a temporal compression of those q that start at a later time τ^* in the time interval Δt in such a way that the thickness oscillators that start later within one cycle finish their cycle earlier. This situation is depicted in Figure 2.57, which shows a temporal sequence of the synchronization functions $p_i(t)$ together with two $q_i(t)$, labeled $q_{i,\tau}$ and q_{i,τ^*}. The former has been assumed to start earlier and the latter sets in later but has a shorter time interval for completion of the oscillation which results in an earlier endpoint (in time) than for $q_{i,\tau}$. Then, in the subsequent cycle, these oscillators start earlier and the consecutive temporal development yields a sustained oscillation. Hence, the *mathematical* feedback mechanism is given by the contraction of the cycle time of the thickness oscillators. Now, the question arises: what could be the physicochemical origin of this contraction? As the oxidation process itself is fast and etching is slow, a valid assumption is that the synchronization is associated with the etching process. One needs at least two different oxide etching rates that differ enough to allow for fast etching of later formed oxides within one oscillatory cycle. Therefore, one considers the oxide growth within an idealized cycle (*i*th) as shown in Figure 2.58. The side view (Figure 2.58a) shows initially formed oxide islands which grow into the Si substrate but also extend towards the exterior of the geometrical surface, due to the

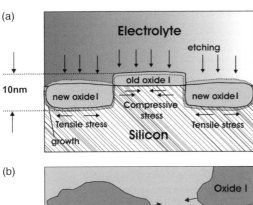

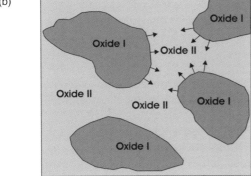

Figure 2.58 Schematic of the generation of stress by earlier formed oxides on later formed ones within an oscillatory cycle. (a) Side view indicating compressive stress in the oxide and tensile stress in the Si substrate. Between oxide islands, the stress is compressive underneath the oxide that is etched in the former cycle (old oxide); underneath the old oxide I, a defect-rich oxide II is formed that etches faster. (b) Top view at an advanced stage where the later formed oxide II is under considerable stress, indicated for a selected region between oxides I by the arrows.

volume mismatch of 2.27 between the oxide and Si. The oxide is thus under compressive stress whereas Si is underneath the oxide and is under tensile stress, a fact that is important in microelectronic devices [234]. Between oxide islands, the stress in the substrate is compressive as indicated. The top surface view (Figure 2.58b) shows also a rather advanced stage of the oxidation process. For simplicity and intelligibility, the oxides are labeled I and II to characterize oxides that have been formed at earlier stages of the process (I) and at later stages (II). It is seen that oxide II forms in Si that is under substantial tensile stress, and thus this oxide will exhibit a considerably enhanced defect density. Such defects could be point and line defects but also small cracks that, in the presence of the electrolyte and under etching conditions, can transform into nanopores.

Etching at nanopores becomes a three-dimensional process whereas the non- or less-defected oxides etch predominantly in a two-dimensional process that is much slower. In Figure 2.58b, the successive oxidation process is indicated by the areas located between the oxide I islands where the arrows symbolize the stress forces. It is known from fractal photocorrosion of Si, for example, that oxidation and dissolution are prevalent at sites with stress-induced distortion of bonds [235].

In Figure 2.59, these processes are condensed in a schematic that shows the temporal evolution of sustained oscillations. Damped oscillations occur if the etching time interval of successively formed oxides does not contract enough. In this case, the bell-type shape of the synchronization function in Figure 2.57 becomes more spread out until the oscillations cease. Under these conditions, the Si–oxide interface will exhibit an increased roughness because the system is characterized by the simultaneous existence of various $p_i(t)$ (e.g., $i - N$, ..., $i - 1$, i). Indeed, such oscillatory phenomena have been used for the preparation of electronically superior Si single-crystal surfaces [236], because the oscillatory behavior is a macroscopic phenomenon and hence synchronization occurs throughout the sample surface. Any structure formed in this process exists on the overall sample surface resulting in the scalability of the related nanostructures which can be used for the preparation of solar energy converting devices (see Sections 2.5 and 2.6).

Another example of nanotopography development, achieved by simultaneous monitoring of photocurrent and reflectance measured by Brewster angle analysis, uses the transformation of the dissolution mechanism from the divalent process to the tetravalent one; the principle of the conditioning is shown in Figure 2.60 where the photocurrent–voltage characteristic for n-Si(111) in dilute ammonium fluoride solution is dipicted in the potential region around the first current maximum where this transition occurs [169]. As seen in Figure 2.60, the I–V characteristic shifts to more cathodic values with increasing light intensity. At a chosen electrode potential (here, -0.1 V), the current shows markedly different values depending on the light intensity. As a consequence, the light intensity can be used to establish current flow either on the cathodic or anodic side of the first current maximum. On the cathodic side, the divalent dissolution mechanism dominates (cf. Figure 2.39) and on the anodic branch, anodic oxide formation occurs. Figure 2.61 shows a simultaneous measurement of the photocurrent–voltage characteristic and the Brewster angle reflectance, measured at the Brewster

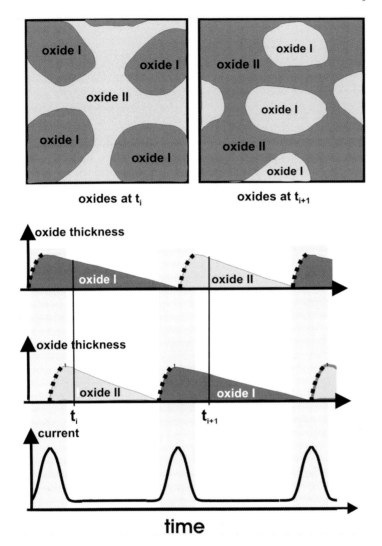

Figure 2.59 Schematic of the temporal development of overlapping oxide formation and etching assuming that regions exist on the surface that are later oxidized but etch faster due to more three-dimensional attack at extended defects (nanopores); the faster oxide formation than the etching results in an asymmetric temporal profile for both oxide types. Also shown is the shorter lifetime of the more defect-rich oxide II resulting in a compression of its lifetime with respect to oxide I (see text).

angle, φ_B, of a smooth (H-terminated) Si substrate. It is seen that the reflectivity reaches its maximum before the current and that $R_p(\varphi_B)$ drops strongly afterwards. The increase of the reflectance signal is attributed to an increase of size and density of nanostructures that are formed in the divalent dissolution regime (see, for instance, Figures 2.35 and 2.38). Before the first photocurrent maximum, R_p

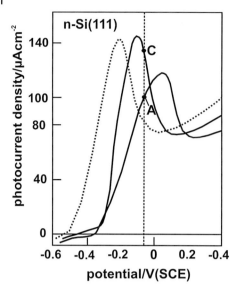

Figure 2.60 Photocurrent–voltage behavior of n-Si(111) in 0.1 M NH$_4$F, pH 4, for increased light intensities; dotted curve, $I = 40\,mW\,cm^{-2}$; full curves, (C) $I = 7\,mW\,cm^{-2}$, (A) $I = 1\,mW\,cm^{-2}$. A and C indicate electrode potentials which are, dependent on the light intensity, connected to a transition from divalent to tetravalent dissolution.

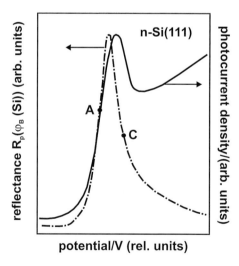

Figure 2.61 Simultaneous reflectance R_P (dashed-dotted curve) and photocurrent (solid curve) measurement near the first photocurrent maximum; the shift of the maximum of R_P relative to the current maximum indicates transition to tetravalent dissolution before the current maximum (see text); points A and C indicate the reflectance behavior and thus the surface condition that can be obtained at fixed potential by varying the light intensity (cf. Figure 2.60).

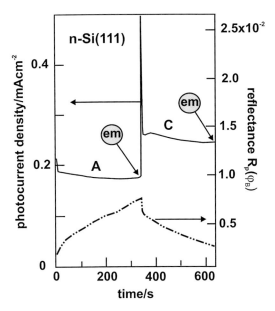

Figure 2.62 Chronoamperometric conditioning of n-Si(111) in 0.1 M NH$_4$F, pH 4; potentials A and C correspond to the respective light intensities in Figure 2.60 and indicate divalent and tetravalent dissolution, respectively; the encircled "em" denote the instant of sample emersion from solution for subsequent surface topographical analysis by AFM and SEM (Figure 2.63).

decreases due to the commencement of nanostructure oxidation and oxide etching in the HF-containing solution. Also indicated in Figure 2.61 are the R_p values for conditions A and C in Figure 2.60, that is, at constant potential. In Figure 2.62, the differences in the surface dissolution processes between conditions A and C are exploited for the preparation of a novel nanostructure. The temporal evolution of the current and the reflectivity signal are displayed simultaneously for a low and a higher light intensity, corresponding to conditions A and C, respectively. A successive roughening of the surface at condition A is seen in the reflectivity increase. Moreover, upon switching the light intensity to 7 mW cm^{-2}, a step in the current is observed, accompanied by a decrease in R_p that results from the different optical properties of the electrolyte–oxide–Si interface and smoothening due to the transition into the electropolishing region where the oxide is etched. The surface after conditioning A for 300 s is shown in the upper left AFM image in Figure 2.63. After 300 s at condition C, the then oxidized surface is characterized by Si nanostructures that are embedded into an SiO$_2$ mask as shown by the upper right AFM image in Figure 2.63, and the corresponding SEM image is shown below. The described treatment serves as a novel nanostructure conditioning step, controlled by Brewster angle reflectance, which results in partially oxidized Si nanostructures that can be used as templates for metal deposition in nanoemitter structures or as catalyst supports in photoelectrocatalysis applications, for example.

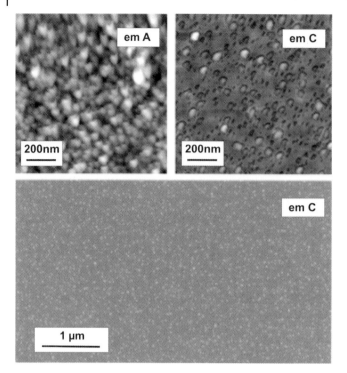

Figure 2.63 AFM and SEM analysis of surface structure induced by switching the Si dissolution process from the divalent mode to the oxide-forming tetravalent one. Top: AFM images upon emersion at condition A and C in Figure 2.62, respectively; bottom: SEM image for emersion at condition C in Figure 2.62 (see text).

Possibilities for the preparation of rod-type Si features are provided by the finding that there is an increased aspect ratio after oxide removal.

2.4.2
Indium Phosphide

Indium phosphide has been a successful material in the preparation of solid-state, photovoltaic, and photoelectrocatalytic electrochemical solar cells [237–240]. Photovoltaic solid-state solar cells reach single-junction efficiencies above 24% [237]. When used as a photocathode in photoelectrochemical solar energy conversion, the material has shown excellent stability [239], related to the unique surface chemistry of the polar InP(111) A-face that exposes In atoms only [240]. The photoelectrochemical conditioning of single-crystalline p-type InP with the aim of preparing efficient and stable photoelectrochemical solar cells for photovoltaic and photoelectrocatalytic operation is described in the following and the induced surface transformations are analyzed employing a variety of surface-sensitive methods.

Two surfaces are conditioned: the aforementioned InP(111) A-face with In atoms protruding a quarter of a monolayer and the In-rich InP(100) (2×4) reconstructed surface of a thin homoepitaxial layer, prepared by metal organic vapor-phase epitaxy (MOVPE) on a crystalline InP wafer substrate [241]. Whereas the former surface has been successfully conditioned for the development of an efficient and stable photovoltaic photoelectrochemical solar cell (see Section 2.5.2), the latter surface has been employed in the development of an efficient photoelectrocatalytic structure for light-induced hydrogen generation (see Section 2.6.2). The functioning of solar energy conversion devices is typically described by energy band diagrams of the respective semiconductor–electrolyte junctions. Such band diagrams can be obtained by a variety of experimental methods in conjunction with theoretical data. Among the particularly well-suited methods is UPS [150] as this method does not only allow the analysis of the valence band structure but also monitors changes in band bending due to surface states and in electron affinity related to surface dipole changes [242]. Such experiments can be performed using the in-system apparatus shown in Figures 2.26 and 2.27.

2.4.2.1 The InP(111) A-face

A schematic of the atomic arrangement of the (111) A-face is shown in Figure 2.64. The In atoms form the topmost surface layer, a quarter of a monolayer on top of the P atoms. Therefore, this polar surface exhibits specific chemical properties which are revealed, for instance, in its etching behavior. Whereas the P-terminated (111) B-face etches in phosphoric acid, for instance, the A-face is only very slowly attacked; HCl etching prepares the (111) A-face [243].

The maximum power point of efficient solar cells is located close to the open circuit voltage (see Figures 2.12 and 2.89). For p-type semiconductors, the open circuit condition is the most anodic potential at which the photocathode is operated and anodic dark currents compensate the cathodic photocurrent at this potential. It is therefore important to stabilize the photocathodes for operation at the maximum power point where anodic as well as cathodic photocorrosion can occur. Among the approaches presented in Figure 2.11, the protection of the semiconductor surface from the reactive electrolyte interface has been considered as most promising. Interfacial film formation can be achieved *in situ* by scanning the

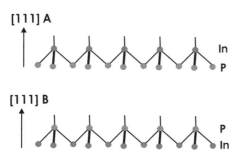

Figure 2.64 Atomic arrangement of the InP(111) A-face (top) and the (111) b-face (bottom).

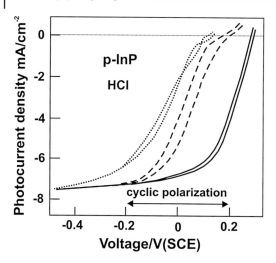

Figure 2.65 Photocurrent–voltage characteristic of p-InP(111) A for different surface conditions: dotted curve, untreated sample; dashed curve, after etching in 1% Br$_2$/CH$_3$OH (condition (a)); solid curve, after 50 cycles in 0.5 M HCl (condition (b)); light intensity, 20 mW cm^{-2}, W–I lamp.

electrode towards anodic potentials where surface oxidation occurs and, in the backward scan, to more cathodic potentials where the photoreduction reaction, which typically is H$_2$ evolution, results in a reduction of the formed anodic layer. In this respect, the repeated scanning between these potentials bears a similarity to electropolishing [244] where simultaneous (photo)electrochemical oxidation and oxide reduction by etching lead to the formation of a thin interfacial oxide and sample smoothening. Such a strategy has been successful with p-InP exposed to an HCl electrolyte [245]. The conditioning procedure, consisting of multiple scans between −0.2 and +0.2 V under illumination, leads to a marked increase in the fill factor, as shown in Figure 2.65. The figure shows also the I–V curve for the untreated sample and the improvement after Br$_2$/methanol etching; the cycling procedure, however, leads to a much more pronounced improvement.

Experiments on the induced changes in surface electronic properties have been preformed using the combined in-system electrochemistry/UHV surface analysis approach (Figure 2.27) where the electrochemical vessel is attached to a laboratory apparatus, performing UPS and standard XPS with a Mg K$_\alpha$ photon source. The influence of photoelectrochemical conditioning on the surface dipole changes can be deduced from the data in Figure 2.66. Here, HeI (photon energy of 21.2 eV) UPS spectra are shown for the sample after etching in methanol/bromine (dashed curve in Figure 2.65), spectrum (a), and after optimization in HCl upon repeated cycling (solid curve in Figure 2.65), spectrum (b). The InP valence band emission features are displayed including the photoemission onset near the Fermi level and the secondary electron cutoff. It is seen that the cutoff E_{co} changes from −16.9 eV (initial surface) to −16.1 eV after cycling in HCl. The valence band is located at −0.6 eV for spectrum (a). Due to surface phase formation during HCl cycling,

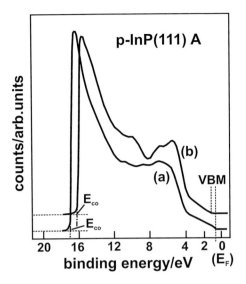

Figure 2.66 UPS of p-InP(111) A for conditions (a) and (b) in Figure 2.65; excitation by He I (21.2 eV) (see text).

assessment of the valence band emission onset is more difficult, yielding an extrapolated value of −0.8 eV, a value that has also been deduced from shifts in those core levels upon treatment which do not undergo chemical shifts. The $p_{3/2}$ level was taken in conjunction with the valence band extrapolation. Before discussing the spectral features, the cutoff data and the valence band position changes will be discussed to derive an energy band schematic that describes the induced changes. To determine the electron affinity change, the kinetic energy, photon energy, doping level (determining the energetic distance of Fermi level and valence band edge), energy gap, and secondary electron cutoff are related by

$$E_{kin} = h\nu - \chi - E_g + (E_F - E_V) - E_{co} \equiv 0 \tag{2.59}$$

With a photon energy of 21.2 eV and an energy gap of 1.35 eV, Equation 2.59 can be rewritten, also taking into account a possible band bending (cf. Figure 2.29), and solved for the electron affinity χ (for kinetic energy zero):

$$\chi = 19.85\,\text{eV} + (\Delta E_{dop} \pm eV_{bb}) - E_{co} \tag{2.60}$$

For the initial condition (a) in Figure 2.66, the overall energetic distance between E_F and E_V is 0.6 eV and the cutoff energy is 16.9 eV. This results in an electron affinity of 3.6 ± 0.1 eV. With $\Delta E_{dop} \pm eV_{bb} \approx -0.8\,\text{eV}$ and $E_{co} = 16.1\,\text{eV}$ after HCl conditioning, one obtains $\chi = 4.5 \pm 0.1\,\text{eV}$ yielding a change in χ by 0.9 eV. The resulting energy schematic which includes electrolyte levels such as the cathodic decomposition levels and the hydrogen redox couple is shown in Figure 2.67. An example for cathodic decomposition in HCl is the reaction

$$\text{p-InP} + 2e_{CB}^-(h\nu) + \text{Cl}_{ad}^- + 3\text{H}_{aq}^+ \rightarrow \text{InCl} + \text{PH}_3 \tag{2.61}$$

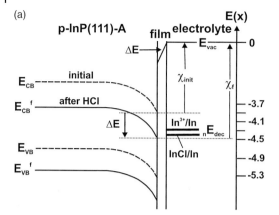

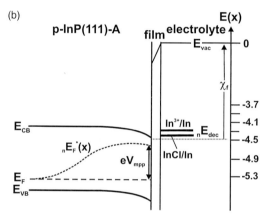

Figure 2.67 Energy scheme deduced from the UPS data shown in Figure 2.66. (a) Schematic for conditions (a) and (b) in Figure 2.65, labeled *initial* and *after HCl*; also shown are cathodic decomposition levels In^{3+}/In^0 and $InP/InCl$; $E^f_{CB,VB}$, conduction and valence band positions after HCl conditioning; χ_{init}, χ_f, electron affinities before and after HCl conditioning; ΔE, energy shift resulting from the treatment, occurring across an interfacial film that has been identified by XPS (Figure 2.68). (b) Energy relation for operation at the maximum power point, that is, close to the open circuit value as indicated by the reduced band bending and the large difference between the quasi-Fermi level and the bulk Fermi level at the surface; eV_{mpp}, voltage (energy) at the maximum power point.

with a potential of −0.5 V (SCE). It will be shown below in the XPS data analysis that Cl is incorporated into an interfacial film which forms during cycling in HCl. Further decomposition reactions are InP/In^0 (−0.4 V (SCE)) and In^{3+}/In^0 at −0.58 V. Translated to the vacuum scale, the levels are located at 4.4 (Equation 2.61), 4.5 and 4.3 eV, respectively. After optimization, the energetic position of the conduction band edge has shifted downward such that the H_2 evolution reaction can still occur with a large contact potential difference but the decomposition is unfavorable even if the interfacial film were absent (Figure 2.67a). Figure 2.67b shows the

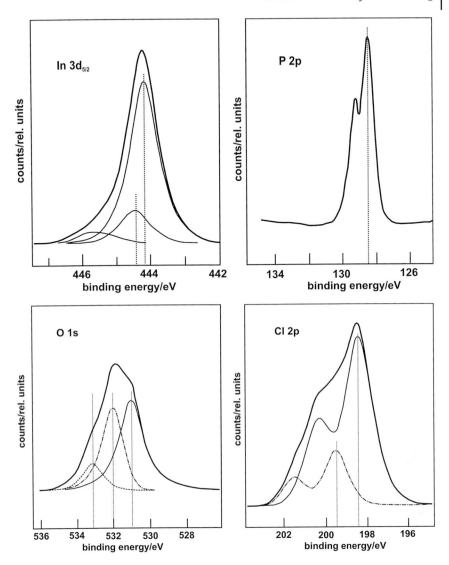

Figure 2.68 XPS analysis of treatment (b) in Figure 2.65. Excitation, Mg K_a (hv = 1256 eV); the In 3d, P 2p, O 1s, and Cl 2p lines are shown (see text); lines obtained from spectrum deconvolution are indicated by dotted vertical lines.

situation near the maximum power point, that is, close to the flatband condition assuming negligible surface/interface recombination as indicated by the good fill factor in Figure 2.65.

The data for surface chemical composition after cycling in HCl, analyzed by XPS, are presented in Figure 2.68. The In 3d line at 444.2 eV is attributed to In in InP but also to In_2O_3. The line at 444.5 eV has been attributed to In in InCl [246]. The weak line at 445.7 eV is related to $In_x(PO_4)_y$-type phosphates, whereas metallic

In would be found at 443.4 eV. The absence of In^0 shows clearly that the cyclic photoreduction process in HCl does not lead to In deposition. The phosphorous line (E_B = 128.3 eV) shows P in InP and possibly the small signal around 133 eV could be assigned to phosphates but should not be overinterpreted. It should be noted that with 1256 eV photon energy, the surface sensitivity is higher for the In 4d line because the kinetic energy of the emitted electrons is smaller; also, the atomic sensitivity factor is much smaller (by about a factor of 10) for P than for In. The O 1s signal at 531 eV belongs to In_2O_3 and the lines at E_B = 532 and 533.2 eV originate from OH and molecular water, respectively. The Cl 2p signal shows the presence of InCl (E_B = 198.3 eV) and of adsorbed Cl^- (E_B = 199.4 eV). The evolving picture is that the cyclic polarization and optimization of the I–V curve in HCl results in the formation of a surface phase that consists of a mixture of indium oxide and indium monochloride. The formation of InCl could be facilitated by adsorption of Cl^- at the film surface. The thickness determination yields a value of about 1.2 nm. It should be noted that, after further optimization as a photoelectrochemical solar cell in a V(II)/V(III) electrolyte, the thickness of the interfacial film decreases to about 0.8 nm. These films allow carrier transfer via tunneling, but also the conduction band of In_2O_3 is energetically located slightly below that of InP. If the InCl–In_2O_3 composite film would have a similar electron affinity, the excess minority transport could also result from a conduction band process as has been postulated for a device prepared with thin-film InP(100) (Figure 2.113; Section 2.6.2). The earlier concept of an MOS structure [237] has therefore to be partially revised as the main constituent of the interphase formed is InCl.

2.4.2.2 The In-Rich InP(100) (2×4) Surface

The atomic arrangement of the (2×4) reconstructed p-InP surface of a thin homoepitaxial layer has been determined from low-energy electron diffraction experiments and is shown in Figure 2.69. Samples with an In-rich surface have been prepared by MOVPE on a (100)-oriented InP wafer that was doped with Zn to 10^{18} cm^{-3}. Thin films were also Zn-doped with a dopant concentration of 4×10^{17} cm^{-3}. As precursors in the deposition process, trimethylindium, *tert*-butylphosphine, and diethylzinc have been used for doping. The ratio of the

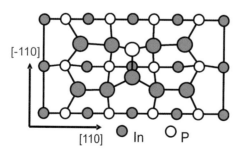

Figure 2.69 Surface arrangement of the InP(100) 2×4 reconstructed In-rich surface (see text).

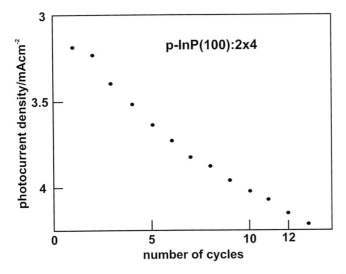

Figure 2.70 Activation procedure of an InP(100) 2×4 surface in 0.5 M HCl.

compounds containing group V and group III elements was 40 and, accordingly, growth occurred with substantial P excess. The In-rich surface was prepared by keeping the surface temperature of the sample above the sublimation temperature of the group V compound [247]. Nine In top-surface atoms (large dark circles in Figure 2.69) exist with only one P atom at the same top-surface layer [248]. Back contact to the wafer supporting the sample was made by the standard Zn–Au procedure established for p-type InP.

TM-AFM images of the samples, measured as obtained from the growth process in ambient atmosphere, show atomic terraces, indicating a substantial stability of the (2×4) (100) surface. Upon immersion into an HCl electrolyte, the cathodic photocurrent was small, indicative of a low activity towards hydrogen evolution. As a so-called activation treatment, a cyclic polarization under illumination was tried that indeed showed pronounced improvement of the cathodic photocurrent in HCl (Figure 2.70). Present investigations point to the formation of an indium oxide phase that contains chlorine, very similar to the findings for the interfacial film on the unreconstructed (111) A-face. It was found that after this treatment, the atomic steps on the surface remain discernible although the terrace widths appear somewhat smaller. The surface after 15 cycles was used as a reproducible starting condition for the preparation of a photoelectrocatalytic device which is described in Section 2.6.2.

2.4.3
Copper Indium Dichalcogenides

The surface and interface conditioning procedures presented here are predominantly related to the development of photovoltaic solid-state and electrochemical solar cells which will be described in Section 2.5. The materials encompass single-crystalline and thin-film polycrystalline $CuInS_2$ and $CuInSe_2$. The sequence follows the time line of developments, discussing (i) those conditioning treatments that lead to efficient and stable solar cells and (ii) procedures that are particularly intriguing regarding the surface chemistry and/or surface electronics.

2.4.3.1 CuInSe$_2$

Whereas the tendency to cathodic protection has resulted in the preparation of stable photocathodes based on InP, n-type semiconductors are generally more susceptible to corrosive attack, due to the deflection of holes as minority carriers towards the reactive electrolyte interface. Therefore, initially, from the experience gained with GaAs, for instance, the chances of achieving stable and efficient operation with an n-type chalcopyrite material were considered small. The chalcopyrite structure resembles that of zinc blende with a sublattice for the In and Cu atoms replacing Zn in zinc blende (Figure 2.71). For the development of a photovoltaic solar cell with an n-type semiconductor, a first estimate for the preparation of a rectifying junction can be made using the semiconductor electron affinity (4.6 eV for $CuInSe_2$) and the vacuum work function of the redox electrolyte. Taking acidic I^-/I_3^- and a vacuum work function of the normal hydrogen electrode of about 4.6 eV yields a value for the I_2/I^- couple of 5.1–5.2 eV (+0.53 V vs. NHE). With the energetic difference between the $CuInSe_2$ conduction band and the Fermi level for

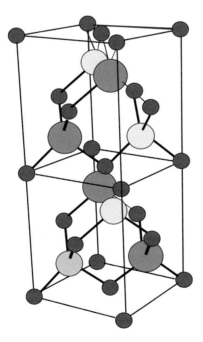

Figure 2.71 Structural arrangement of Cu, In, and S or Se atoms in ternary chalcopyrites; large gray circles, In; smaller light gray circles, Cu; small dark gray circles: S, Se.

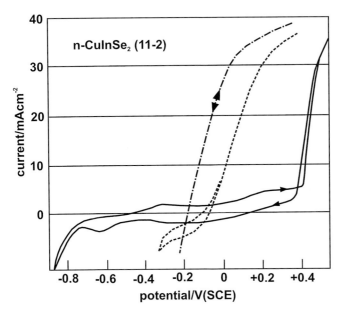

Figure 2.72 Electrochemical behavior and conditioning procedure for n-CuInSe$_2$ (112) in various electrolytes; solid curve, 1 M HCl; dashed curve, 1 M CaI$_2$ + 50 mM I$_2$; dashed-dotted curve, after addition of 1 M HI to the I$^-$/I$_3^-$ electrolyte; note the occurrence of anodic photocurrents in the redox electrolytes.

n-type doping of 0.1–0.2 eV, the contact potential difference is of the order of 0.4–0.5 eV. Surface treatments, however, can alter the energetic positions of band edges markedly (see Figures 2.66 and 2.67). Figure 2.72 shows photocurrent–voltage curves of (11-2)-oriented n-CuInSe$_2$ single-crystal samples immersed into supporting (HCl) and I$_2$/I$^-$-containing electrolytes. The solid curve shows an extended potential range where repeated cycling between −0.7 and +0.4 V results in pronounced current decrease with or without illumination. In this range, the semiconductor appears to be passivated. The breakdown of the passivity at potentials positive from +0.4 V results in photocurrents that are attributed to the oxidation of the CuInSe$_2$ absorber. The electrochemical passivity is lifted by addition of the I$_2$/I$^-$ redox couple. The dotted curve shows that the photoeffect is observed within the otherwise passive range of the sample. An improvement of the photocurrent–voltage characteristic is found when HI is added to the solution showing a photovoltage of about 0.4 V (dot-dashed curve). The system does, however, corrode and CuI crystallites are found after extended operation which results in a gradual decrease of the photocurrent output. It was found that the addition of Cu$^+$ ions altered the surface chemistry such that long-term stable operation was achieved. It was found that Cu$^+$ addition resulted in the formation of a CuISe$_3$ interphase that also contained elemental Se. The film was p-type and actually located the rectifying junction at the n-CuInSe$_2$/p-CuISe$_3$–Se^0 contact [249]. The contact potential difference was determined to be 0.65 eV by UPS He II

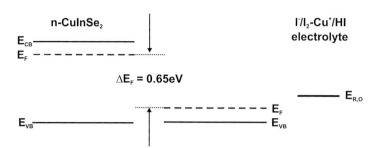

Figure 2.73 Energy band alignment between n-CuInSe$_2$, p-CuSe$_3$, and I$^-$/I$_3^-$–HI redox electrolyte (Cu$^+$ addition did not change the redox potential) before contact, based on UPS He II valence band spectra and measurement of the redox potential; the contact potential between the solid phases is 0.65 eV; the maximum band bending in n-CuInSe$_2$ is $eV_{bb} = E_{R,O} - {}^nE_F \approx 0.5$ eV.

($h\nu = 41.8$ eV) measurements of the valence band onset of photoemission. Also indicated was a rather high doping level that resulted in the asymmetry of the junction necessary to induce a large band bending in the photoactive CuInSe$_2$. In this sense, the cell was a solid-state n–p heterojunction device, prepared by *in situ* photoelectrochemical conditioning, with a liquid front contact. The energy band alignment before contact is shown in Figure 2.73 and includes the measured redox potential. Upon contact, the system will equilibrate with the solution level yielding a contact potential close to the photovoltage observed in the solar cell efficiency curve (Figure 2.78) of $V_{oc} = 0.42$ V.

2.4.3.2 CuInS$_2$

The preparation of thin-film polycrystalline CuInS$_2$ (CIS) solar cells with efficiency larger 10% involves the use of Cu-rich material. Cu enrichment beyond the homogeneity range of the compound results in phase segregation and leads to the formation of Cu–S phases on the surface. Highly p-type CuS (covellite) and Cu$_2$S (djurleite) are formed and their presence makes these cells photoinactive. To remove the Cu–S phases, a cyanide etch has been developed. Since present CIS solar cells are characterized by toxic components such as a CdS buffer layer and the toxic KCN processing step, wider application is problematic. Therefore, attempts are made to replace the toxic factors by nontoxic or at least less toxic ones. A promising route is the replacement of the KCN etchant by an electrochemical treatment since electrochemistry allows the well-defined control of processes. Several electrochemical treatments have been developed; most of them, however, yield very low or only moderate (3%) solar cell efficiencies. In an attempt to understand the origin of these restrictions, surface analysis using XPS was difficult

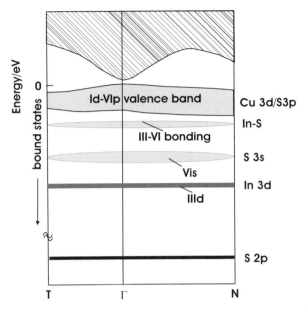

Figure 2.74 Generic energy band diagram for I–III–VI$_2$ (ABX$_2$) chalcopyrite semiconductors including symmetry points of the I–III–VI$_2$ Brillouin zone; on the right-hand side, the occupied states for CuInS$_2$ are indicated.

because of the rather small changes of the Cu 2p$_{3/2}$ core level (932.7 eV) and the LMM Auger line (337 eV) for the notorious Cu$_2$S compound and the host, CIS. Cu$_2$S is a remnant after the removal of covellite (CuS) and resisted various corrosion treatments. Its removal was considered pivotal for attaining higher solar cell efficiencies.

A clearer picture of the processing with regard to Cu$_2$S can be obtained from XES experiments (see Figures 2.22–2.24) where the spectator decay from the S 3s bands to the S 2p core levels was analyzed [250]. Figure 2.74 shows the principal energy band structure for ABX$_2$ chalcopyrite semiconductors (X = S, Se), where the X s-band (3s for the sulfide compounds, 4s for the selenide compounds) is about 2 eV wide and energetically separated by about 6 eV from the In–X bonding orbitals (B^{III}–X^{VI} band).

Using an excitation energy of 200 eV, the spectral range of interest is given by the valence band maximum and the core-level energy for S 2p (about 162 eV for the 3/2 peak). The lowest energy of the L$_{2,3}$ transition is about 146 eV expected from band structure analysis [251]. The valence band maximum is at 162 eV and the 3s → 2p decay peaks at the allowed $\Delta m = 1$ transition at 148 eV. Figure 2.75 shows an XES dataset for various treatments also comparing the effect of the cyanide etch with the electrochemical procedures. The most marked differences are found in the peak position around 148 eV and in the upper valence band features displayed on the right-hand side of Figure 2.74. One sees that the unetched CIS sample, the Cu$_2$S reference and the CIS crystal show pronounced differences

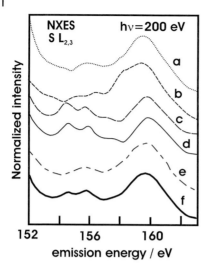

Figure 2.75 Comparison of nonresonant X-ray emission spectroscopy (NXES) data for standards and after chemical/electrochemical treatments; the energy region of transitions from the upper valence band of the S $L_{2,3}$ XES spectra for as-deposited $CuInS_2$ (a), a CuS reference (b), a $CuInS_2$ crystal (c), after KCN etching of the thin film (d), and for two (photo)electrochemical etching procedures (e, f) where the former was an alkaline treatment that did not completely remove the Cu–S surface phases and the latter was an advanced treatment in a vanadium redox electrolyte that resulted in a Cu–S-free surface (see also text) as seen by comparison with the crystal signature in (c).

in the region between 158 and 153 eV. The KCN-etched sample shows striking similarity with the signal from the crystal and that after an advanced electrochemical two-step treatment that will be described below [252]. It gives a similar result to the KCN etch but the feature at 157.5 eV is missing. The first step of the electrochemical etching procedure (curve (e) in Figure 2.75) shows similarity with the signature of the untreated sample indicating that Cu–S phases have not been sufficiently removed. This conclusion is supported by the observation that solar cells fabricated after the one-step etching procedure showed negligible efficiencies. The closest similarity with the signal for the CIS crystal is seen for the KCN etch. The differences between curves (c) and (f) point to presently difficult-to-assess electrochemically induced surface conditions which deviate from those of the crystal as also reflected by the lower solar cell efficiency after the two-step electrochemical treatment (see Figure 2.79 in Section 2.5.1).

The basic concept of the (photo)electrochemical removal of Cu–S phases is shown in Figure 2.76. Here, the current–voltage behavior of polycrystalline Cu-rich CIS is plotted. From comparison with thermodynamic stability data, the indicated dissolution regimes for the Cu–S phases and for CIS were assigned. Therefore, scanning the electrode between −0.2 and +0.35 V should induce dissolution of CuS; Figure 2.77a shows the processing protocol. Repeated scanning between the potential limits leads to the disappearance of an anodic peak and, simultaneously,

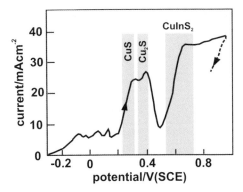

Figure 2.76 Dark I–V characteristic of a polycrystalline Cu-rich CuInS$_2$ film in V$^{2+/3+}$–HCl electrolyte (see text).

a cathodic photocurrent develops. The second step of the procedure is shown in Figure 2.77b where the potential scan was extended into the dissolution region of the CIS substrate, that is, to +0.8 V. As this will result in sample destruction, only short potential pulses were applied. The goal was (i) to dissolve the strongly attached Cu$_2$S phase and (ii) to leave the substrate absorber mostly unaffected by the treatment. Three fast scans were made (Figure 2.77b) and, finally, the potential was kept at +0.45 V where a small constant dissolution current is seen. The resulting conversion efficiency curves are shown in Figure 2.79 discussed below.

2.5
Photovoltaic, Photoelectrochemical Devices

In this section, photoelectrochemical solar cells are presented that convert sunlight into electricity. Also, a solid-state photovoltaic cell with a polycrystalline CIS absorber has been included that is prepared by a photoelectrochemical conditioning procedure and by KCN etching.

Generally, in photoelectrochemical systems purely regenerative operation is difficult to achieve and there are only a few examples where the issue of stability against photocorrosion has satisfactorily been solved. The main achievements have been made using passivating interfacial films [253–256] and using absorber materials with specific band structure properties as shown in Figure 2.11b where the absorption process does not affect the bonding in compound semiconductors [257, 258]. Unfortunately, the group VIb transition metal dichalcogenides cannot yet be prepared in a scalable manner [259–261]. Recent preparation using reactive magnetron sputtering might represent a promising alternative for large-scale preparation [262]. The dye-sensitized solar cell has shown considerable kinetic stability (cf. Figure 2.11c) [263, 264] and provides an example where kinetic

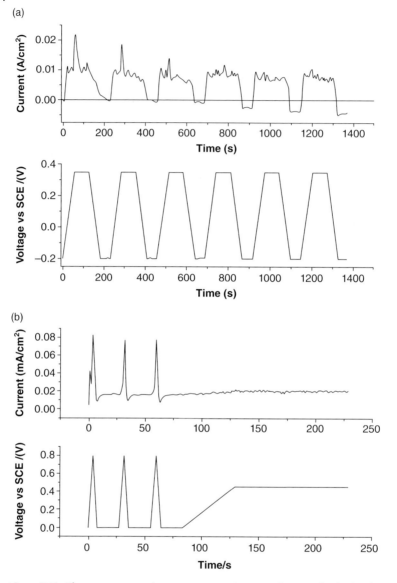

Figure 2.77 Chronoamperometric two-step conditioning of polycrystalline $CuInS_2$ films for removal of Cu–S surface phases; solution, $V^{2+/3+}$–HCl. (a) Step I: scan between −0.2 and +0.35 V until onset of cathodic photocurrents; (b) fast scans to +0.8 V for removal of strongly bound djurleite (Cu_2S) layer; illumination, 50 mW cm^{-2}, W–I lamp.

stabilization resulted in the first *technical* realization of a photoelectrochemical solar cell. This advance also points to the importance of device architectures that allow efficient light harvesting, minority carrier scavenging, and carrier transport in complex nanostructures.

2.5 Photovoltaic, Photoelectrochemical Devices

Here, the preparation of stabilizing interfacial films by *in situ* electrochemical and photoelectrochemical conditioning, described in Section 2.4, is emphasized. Such films can either be of tunneling thickness (Section 2.4.2) to allow efficient charge transfer or, for larger thickness, have to provide an efficient charge transport channel (Section 2.4.1). The advantages of using interphases, prepared *in situ*, are low-temperature processing, scalability, and conformal attachment to the absorber surface. In addition, (photo)electrochemical conditioning provides experimental parameters such as current, potential, and illumination intensity which allow fine tuning of the processes in the submonolayer range.

The selection made covers the first efficient and stable system based on the ternary chalcopyrite $CuInSe_2$, an electrochemical treatment to avoid a toxic etching step in solid-state CIS device fabrication, the first stable and efficient liquid-junction solar cell (InP), and a novel concept where nanoemitters, interspersed in a nanoporous passivating film, are used to scavenge excess minority carriers.

2.5.1
Ternary Chalcopyrites

Figure 2.78 shows the output power characteristic, yielding 9.5% efficiency under simulated solar light under simulated AM 1 conditions. Due to the small energy

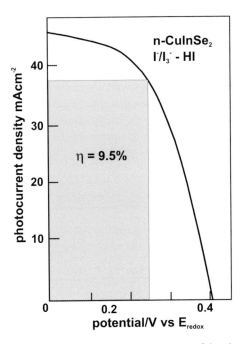

Figure 2.78 Output power characteristic of the photoelectrochemical solar cell n-$CuInSe_2$ (112) in acidic iodine–iodide electrolyte (see Figure 2.72) with added Cu(I) ions; illumination, simulated AM 1, $I = 99\,mW\,cm^{-2}$.

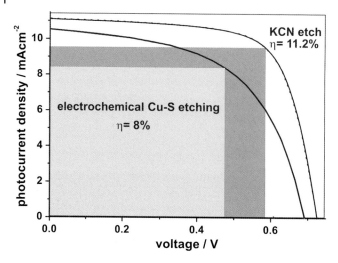

Figure 2.79 Output power characteristic of a solid-state photovoltaic solar cell with polycrystalline thin-film CuInS$_2$ for a chemical etching and an electrochemical process as indicated; illumination, solar simulation AM 2, $I = 50\,\text{mW cm}^{-2}$.

gap of 1.05 eV, the photovoltage is reduced but the short circuit current is rather large. The device has subsequently been improved in its efficiency to 12% [265] by modifying the growth conditions of the p-CuISe$_3$ surface phase. Treatments involved repeated etching–film formation sequences, air annealing of the film, and the use of higher concentrations of HI and Cu$^+$ ions.

The output power characteristic of a solid-state CIS solar cell where the deleterious Cu–S phases have been removed by an electrochemical process instead of the toxic KCN etch step is presented in Figure 2.79. A conversion efficiency of 8% has been reached. The figure also shows a device from the same sample batch prepared using the KCN etch. It is seen that the efficiency is higher after KCN etching. This is attributed to a near-surface depletion of Cu from the originally Cu-rich compound, resulting in the formation of a graded and buried junction already within the CIS surface.

2.5.2
InP Solar Cells

Based on the processing of InP(111) A, described in Section 2.4.2.1, a stable photoelectrochemical solar cell has been developed. After the aforementioned cyclic polarization in HCl, an additional cyclic conditioning in $V^{2+/3+}$/HCl redox electrolyte was performed. This treatment resulted in a further increase of the fill factor as shown in Figure 2.80. The solar-to-electrical conversion efficiency, measured in natural sunlight, was 11.6% in a two-electrode configuration and the result is shown in Figure 2.81.

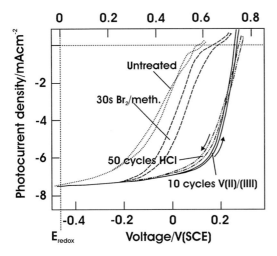

Figure 2.80 Cyclic conditioning procedures of InP(111) A in HCl (cf. Figure 2.65) and subsequently in 0.35 M VCl$_2$–VCl$_3$–4 M HCl redox electrolyte; the treatment in vanadium solution is labeled V(II)/V(III); E_{redox} denotes the potential of the V(II)/V(III) redox couple and represents the short circuit condition.

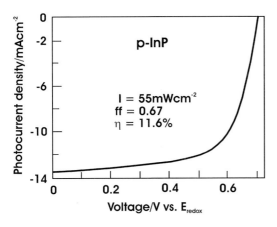

Figure 2.81 Output power characteristic of the photovoltaic photoelectrochemical solar cell p-InP(111)/film/V(II)/V(III)–HCl–C (C, carbon counter electrode) measured in natural sunlight, $I = 55$ mW cm^{-2}.

2.5.3
Nanoemitter Structures with Silicon

2.5.3.1 Device Development

The photocurrent (n-Si) and dark current (p-Si) oscillations, described in Section 2.4.1, result in nanoporous oxide films as can be seen in Figure 2.55. Those pores that are deep enough to reach the underlying Si have electrical contact and

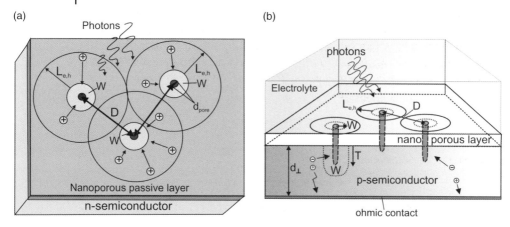

Figure 2.82 Schematic of the operational principle of a nanoemitter solar cell. (a) Top view showing the nanocontacts, their diameter d_{pore}, the space charge region W, minority carrier diffusion length $L_{e,h}$ for electrons or holes, respectively, and the distance D between the nanoemitters; the surface is covered by a nanoporous passivating layer where electrical contact with Si is only possible at metallized pores. (b) Side view including the electrolyte phase indicating conditions for efficient collection of excess carriers generated near the back contact with weakly absorbed light (see text).

the light-induced minority carriers (n-Si) or the potential-induced majority carriers (p-Si) are collected at the bottom of the pores where they react to from new oxide. This carrier scavenging effect can be used more generally if one deposits material into an otherwise passivating film that forms a rectifying contact with the absorber. Such nanostructure systems can be applied in solid-state photovoltaics but they have a specific stability advantage in photoelectrochemistry where they can be used for photovoltaic but also for photoelectrocatalytic operation. The principle is shown in Figure 2.82. The top view in Figure 2.82a correlates the distance D between nanodimensioned emitters with diameter d_{pore}, the space charge layer width W, and the minority carrier diffusion length $L_{h,e}$. If the inequality

$$D \leq \sqrt{2}(L_{h,e} + W + r_{em}) \tag{2.62}$$

is fulfilled, all light-generated excess minority carriers will be collected, thus allowing the preparation of efficient solar cells with nanoemitter front contacts that typically are too small in number and size to block an appreciable part of the sunlight. With D depending, for not too small diffusion lengths, mostly on L, the concept allows tailoring of the emitter patterns to specific absorbers. Diffusion lengths show an extreme spread in values for presently considered photovoltaic materials; whereas for copper and zinc phthalocyanides, $L \approx 15\,nm$, the value is in the range 30–250 nm for microcrystalline Si, and for the technologically advanced semiconductors GaAs and InP, $L \approx 8$–$10\,\mu m$. The largest diffusion lengths occur in crystalline Si where $L > 100\,\mu m$ for Cochralski Si and $L \approx 1200\,\mu m$ for float-zone Si, the latter being used in concentrator applications [266].

Although external patterning is possible for the realization of nanoemitter solar cells, hitherto only self-organized processes have been used to provide the templates for deposition of nanoemitter material [267]. The concept can be extended to increase the red sensitivity of a solar cell by deepening of pores into the absorber as shown in Figure 2.82b. The minority carriers generated close to the back contact can be collected if the inequality

$$d_\perp \leq T_{\text{pore}} + W + \frac{L_{h,e}}{2} \qquad (2.63)$$

holds. Here, the pore depth T_{pore} and the absorber thickness $d_\perp$ are related via the diffusion length and the space charge layer width.

For realization of a photovoltaic nanoemitter structure with an n-Si(100) absorber, a nanoporous oxide matrix is prepared using the oscillatory photocurrent described in Section 2.4.1. Sample emersion was at the increasing branch of the photocurrent where the pores are comparably large and the oxide coverage is smaller than in the other photocurrent phases (cf. Figures 2.54 and 2.55) to ensure that most nanopores provide contact between Si and the electrolyte, thus facilitating pore deepening.

The second step of the processing sequence is visualized in Figure 2.83. After pore formation by current oscillation, the pores are deepened into the substrate and, subsequently, site-selective metal deposition into the pores is carried out

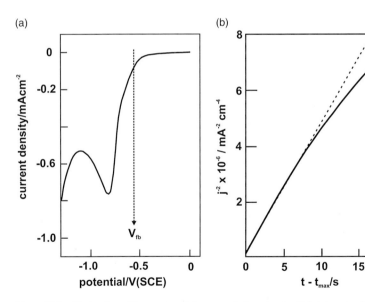

Figure 2.83 Electrodeposition protocol for site-selective Pt deposition into the pores of an anodic oxide matrix prepared by current oscillations on a Si(100) surface. (a) Current–voltage characteristic (V_{fb} is the flatband potential determined by Mott–Schottky plots); (b) plot of j^{-2} versus time for evaluation of the diffusion constant (see text); solution, 1 mM H_2PtCl_6, 0.1 M K_2SO_4.

(Figure 2.83a). Because chemical etching of the Si/anodic nanoporous oxide structure in alkaline solution (KOH, NaOH) resulted in the formation of inverted pyramids that were spatially extended, an electrochemical method was developed which allowed confined pore deepening. Pt deposition takes place at the cathodic current maximum, negative from flatband situation at −0.55 V. The n-Si sample is then in accumulation and the peak indicates proton (hydronium) reduction via electrons from the conduction band. Under these conditions, the majority carrier concentration easily exceeds values of $10^{18} \, cm^{-3}$. Figure 2.83b shows a plot of j^{-2} versus time which is linear in the initial regime, indicating hydrogen diffusion with a diffusion coefficient of $D = 4.4 \times 10^{-5} \, cm^2 \, s^{-1}$, determined from the Cottrell equation

$$j = nF \frac{A c_R D^{1/2}}{\pi^{1/2} t^{1/2}} \tag{2.64}$$

where area A, concentration c_R, diffusion coefficient, and inverse square root of the time are related. Hydrogen evolution is superimposed because Pt is already deposited by polarizing the sample from 0 to −0.8 V. The origin of the behavior can be identified by inspection of the energy relations at the n-Si–electrolyte contact (Figure 2.84). The Pt deposition levels are located energetically below the Si valence band edge, and also deposition from occupied surface states can occur (for details, see below). Once Pt has been deposited, which, due to the presence of surface states, occurs at rather anodic potentials (−0.3 V), H_2 evolution sets in characterized by a low reaction overpotential and the processes related to the actual electrodeposition process cannot be directly examined. A detailed discussion will be presented in Section 2.5.3.2.

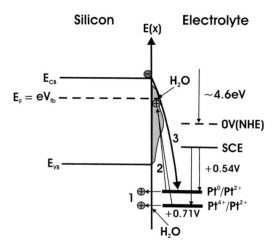

Figure 2.84 Schematic of the Pt electrodeposition processes at the Si–electrolyte contact: (1) hole injection into the valence band; (2) hole injection into surface states; (3) Pt ion reduction by electrons from the conduction band.

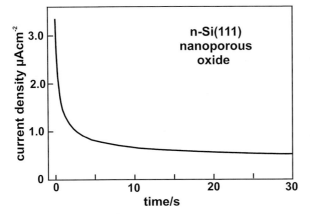

Figure 2.85 Pore deepening protocol (current versus time) for an oxide matrix prepared by photocurrent oscillations after sample emersion at the increasing branch of the oscillatory current; solution, 4 M NaOH; potential held at −0.85 V (SCE).

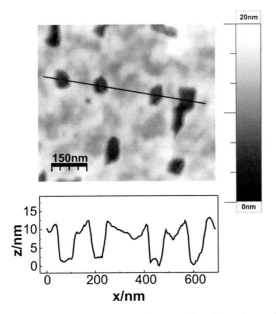

Figure 2.86 CM-AFM image of Si(100) after electrochemical pore deepening; oxide removal by a 10 s dip of the sample in 25% HF.

The processing protocol for the pore deepening is shown in Figure 2.85, and Figure 2.86 shows a CM-AFM image of a sample after the pore deepening procedure where the anodic oxide mask has been etched off by an HF dip, leaving an H-terminated Si surface. The image in Figure 2.86 shows rectangular depressions that, with the twofold symmetry of the (100) orientation and the cross-sectional view, shows that the pits formed are about 10 nm deep. Although this is small in

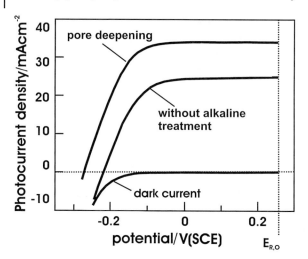

Figure 2.87 Output power characteristic in 2 M HI/50 mM I$_2$ redox electrolyte showing the influence of the pore deepening procedure on the current collection efficiency and the attainable photovoltage; illumination, W–I lamp, 100 mW cm^{-2}; three-electrode potentiostatic arrangement.

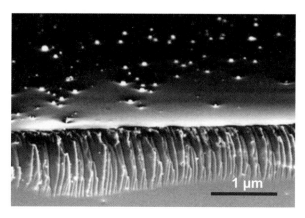

Figure 2.88 Edge-view scanning electron micrograph of n-Si after Pt deposition; note the Pt islands visible in the material contrast.

regard to the condition in Equation 2.63 for the large diffusion lengths in c-Si, it turns out that the process increases the solar conversion efficiency as shown in Figure 2.87.

Figure 2.88 shows a scanning electron micrograph that shows a side view of a sample after deposition of Pt particles. The site-selective deposition of Pt is clearly visible. It should be noted that not all nanopores are necessarily filled because, in the progressing oscillation, pores at different phases of their development exist on the surfaces and, depending on the process, some do not make contact with the

Si substrate. Then, electrodeposition does not take place. It should further be considered that the processing sequence in Figure 2.85 uses the fact that Si etches in alkaline solution but that its oxide remains almost unaffected by these etchants. On the other hand, silicon oxide can be etched by HF-containing solutions that leave the silicon substrate mostly unaffected. This binary etching behavior of Si and its oxide has resulted in a wealth of applications in micro- and nanoelectronics [94].

For application in solid-state devices, the structure would be covered by a transparent conducting oxide (TCO) such as ZnO or indium tin oxide (ITO), followed by the contact finger system. For electrochemical solar cells, a redox electrolyte is used as contact. In photovoltaic systems, typical redox couples such as Fe(II)/(III), I_3^-/I^- (n-type), or V(II)/(III) and S_2^{2-}/S_x^{2-} (p-type) can be used. For photoelectrocatalytic devices, a first approach is to use noble metal catalysts as nanoemitters such as Pt, Ir, and Rh (H_2 generation, CO_2 reduction) on p-type samples or Ru, RuIr, RuO_2, and mixed oxides on n-type samples (O_2 generation). It should be noted that the rectifying contact behavior is dominated in electrochemistry by the redox energy of the respective redox couple in solution, whereas for solid-state devices the work function of the nanoemitter with respect to the semiconductor Fermi level defines the barrier.

The photoelectrochemical solar cell is completed by using an acidic I_2/I^- electrolyte as redox couple with a carbon counter electrode and calomel reference electrode. The solar-to-electrical conversion efficiency characteristic is shown in Figure 2.89. Upon illumination with a W–I lamp, an efficiency of 11.2% is reached. This is the hitherto best value out of a batch of five samples which all showed efficiencies above 10%.

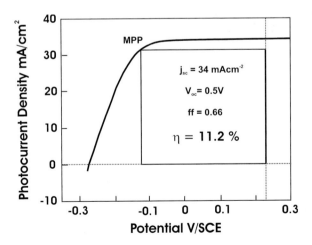

Figure 2.89 Output power characteristic of the hitherto best photovoltaic electrochemical nanoemitter solar cell; redox electrolyte, 2 M HI/50 mM I_2; illumination, 100 mW cm^{-2}, W–I lamp; three-electrode potentiostatic arrangement.

2.5.3.2 Surface Chemical Analysis of the Electrodeposition Process

Electrodeposition has been analyzed using SRPES at Bessy II, where the experiments were performed at the SoLiAS of the collaborating research group with members from the Brandenburgisch-Technical University Cottbus, the Technical University Darmstadt, and the Helmholtz Center Berlin (see Figures 2.26 and 2.27). In these *model experiments*, H-terminated n-Si(111) was used and electrodeposition was done following the protocol shown in Figure 2.90. The cathodic current shows two peaks, located at −0.3 and −0.55 V. The flatband potential has shifted anodically compared to that of the oxide-covered (100) sample by about 70 mV. This could in part be related to the higher electron affinity of the (111) surface compared to the (100) surface where the surface corrugation dipole is larger [268] and also to differences in the surface coverage (H versus oxide) which can influence surface dipoles.

SRPES measurements have been performed for three conditions along the polarization curve in Figure 2.90, indicated by C_1, C_2, and at 0 V (SCE). First, the energetic situation at the Si–electrolyte contact is reviewed (Figure 2.84) as mentioned above: for alignment between electrolyte levels and the Si band structure, the measured flatband potential of −0.48 V vs. SCE is used. The Pt deposition occurs in two major steps, for example, Pt(IV)/(II) and Pt(II)/(0). The redox levels are indicated in the figure and the reaction scheme is

$$PtCl_6^{2-} + 2e^-_{CB,VB} \rightarrow PtCl_4^{2-} + 2Cl^- \quad (+0.71 \text{ V vs. SCE}) \tag{2.65}$$

$$PtCl_4^{2-} + 2e^-_{CB,VB} \rightarrow Pt^0 + 4Cl^- \quad (2. + 0.54 \text{ V}) \tag{2.65a}$$

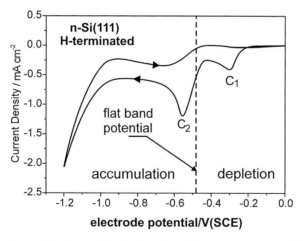

Figure 2.90 Protocol for Pt electrodeposition on chemically H-terminated n-Si(111) for model experiments on surface chemistry using the SoLiAS at undulator U49/2 at Bessy II, recorded in the in-system apparatus shown in Figure 2.27. Three conditions have been analyzed by SRPES: deposition under depletion (peak C_1), and accumulation (C_2) conditions, and deposition near 0 V where the open circuit potential is located; note the anodic shift in flatband potential compared to that of the device (using Si(100)) that can be related to the differences in electron affinities (surface dipole contributions) of the two faces ($\chi(100) < \chi(111)$).

The first step in Pt metal deposition is the reduction to Pt(II), followed by the actual metal deposition. Also shown in Figure 2.84 are Si surface states. Three principal charge transfer processes have been drawn: 1, hole injection into the valence band which is a potential-independent process; 2, hole injection into occupied surface states; and 3, reduction of the Pt complexes by electrons from the conduction band. The latter two processes are potential dependent. At potentials for which the semiconductor is in depletion or in a normal (compared to strong) inversion condition (between 0 V and flatband), the structure under the peak C_1 is attributed to charging and discharging of surface states and at $V < -0.48$ V (V_{fb}) reduction occurs by electrons from the conduction band.

A complete set of SRPES data on the relevant core levels Si 2p, O 1s, and Pt 4f is shown in Figures 2.91–2.93 for that situation which was also used in the fabrication of the devices, for example, process 3 in Figure 2.84. For the measurements, the photon energy has been adjusted such that the kinetic energy of the photoelectrons is in the region of 50 eV. This is the minimum region of the inelastic mean free electron scattering length of about 0.3–0.4 nm and hence the surface sensitivity is in the monolayer range. This high surface sensitivity is visible in Figure 2.91 where the Si 2p core-level signal is compared for an H-terminated surface and after cathodic Pt deposition via electrons from the conduction band (process 3). After Pt deposition, a pronounced additional peak at a binding energy

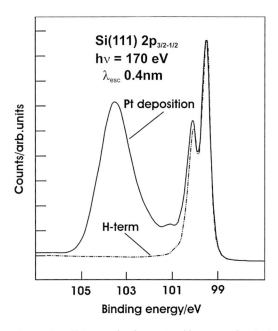

Figure 2.91 Si 2p core level, measured by SRPES at a photon energy of 170 eV where the inelastic mean free scattering length for photoelectrons, λ_{esc}, has a minimum resulting in monolayer surface sensitivity; after Pt deposition under accumulation condition (peak C_2); the result is compared to that of a smooth H-terminated sample (see text).

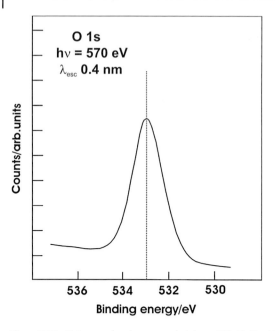

Figure 2.92 O 1s core level measured at $h\nu = 570\,eV$; the kinetic energy of the photoelectrons that determines the escape depth of elastic electrons is similar to that in Figure 2.91; therefore, the surface sensitivity is here in the monolyer range, too.

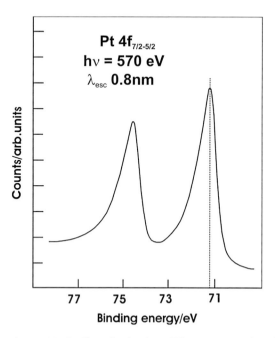

Figure 2.93 Pt 4f core level at $h\nu = 570\,eV$; the surface sensitivity had to be reduced because of a Cooper minimum in the excitation cross section at lower photon energies where the surface sensitivity would be higher; note the asymmetric, so-called skew lines in the signals from the spin orbit split $4f_{7/2}$ and $4f_{5/2}$ signal.

of about 103.6 eV is seen which is typical for silicon oxide and, as already discussed, corresponds to about one monolayer of oxide. The O 1s line position (Figure 2.92) is also in accordance with silicon oxide formation and the Pt 4f line at 71.2 eV (Figure 2.93) indicates elemental Pt. The Pt line has been measured at higher photon energy compared to the above discussion regarding the kinetic energy of photoelectrons. This is due to the existence of a Cooper minimum in the excitation cross section at lower photon energies [269]. Also, it is noted that the $4f_{7/2\text{-}5/2}$ signal shows so-called skew lines, that is, signals that are asymmetric [270, 271]. The asymmetry is attributed to reduced screening of photoelectrons by the Pt conduction electrons at the Fermi level which is typically described an expression derived by Doniach and Sunjic that contains an asymmetry parameter [129]. Figure 2.94 shows a set of Pt core levels measured for different Pt deposition conditions; the most symmetric curve is obtained for Pt films, the second one for Pt deposition as in Figure 2.93, and the most asymmetric curve is recorded for Pt islands deposited on step bunched Si surfaces (see also Section 2.4.1). On step bunched Si, typical sizes of the Pt nanoislands are 30 nm in width and about 3 nm in height, whereas for Pt deposition on H-terminated Si, the size is about 100 nm in width and 30 nm in height. Therefore, the result in Figure 2.94 indicates a size effect in the screening of the photoelectrons.

The SRPES results show that silicon oxide is formed during Pt deposition due to the scanning conditioning procedure (peak C_2). The oxide thickness on H-terminated

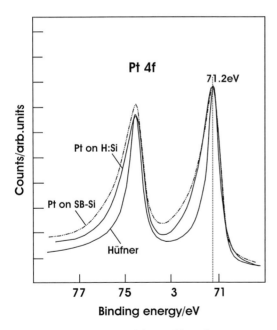

Figure 2.94 Comparison of the Pt 4f lines for different samples: solid curve 1, Pt film measured by Hüfner as indicated; solid curve 2, signal from Pt islands deposited on H-terminated surface; dashed-dotted curve, signal from Pt islands deposited onto a step bunched Si surface; the latter are substantially smaller (see text).

Si amounts to approximately 1 ML. Under accumulation conditions, Si oxidation by hole injection processes is difficult because of the large concentration of conduction band electrons in an accumulation layer that leads an increased surface recombination rate. Hole injection into occupied surface states would only be possible if the reaction rate of an oxidized surface defect with the surrounding water were faster than the recombination via conduction band electrons which is unlikely. It is therefore concluded that oxide formation is likely to have occurred by the procedure where the sample was scanned from open circuit potential (about −0 V) to the peak C_2 thus allowing for oxidation under transient depletion conditions.

The basic results from the measurements of processes 1 and 2 in Figure 2.84 are condensed in Figures 2.95 and 2.96 which show the O 1s and the Pt 4f lines, respectively. It is seen that for deposition under depletion conditions at C_1, silicon oxide is formed and Pt is deposited as expected in a simultaneous hole injection

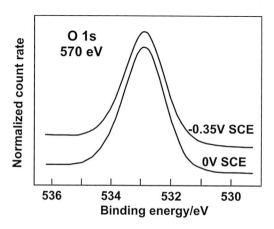

Figure 2.95 O 1s core level corresponding to peak C_1 in Figure 2.90, indicated by the potential at which the process was interrupted (−0.35 V), and to open circuit conditions (0 V).

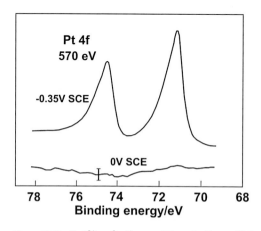

Figure 2.96 Pt 4f line for the conditions in Figure Z14; note the absence of a Pt signal for 0 V.

process that results in bond weakening at Si surface atoms in the presence of water and in Pt deposition from the tetrachloride Pt(II) complex. The oxide formation process will be discussed below (see Figure 2.98). It is surprising, however, that for sample emersion at open circuit potential, oxide formation is observed but the Pt signal is absent.

Inspection of the electronic situation of the semiconductor surface shows a distinct difference between C_1 and open circuit potential; whereas for emersion at −0.35 V (C_1), the semiconductor is in depletion, at open circuit potential the inversion condition has been reached. In this situation, the concentration of holes at the surface is larger than that of the electrons which are majority carriers from the doping. The band bending can be determined from the difference to the flatband potential yielding $eV_{bb} \approx 0.5\,eV$. With the Fermi level due to the doping located 0.25 eV below the conduction band edge, the energetic distance $E_{CB}-E_F$ (surface) is 0.75 eV. Inversion sets in at values above half of the energy gap (1.12 eV at room temperature), not taking into account, for simplicity, the slight deviation from the midgap position due to different effective DOS at the conduction and valence band edges. Recent data for the surface states at Si–silicon oxide interfaces on (111) surfaces show that levels exist that are related to the protrusion of an unpaired electron in an sp^3-like hybrid into an oxide microvoid [272]. Such defects appear likely to exist at a Si–anodic oxide interface. The energetic positions of the corresponding P_b^0 levels are 0.35 and 0.85 eV above the valence band edge as shown in Figure 2.97. At open circuit potential, the lower lying surface state becomes unoccupied as the Fermi level lies at or below the respective energy as indicated in the figure. In addition, the inversion region extends into the Si surface up to the perpendicular line

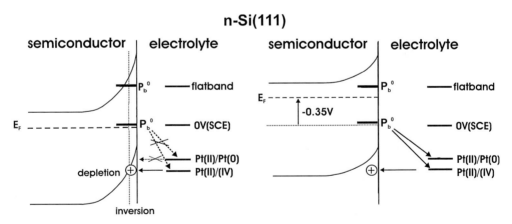

Figure 2.97 Energy scheme for interpretation of the SRPES data obtained at C_1 and open circuit potential in Figure Z9. The left-hand side shows the semiconductor band bending at open circuit potential, the P_b^0 surface states, and the Pt deposition levels; also shown is the inversion layer extension; the inhibited Pt deposition processes are indicated by crossed dotted lines. The right-hand side shows the situation under 0.35 V forward bias compared to open circuit potential; possible deposition channels are indicated by solid lines (see text).

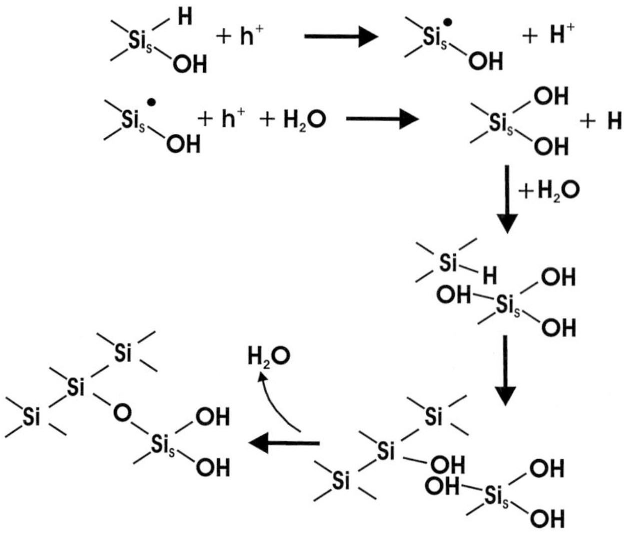

Figure 2.98 Schematic for oxide formation via a kink site (surface state) at which a hole from the Pt levels is injected; further oxidation of the radical results in OH termination, which is followed by solvolytic backbond splitting; in the subsequent process (step 4), the H-termination of the second-layer Si atom is replaced by an OH group by two holes from the Pt complex as shown in steps 1 and 2, followed by the formation of an oxo-bridge (see text).

drawn beneath the outmost surface layer. Hole accumulation in this region can inhibit hole injection from the Pt(II)/(0) level that is energetically located above the Pt(IV)/(II) redox energy (see Figure 2.90). As a consequence, the processes for hole injection into surface states that would result in reduction of Pt (IV) and Pt(II) and the hole injection into the valence band for Pt(II) reduction are inhibited and only hole injection into the valence band from the lower lying Pt(IV)/(II) level is possible. The latter results in oxide formation, but Pt deposition is inhibited as has been found in the corresponding SRPES experiment. The experiment shows how surface chemistry can be "switched" by surface electronics. Figure 2.98 shows a schematic of oxide formation that starts at a kink site atom which acts as a surface state is presented. The process involves a four-electron transfer. The initial step is hole capture (h^+) that leads to surface radical formation. Subsequently, the capture of the second hole results in the OH termination of the surface atom. After solvolytic splitting of the backbond that results in the H-termination of the underlying Si atom, the initial two-hole process is repeated to oxidize this atom and to form a hydroxyl group. The recombination of the hydroxyl groups in the backbond leads to a localized oxo-bridge as precursor of oxide formation. The formation of localized metal–oxide–semiconductor junctions can be seen in Figure 2.99 where TEM cross-sectional views show an amorphous interfacial film between Pt islands and the n-Si substrate. This film is about 1–2 nm thick and extends laterally across the surface. The origin of this delocalized film formation has not yet been identified;

2.5 Photovoltaic, Photoelectrochemical Devices

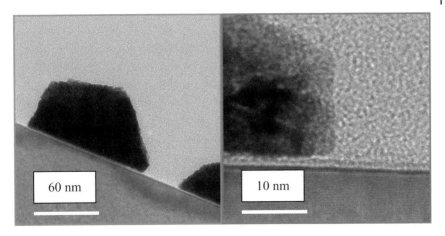

Figure 2.99 TEM side view of Pt islands on an n-Si(111) H-terminated sample (see text).

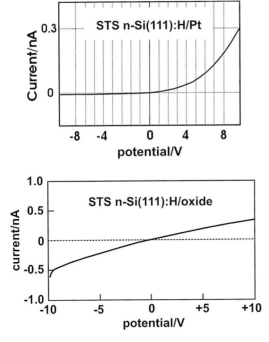

Figure 2.100 Scanning tunneling spectroscopy data of samples as in Figure 2.99 recorded with the tip on top of a Pt island (top) and on top of the interfacial oxide film (bottom).

it could be related to surface states with delocalized character or the influence of the hole injection into the valence band, leading to oxide formation that should occur throughout the sample surface. The results of STS experiments, performed on top of the Pt islands and above the oxide layer, are shown in Figure 2.100.

Rectifying behavior is found at the tip–air gap–Pt–oxide–Si junction and a more ohmic behavior is seen above the oxide where the current–voltage curve is mostly linear. This shows that nanodimensioned rectifying junctions between Pt islands and n-Si have been formed that dominate the properties of the device with regard to its solar-to-electricity conversion efficiency. A more detailed analysis of the current–voltage behavior is presently difficult due to the two tunneling mechanisms (through the oxide and through the air gap to the tip) and the unknown tip area and shape. Further insights into the functioning of such devices (also in the photoelectrocatalytic mode; see below) can be gained by *in situ* STM and scanning electrochemical microscopy (SECM) experiments that are presently being pursued.

2.6
Photoelectrocatalytic Devices

In the pursuit of the multiple-photon electrolysis route using tandem structures (see Figures 2.13 and 2.15) to capture the photonic excess energy, separate investigation of the half-cell components for the photocathode and photoanode is mandatory. This approach demands the use of technologically advanced semiconductors where doping and interface behavior can be well controlled, which is essential for the future preparation of monolithic tandem structures. Here, the development of photocathodes for light-induced H_2 generation will be described, beginning with p-Si in a nanoemitter configuration. Then, a device made using a thin epitaxial p-InP film, designed for photoelectrocatalytic application, is presented.

2.6.1
Nanoemitter Structures with p-Si

Figure 2.101a shows a HRSEM image of a nanoporous oxide matrix obtained by current oscillations of p-Si(100). The pores exhibit rectangular shapes with the symmetry of the underlying (100) substrate and sizes range from larger pores of about 120 nm basis length to 10–15 nm pores. The photoelectrodeposition procedure (Figure 2.101b) results in the filling of nanopores by Pt. Site-selective photodeposition is seen (Figure 2.101c) and some pores obviously have been overgrown, exhibiting a mushroom-type appearance. The Pt deposits range in size from about 250 nm (overgrown pores) down to 15 nm where rectangular shapes are observed. For comparison of the influence of the noble metal on the photoelectrocatalytic behavior, Rh has also been used (not shown).

The resulting output characteristic for Pt and Rh nanoisland catalysts (Figure 2.102) shows a rather sluggish increase of the cathodic photocurrent, indicative of surface recombination, and, at the potential of the hydrogen redox couple, current saturation has not yet been reached. Even at 0.2 V cathodic from $E_{R,O}$ (H_2/H^+), where band bending is increased, carrier collection does not yet reach a saturation value. The low observed efficiencies are attributed to pronounced surface/interface recombination and to the reduced contact potential difference between p-Si and

(a)

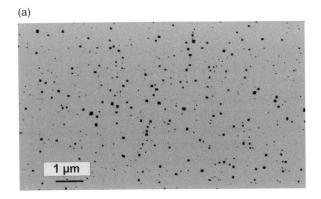

(b)

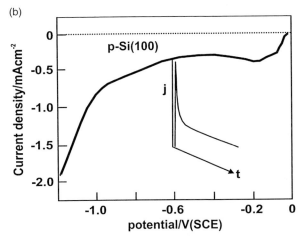

(c)

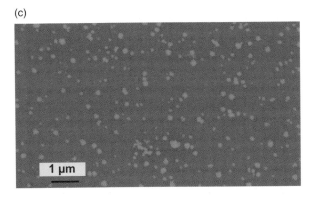

Figure 2.101 Photoelectrodeposition of Pt for preparation of a nanoemitter photocathode: (a) SEM image of porous oxide matrix on Si(100); (b) deposition protocol: solution, 1 mM $PtCl_6^{2-}$, 0.1 M K_2SO_4, pH 3.8; light intensity, 130 mW cm^{-2}, W–I lamp; (c) SEM image after Pt deposition.

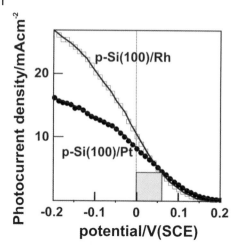

Figure 2.102 Output characteristic of nanoemitter devices made with p-Si–nanoporous oxide–Pt or Rh contacts in 1 M H_2SO_4; light intensity, $100\,mW\,cm^{-2}$, W–I lamp.

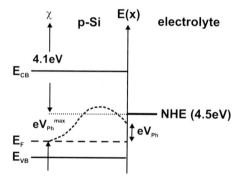

Figure 2.103 Energy scheme for the p-Si–H_2/H^+ contact; the metals (Pt and Rh) have been neglected here for simplicity because the structure forms an equilibrium with the redox solution; eV_{Ph}, achieved photovoltage; eV_{Ph}^{max}, maximum photovoltage attainable from the energetic difference $E_{R,O}(H_2/H^+) - E_F(p\text{-Si})$; χ, semiconductor electron affinity; the minority electron quasi-Fermi level (dashed curve) indicates surface recombination.

the hydrogen redox couple. The energetic situation is presented in Figure 2.103: the Fermi level of p-Si is located at 0.3–0.4 eV positive from the hydrogen redox couple. This difference gives the maximum attainable photovoltage. Surface recombination is typically considered by a downward bending of the minority carrier (electrons) quasi-Fermi level at the surface, indicating the reduced excess carrier concentration at the interface as a result of the surface recombination processes (see Figure 2.7). The minority carrier recombination is unlikely to result from the action of so-called metal-induced gap states [180] because the spatial separation of the metal from the semiconductor surface by the interfacial oxide

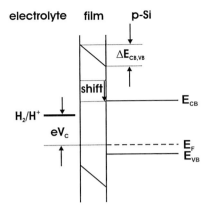

Figure 2.104 Envisaged influence of an interfacial film with dipole character that results in a downward shift of the Si band edges ($\Delta E_{CB,VB}$) which leads to a larger contact potential difference, eV_c (see text).

inhibits penetration of the evanescent metal states into the complex band structure within the semiconductor energy gap.

Strategies for increasing the contact potential difference involve changes of surface dipoles. For an increase of the electron affinity, the surface dipoles must be oriented with their negative charge towards the exterior as shown in Figure 2.104. Presently, the deposition of an atomic layer deposition film [273] is envisaged for this purpose. The dipole character for photoelectrocatalytic application, however, has yet to be established.

2.6.2
Thin-Film InP Metal–Interphase–Semiconductor Structure

2.6.2.1 Basic Considerations

The electron affinity and the energy gap of InP make this material a well-suited candidate for light-induced H_2 evolution. With $\chi \approx 4.4$ eV and $E_g = 1.35$ eV, the valence band maximum is located at 5.75 eV below vacuum. Taking a value of 4.5–4.6 eV as the work function of the normal hydrogen electrode, and with the Fermi level of p-InP about 0.25 eV above E_{VB}, the contact potential difference is about 1 eV. Accordingly, much larger photovoltages are expected with this material and have indeed been realized [67, 69]. An energy scheme is shown in Figure 2.105.

Earlier work was carried out with crystalline bulk InP samples prepared by the gradient freeze method [274]. As In could become scarce due to its present use in electronics [275], the use of bulk material appears too expensive for applications. Therefore, homoepitaxial thin films have been prepared on InP wafers which can be removed by established lift-off techniques [276]. The preparation by MOVPE allows scaling up and fine tuning of the growth process, enabling fabrication of films with high electronic quality [277]. Besides the hitherto used (111) A-face (see Section 2.4.2.1), the surprisingly stable In-rich (100) surface is considered here.

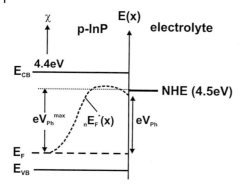

Figure 2.105 Energy scheme of the p-InP–H$_2$/H$^+$ redox electrolyte contact. The larger electron affinity compared to Si (by about 0.3 eV) and the larger energy gap (1.35 eV) yield a contact potential increase of >0.5 V relative to the p-Si–acidic electrolyte contact; the maximum photovoltage is about 1 V, depending on the InP doping level.

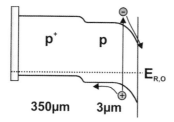

Figure 2.106 Energy band diagram of the structure of p$^+$-InP substrate (doped in the 10^{18} cm^{-3} range) and the homoepitaxial p-InP film (doped in the low 10^{17} cm^{-3} range) showing a back surface field band alignment.

2.6.2.2 Device Preparation and Properties

The energy scheme in Figure 2.106 shows the band alignment between a 10^{18} cm^{-3} Zn-doped substrate and a less doped (low 10^{17} cm^{-3} range) homoepitaxial 3 μm thick film. The doping profile ensures electronic reflection of excess minority electrons thus providing a back surface field [278]. Photoelectrodeposition of Rh from a Rh(III) chloride solution has been performed at −0.2 V (SCE) where the InP band bending is large (Figure 2.107). The temporal profile is a superposition of Rh deposition and H$_2$ evolution and therefore, as in the case of Si, does not allow the determination of the deposition charge from the chronoamperometric profiles. A TM-AFM image (Figure 2.108) shows that the procedure has resulted in a complete coverage of the surface by Rh and a TEM side-view image (Figure 2.109) reveals a vertical nanostructure consisting of InP, an interfacial film, and the Rh deposits. The lattice planes of InP are clearly seen and the amorphous interphase has a thickness of 4–5 nm. The Rh deposits are between 5 and 15 nm thick and form an irregular structure that appears to consist of almost spherical subunits of about 5 nm in size. The optical properties with regard to reflectivity of the structure are presently being investigated and preliminary data indicate a

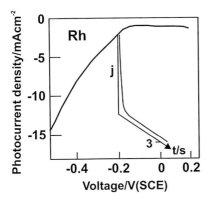

Figure 2.107 Protocol for Rh photoelectrodeposition onto (100)In-rich 2×4 reconstructed p-InP; solution, 5 mM RhCl$_3$, 0.5 M NaCl, 5% isopropanol; light intensity, 100 mW cm^{-2}.

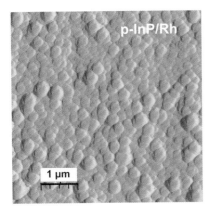

Figure 2.108 TM-AFM image of p-InP after photoelectrodepositon of Rh; note that a closed film is observed.

Figure 2.109 TEM cross-sectional view showing Rh deposits, an amorphous interfacial film, and the epitaxial structure of InP substrate.

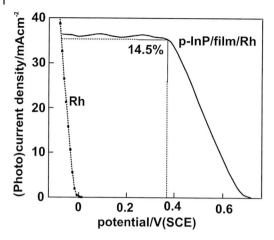

Figure 2.110 Output power characteristic for light-induced H_2 evolution of a Rh/p-InP device; electrolyte, 1 M $HClO_4$; light intensity, 100 mW cm^{-2} (W–I lamp); three-electrode potentiostatic arrangement with carbon counter and calomel reference electrode.

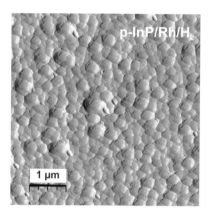

Figure 2.111 TM-AFM image of the device shown in Figure 2.110 after passage of a charge of 45 C cm^{-2} (see text).

reduced reflectivity in the visible spectral range [183]. As a result, the solar conversion efficiency reaches values around 14% as seen in Figure 2.110. Figure 2.111 shows a (top view) TM-AFM image after passage of a 45 C cm^{-2} photocurrent. The basic appearance in comparison to Figure 2.108 remains unaltered and the side-view TEM image in Figure 2.112 demonstrates that the interface between the amorphous film and the InP surface leaves the atomic lattice plane arrangement unaffected. Photocorrosion would result in the transformation of about 15 μm of InP either by dissolution or by deposit formation, and hence this structure exhibits pronounced stability and efficiency. The Rh particles are clearly nanodimensioned with most sizes in the range 3–6 nm. High-resolution TEM images show crystal

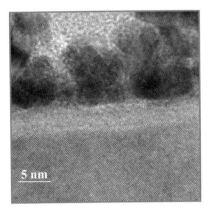

Figure 2.112 TEM cross-sectional view of the device in Figure 2.110 after passage of a charge of 45 C cm^{-2} (see text).

planes within the particles which therefore consist at least of a crystalline core. The solid-state properties of the nanocrystalline particles, in particular with regard to charge transfer, Fermi level position, and energetic states, have yet to be determined.

The nature of the interphase can be deduced from recent XPS analyses that indicate the presence of indium oxide with a smaller admixture of phosphates; contrary to the observation of an InCl phase on the (111) A face, chloride has not been found here, indicating a different surface chemistry of the (100) In-rich surface that is presently investigated. The film formed on InP(100), however, is too thick to enable efficient tunneling; therefore, charge transport can take place via a defect band or by appropriate energy band alignment. Taking the data for the electron affinity of In_2O_3, for example, 4.5 eV [279], the energy band alignment between p-InP, the interphase, Rh, and the redox couple shows that a conduction band cliff would allow efficient electron transport across the film to the Rh film where the reduction reaction occurs (Figure 2.113). It is presently not known how the incorporation of indium phosphates into indium oxide affects the electron affinity. UPS experiments can provide information on the band offset between InP and the film if the energy gap of the film is known. Because the deposition steps in the Rh(III) reduction to Rh have not been as well investigated as those for Pt reduction, where further oxide growth has been observed (see Figures 2.91, 2.92, and 2.95), a UPS measurement on the film after the conditioning procedure in HCl (Section 2.4.2.2) might not yield conclusive information. Rh deposition could result in an increase of the film thickness if it is accompanied by hole injection processes. Therefore, model experiments on the Rh deposition onto well-defined InP surfaces before and after the optimization procedure in HCl are necessary. The absorption behavior of the structure can be assessed by BAS, and first experiments show specific optical properties induced by the Rh film, consisting of nanoparticles and an influence of the highly doped substrate on the red part of the spectral sensitivity.

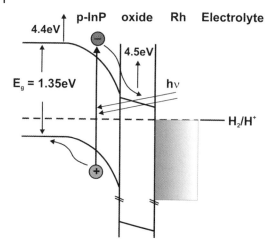

Figure 2.113 Energy schematic of the p-InP–interfacial film–Rh–electrolyte junction including the transport processes of light-induced minority electrons (see text).

2.7
Synopsis

2.7.1
Summary

The use of surface-sensitive techniques for the development of photoelectrochemical devices that convert solar energy has been described. The essence of this approach is the control of interfacial properties. This is achieved by a combination of empirical procedures that are developed into directed approaches of interface modification for desired electronic, chemical, and structural properties by a feedback between preparation and analysis. Besides the multitude of commercially available surface analytical techniques employed (AFM, STM, TEM, HRSEM, HREELS, SRPES, FTIR), novel methods have been developed such as Brewster angle reflectometry and stationary microwave reflectivity. The detailed highly surface-sensitive analysis of the surface chemistry of samples where electrochemical currents have passed has become possible by the development of in-system photoelectron spectroscopy and HREELS.

Several devices presented here are based on the novel nanoemitter concept that allows realization of efficient solar energy conversion structures. The device architecture also demonstrates that arguments regarding costs and abundance of materials have to be considered carefully; cost-limiting factors in terrestrial applications can be drastically reduced if only trace amounts of precious metals have to be incorporated into the structures. For the photovoltaic nanoemitter cell shown in Figures 2.82 and 2.89, for a distance of 5 μm between the emitters, the cost for a square kilometer would be negligible (about €2500). In addition, noble metal consumption can be further reduced by capping of more abundant materials with a few atomic layers of the noble metals. Also, recycling is definitely a consideration, as has been demonstrated clearly in paper photography where huge areas have been processed over the decades using AgI [280]. Nevertheless, it is of course important to pursue research directions that are based on the use of more abundant materials, inspired, for example, by the photocatalytic processes in photosynthesis.

2.7.2
Reflections on Future Development Routes

Presently, the advent of nanoscience, merging disciplines such as physics, chemistry, material science, biology, and informatics, has led to a large number of novel approaches and it is difficult to foresee the development routes. It is, however, possible to give an overview on the innovation potential; this involves processes, materials, specifically designed architectures (an example is the nanoemitter concept that has been presented in this chapter), and more general approaches such as "learning from photosynthesis," combinatorial materials development, and the application of micro- and nanostructures prepared in physics and materials science (photonic crystals). Novel (advanced) architectures could be based on either self-organization or on processes in physics and materials science and allow the use of scarce materials and/or the funneling of excitation energy to emitters or reactive sites. The issue of alternative excitation energy transport compared to direct charge transfer is definitely a promising route to follow. Here, processes such as Förster and Dexter transfer [281, 282] and possibly the mimicking of multichromic Förster transfer [283] appear interesting.

Another aspect is the plasmonic field enhancement by nanodimensioned noble metal particles that exhibit localized surface plasmon resonances in the visible spectral range [284, 285]. The preparation of structures and designs that incorporate plasmonic nanoparticles can increase solar conversion efficiencies by better adapting the incoming light to the device features. The plasmonic enhancement of the absorption of indirect semiconductors (crystalline Si) has already been demonstrated [286]. Photonic crystals with tailored reflectivity properties could be used as light scavengers or as spectral selective optical mirrors near the back contact of a solar converter. The enhancement of spontaneous photon emission from atoms in a quantum electrodynamic cavity [287] might be of use in molecular solar energy converting systems.

Among hitherto less-well-known photophysical processes is the exploitation of multiple exciton generation [288], resulting in multiple low-energy photon generation from a higher energetic one, and singlet fission [289] where the decomposition of a molecular singlet state into (two) lower energetic triplet states takes place.

In artificial photosynthesis, the use of macromolecules that contain a catalytic center consisting of, for example, Ru [290] or Mn [291] provides a promising route for more robust molecular catalysts. Similarly, the use of genetically altered mutants of the enzymes of photosynthesis might be an avenue worth pursuing [292]. The chemical modification of PS I for more efficient H_2 evolution [293] is another promising route, particularly if this much more robust system can be attached to a conducting electrode such as the step bunched Si surface described in Section 2.4.1.

Acknowledgments

The author wishes to thank his research team and, in particular, Helmut Jungblut, Mohammed Aggour, Katarzyna Skorupska, Michael Lublow, Andres Munoz, Jürgen Grzanna, and Michael Kanis. The collaboration with Karl Jacobi (Fritz-Haber-Institut der Max-Planck-Gesellschaft) and Sheelagh Campbell (Portsmouth University) was extremely stimulating. From the "earlier days," thanks go to Ulrich Störkel, Baptiste Berenguier, Nikolaus Dietz, now at Georgia State University, Stefan Rauscher, and Thomas Bitzer (Infineon). Finally, I would like to thank my daughter Sophie Antonia for her patience.

Appendix 2.A

The symbols used in Equation 2.7 are defined as follows:

$$A^* = \frac{\alpha L_p^2 P_0}{[1 - \alpha^2 L_p^2 - (q/kT)\alpha L_p^2 E_x]D_p} \tag{2.A.1}$$

$$C = \frac{e^{-\alpha W} - e^{-\beta}}{e^{-\kappa} - e^{-\delta}} \tag{2.A.2}$$

$$\beta = \alpha d - \kappa \tag{2.A.2a}$$

$$\kappa = \frac{d - W}{L_p} \tag{2.A.2b}$$

$$\delta = \frac{W}{L_p} \tag{2.A.2c}$$

$$D = e^{-\delta} - e^{-2\kappa} \tag{2.A.3}$$

$$C' = \frac{e^{-\alpha W} - e^{-\beta}}{e^{2\kappa} - e^{\delta}} \tag{2.A.4}$$

$$D' = e^{\delta} - e^{2\kappa} \tag{2.A.5}$$

Appendix 2.B

The symbols used in Equation 2.11 are defined as follows:

$$s = \frac{x - W}{\sqrt{2}L_d} \tag{2.B.1}$$

$$u = \frac{W}{\sqrt{2L_d}} \tag{2.B.2}$$

$$\beta' = \frac{\alpha L_d}{2} \tag{2.B.3}$$

$$k = \sqrt{\frac{2}{\pi} \frac{L_d}{D_p}} P_0 e^{-\alpha W} \tag{2.B.4}$$

$$a = \frac{1 + \alpha L_p R_j}{1 + \alpha L_p}; \quad R_j = f(\alpha, d, W, L_p) \tag{2.B.5}$$

Appendix 2.C

The dependence of $\Delta p(0)$ on surface recombination velocity and charge transfer rate:

$$\Delta p(0) = \frac{P_0 \left(1 - \frac{(1 + \alpha L_p R_j) e^{-\alpha W}}{1 + \alpha L_p}\right)}{(1 + S_j)\frac{D_p}{L_p} \exp\left(-\frac{W^2}{2L_d^2}\right) + k_r + S_r} \tag{2.C.1}$$

$$S_j = \frac{2e^{-2\kappa}}{1 - 2e^{-2\kappa}} \tag{2.C.2}$$

References

1 Rajeshwar, K. (1995) *J. Appl. Electrochem.*, **25**, 1067.
2 Alkire, R.C. and Kolb, D.M. (eds) (2003) *Advances in Electrochemical Science and Engineering*, vol. 8, Wiley-VCH Verlag GmbH, Weinheim.
3 Lewerenz, H.J. (1992) *Electrochim. Acta*, **37**, 847.
4 Lewerenz, H.J. and Schulte, K. (2002) *Electrochim. Acta*, **47**, 2633.
5 Parkinson, B.A., Heller, A., and Miller, B. (1978) *Appl. Phys. Lett.*, **33**, 521.
6 Heller, A., Lewerenz, H.J., and Miller, B. (1980) *Ber. Bunsenges. Phys. Chem.*, **84**, 592.
7 Fornarini, L. and Scrosati, B. (1983) *Electrochim. Acta*, **28**, 667.
8 Wrighton, M.S. (1979) *Acc. Chem. Res.*, **12**, 303.
9 Cahen, D., Hodes, G., Grätzel, M., Guillemoles, J.F., and Riess, I. (2000) *J. Phys. Chem. B*, **104**, 2053.
10 Bockris, J.O.M. and Szklarczyk, M. (1983) *Appl. Phys. Lett.*, **42**, 1035.
11 Inoue, T., Fujishima, A., Konishi, S., and Honda, K. (1979) *Nature*, **277**, 637.
12 Khaselev, O. and Turner, J.O. (1998) *Science*, **280**, 425.
13 Tributsch, H. (1997) *Catal. Today*, **39**, 177.
14 Aroutiounian, V.M., Arakelyan, V.M., and Shahnazaryan, G.E. (2005) *Sol. Energy*, **78**, 581.
15 (2009) BP Statistical Review of World Energy.
16 Förster, Th.v. (1948) *Ann. Phys.*, **437**, 55.
17 Dexter, D.L. (1953) *J. Chem. Phys.*, **21**, 836.

18 Becker, K., Lupton, J.M., Müller, J., Rogach, A.L., Talapin, D.V., Weller, H., and Feldmann, J. (2006) *Nat. Mater.*, **5**, 777.
19 Nozik, A.J. (2002) *Physica E*, **14**, 115.
20 Frank, A., Kopidakis, N., and van de Lagemaat, J. (2004) *Coordin. Chem. Rev.*, **248**, 1165.
21 Blackburn, J.L., Selmarten, D.C., Ellingson, R.J., Jones, M., Micic, O., and Nozik, A.J. (2005) *J. Phys. Chem. B*, **109**, 2625.
22 Arney, D., Porter, B., Greve, B., and Maggard, P.A. (2008) *J. Photochem. Photobiol. A*, **199**, 230.
23 Ma, L.Q., Lee, J.Y., Li, J., and Lin, W.B. (2008) *Inorg. Chem.*, **47**, 3955.
24 Solares, S.D., Michalak, D.J., Goddard, W.A., and Lewis, N.S. (2006) *J. Phys. Chem. B*, **110**, 8171.
25 Matson, R.J., Noufi, R., Bachmann, K.J., and Cahen, D. (1987) *Appl. Phys. Lett.*, **50**, 158.
26 Jang, E., Jun, S., Chung, Y.S., and Pu, L.S. (2004) *J. Phys. Chem. B*, **108**, 4597.
27 Intergovernmental report on climate change (IPCC) (2007) Fourth Assessment Report: Climate Change.
28 Woodhouse, M. and Parkinson, B.A. (2008) *Chem. Mater.*, **20**, 2495.
29 Baeck, S.H., Jaramillo, T.F., Brändli, C., and McFarland, E.W. (2002) *J. Comb. Chem.*, **4**, 563.
30 Katz, J.E., Gingrich, T.R., Santori, E.A., and Lewis, N.S. (2009) *Energy Environ. Sci.*, **2**, 103–112.
31 Sato, N. (1998) *Electrochemistry at Metal and Semiconductor Electrodes*, Elsevier Science, Amsterdam.
32 Schottky, W. (1939) *Z. Physik*, **113**, 367; Schottky, W. (1942) *Z. Physik*, **118**, 539.
33 Crowell, C.R. (1965) *Solid State Electron.*, **8**, 395.
34 Shockley, W., Sparks, M., and Teal, G.K. (1951) *Phys. Rev.*, **83**, 151.
35 Gerischer, H. (1960) *Z. Phys. Chem. N.F.*, **26**, 223.
36 Gerischer, H. (1961) *Z. Phys. Chem. N.F.*, **27**, 48.
37 Marcus, R.A. (1956) *J. Chem. Phys.*, **24**, 966.
38 Marcus, R.A. (1997) *Nobel Lectures, Chemistry 1991–1995*, World Scientific, Singapore.
39 Boltzmann, L. (1885) *Ann. Phys.*, **260**, 37.
40 Götzberger, A., Hebling, C., and Schock, H.-W. (2003) *Mater. Sci. Eng.*, **R40**, 1.
41 Guibaly, F.E. and Colbow, K. (1982) *J. Appl. Phys.*, **53**, 1737.
42 Wilson, R.H. (1977) *J. Appl. Phys.*, **48**, 4292.
43 Lewerenz, H.J. and Schlichthörl, G. (1994) *J. Appl. Phys.*, **75**, 3544.
44 Reichman, J. (1980) *Appl. Phys. Lett.*, **36**, 574.
45 Schlichthörl, G. and Lewerenz, H.J. (1998) *J. Electroanal. Chem.*, **443**, 9.
46 Sze, S.M. (1981) *Physics of Semiconductor Devices*, John Wiley & Sons, Inc., New York.
47 Hovel, H.J. (1975) *Semiconductors and Semimetals, Vol. II: Solar Cells*, Academic Press, New York.
48 Wünsch, F., Schlichthörl, G., and Tributsch, H. (1993) *J. Appl. Phys.*, **26**, 2041.
49 Cass, M.J., Duffy, N.W., Kirah, K., Peter, L.M., Pennock, S.R., Ushiroda, S., and Walker, A.B. (2002) *J. Electroanal. Chem.*, **538**, 191.
50 Hamann, C.H. and Vielstich, W. (2005) *Elektrochemie*, Wiley-VCH Verlag GmbH, Weinheim.
51 Bard, A.J. and Faulkner, L.R. (2001) *Electrochemical Methods*, John Wiley & Sons, Inc., New York.
52 Lettieri, S. (2004) *J. Appl. Phys.*, **95**, 5419.
53 Tan, X.M., Kenyon, C.N., and Lewis, N.S. (1994) *J. Phys. Chem.*, **98**, 4959.
54 Kunst, M., Wünsch, F., and Jokisch, D. (2003) *Mater. Sci. Eng. B*, **102**, 173.
55 Ross, R.T. and Hsiao, T.-A. (1977) *J. Appl. Phys.*, **48**, 4783.
56 Gerischer, H. and Mindt, W. (1968) *Electrochim. Acta*, **13**, 1329.
57 Bard, A.J. and Wrighton, M.S. (1977) *J. Electrochem. Soc.*, **124**, 1706.
58 Jungblut, H., Jakubowicz, J., and Lewerenz, H.J. (2005) *Surf. Sci.*, **597**, 93.
59 Yu, H., Webb, L.J., Ries, R.S., Solares, S.D., Goddard, III, W.A., Heath, J.R., and Lewis, N.S. (2005) *J. Phys. Chem. B*, **109**, 671.

60 Morita, Y., Miki, K., and Tokumoto, H. (1992) *Appl. Surf. Sci.*, **60/61**, 466.
61 Dumas, P., Chabal, Y.J., and Higashi, G.S. (1990) *Phys. Rev. Lett.*, **65**, 1124.
62 Kuruvilla, A.B. and Kulkarni, S.K. (1993) *J. Appl. Phys.*, **73**, 4384.
63 Mattheis, L.F. (1973) *Phys. Rev. B*, **8**, 3719.
64 Kato, Y., Kurita, S., and Suda, P. (1987) *J. Appl. Phys.*, **62**, 3737.
65 Hankare, P.P., Chate, P.A., and Sathe, D.J. (2009) *J. Phys. Chem. Solids*, **70**, 655.
66 Archer, M.D. and Nozik, A.J. (2008) *Nanostructured and Photoelectrochemical Systems for Solar Photon Conversion*, Imperial College Press, London.
67 Heller, A. (1984) *Science*, **223**, 114.
68 Tan, M.X., Laibinis, P.E., Nguyen, S.T., Kesselman, J.M., Stanton, C.E., Lewis, N.S., and Kenneth, D. (1994) Principles and applications of semiconductor photoelectrochemistry, in *Progress in Inorganic Chemistry*, vol. 41 (ed. K.D. Karlin), John Wiley & Sons, Inc., New York, pp. 21–144.
69 Heller, A., Miller, B., Lewerenz, H.J., and Bachmann, K.J. (1980) *J. Am. Chem. Soc.*, **102**, 6555.
70 Cahen, D. and Chen, Y.W. (1984) *Appl. Phys. Lett.*, **45**, 746.
71 Schulte, K.H. and Lewerenz, H.J. (2002) *Electrochim. Acta*, **47**, 2633.
72 Noufi, R., Shafarman, W.N., Cahen, D., and Stolt, L. (eds) (2003) *Compound Semiconductor Photovoltaics, MRS Symposium Proceedings*, vol. 763, Materials Research Society, Warrendale, PA.
73 Lewerenz, H.J., Heine, C., Skorupska, K., Szabo, N., Hannappel, T., Vo-Dinh, T., Campbell, S.A., Klemm, H.W., and Munoz, A.G. (2010) *Energy and Environmental Science*, **3**, 748.
74 Miyazaki, S., Maruyama, T., Kohno, A., and Hirose, M. (1999) *Mater. Sci. Semicond. Process.*, **2**, 185.
75 Lublow, M., Stempel, T., Skorupska, K., Muñoz, A.G., Kanis, M., and Lewerenz, H.J. (2008) *Appl. Phys. Lett.*, **93**, 062112.
76 Fermin, D.J., Ponomarev, E.A., and Peter, L.M. (1999) *J. Electroanal. Chem.*, **473**, 192.
77 Parkinson, B.A., Furtak, T.E., Canfield, D., Kam, K., and Kline, G. (1980) *Faraday Discuss. Chem. Soc.*, **70**, 282.
78 Tributsch, H. (1994) *Sol. Energy Mater. Sol. Cells*, **31**, 548.
79 Wilson, J.A. and Yoffe, A.D. (1969) *Adv. Phys.*, **18**, 193.
80 Kulyuk, L., Bucher, E., Charron, L., Fortin, E., Nateprov, A., and Schenker, O. (2002) *Nonlinear Opt.*, **29**, 501.
81 Colev, A., Gherman, C., Mirovitskii, V., Kulyuk, L., and Fortin, E. (2009) *J. Luminescence*, **129**, 1945.
82 Lewerenz, H.J., Lübke, M., Menezes, S., and Bachmann, K.J. (1981) *Appl. Phys. Lett.*, **39**, 798.
83 Nishida, M. (1978) *Nature*, **277**, 202.
84 Grätzel, M. and O'Reagan, B. (1991) *Nature*, **353**, 737.
85 Ito, S., Zakeeruddin, S.M., Humphry-Baker, R., Liska, P., Charvet, R., Comte, P., Nazeeruddin, M.K., Péchy, P., Takata, M., Miura, H., Uchida, S., and Grätzel, M. (2006) *Adv. Mater.*, **18**, 1202.
86 Kambili, A., Walker, A.B., Qiu, F.L., Fisher, A.C., Savin, A.D., and Peter, L.M. (2002) *Physica E*, **14**, 203.
87 Nakato, Y., Shioji, M., and Tsubomura, H. (1981) *J. Phys. Chem.*, **85**, 1670.
88 Bard, A.J. (1982) *J. Phys. Chem.*, **86**, 172.
89 Bakkers, E.P.A.M., Roest, A.L., Marsman, A.W., Jenneskens, L.W., de Jong-van Steensel, L.I., Kelly, J.J., and Vanmaekelbergh, D. (2000) *J. Phys. Chem. B*, **104**, 7266.
90 Gerischer, H. (1977) *J. Electroanal. Chem.*, **82**, 133.
91 Bouroushian, M., Karoussos, D., and Kosanovic, T. (2006) *Solid State Ionics*, **177**, 1855.
92 Hilal, H.S., Masoud, M., Shakhshir, S., and Jisrawi, N. (2002) *J. Electroanal. Chem.*, **527**, 47.
93 Aggour, M., Skorupska, K., Stempel Pereira, T., Jungblut, H., Grzanna, J., and Lewerenz, H.J. (2007) *J. Electrochem. Soc.*, **154**, H794.
94 Campbell, S.A., Port, S.N., and Schiffrin, D.J. (1998) Anisotropy and the micromachining of silicon, in *Semiconductor Micromachining: vol. 2. Techniques and Industrial Applications* (eds S.A. Campbell and H.J. Lewerenz), John Wiley & Sons, Inc., New York, pp. 1–52.

95 Stempel, T., Aggour, M., Skorupska, K., Muñoz, A.G., and Lewerenz, H.J. (2008) *Electrochem. Commun.*, **10**, 1184.

96 Hsyi-En, C. and Chia-Chuan, C. (2008) *J. Electrochem. Soc.*, **155**, D 604.

97 Valderrama, R.C., Sebastian, P.J., Pantoja Enriquez, J., and Gamboa, S.A. (2005) *Sol. Energy Mater. Sol. Cells*, **88**, 145.

98 Sherman, A. (2008) *Atomic Layer Deposition for Nanotechnology*, Ivoryton Press, Ivoryton, CT.

99 Klaus, J.W., Sneh, O., and George, S.M. (1997) *Science*, **278**, 1934.

100 Bisquert, J., Cahen, D., Hodes, G., Rühle, S., and Zaban, A. (2004) *J. Phys. Chem. B*, **108**, 8106.

101 Gregg, B.A. (2003) *J. Phys. Chem. B*, **107**, 4688.

102 Kostedt, W.L., Drwiega, J., Mazyck, D.W., Lee, S.-W., Sigmund, W., Wu, C.-Y., and Chadik, P. (2005) *Environ. Sci. Technol.*, **39**, 8052.

103 Kaneko, M. and Okura, I. (2002) *Photocatalysis: Science and Technology*, Springer-Verlag, Heidelberg.

104 Hussein, F.H. and Rudham, R. (1987) *J. Chem. Soc. Faraday Trans.*, **83**, 1631.

105 Lojou, É. and Bianco, P. (2004) *Electroanalysis*, **16**, 1093.

106 Hickling, A. and Salt, F.W. (1941) *Trans. Faraday Soc.*, **37**, 224.

107 Bockris, J.O.M. (1947) *Nature*, **159**, 539.

108 Grimes, C.A., Varghese, O.K., and Ranjan, S. (2007) *Light, Water, Hydrogen: The Solar Generation of Hydrogen by Water Photoelectrolysis*, Springer-Verlag, Berlin.

109 Hatnean, J.A., Raturi, R., Lefebvre, J., Leznoff, D.B., Lawes, G., and Johnson, S.A. (2006) *J. Am. Chem. Soc.*, **128**, 14992.

110 Carraher, C.E., Jr. and Murphy, A.T. (2005) *Macromolecules Containing Metal and Metal-Like Elements 5*, John Wiley & Sons, Ltd, Chichester.

111 Brecht, M., Radics, V., Nieder, J.B., and Bittl, R. (2009) *Proc. Natl Acad. Sci. USA*, **106**, 11857.

112 Werner, J.H., Kolodinski, S., and Queisser, H.J. (1994) *Phys. Rev. Lett.*, **72**, 3851.

113 Shinkarev, V., Zybailov, B., Vassiliev, I., and Golbeck, J.H. (2002) *Biophys. J.*, **83**, 2885.

114 Grätzel, M. (2001) *Nature*, **414**, 338.

115 Park, H., Vecitis, C.D., and Hoffmann, M.R. (2009) *J. Phys. Chem. C*, **113**, 7935.

116 Kunst, M., and Beck, G. (1986) *J. Appl. Phys.*, **60**, 3558.

117 Jakubowicz, J., Jungblut, H., and Lewerenz, H.J. (2003) *Electrochim. Acta*, **49**, 137.

118 Stumper, J., Lewerenz, H.J., and Pettenkofer, C. (1990) *Phys. Rev. B*, **41**, 1592.

119 Lewerenz, H.J., Stumper, J., and Peter, L.M. (1988) *Phys. Rev. Lett.*, **61**, 1989.

120 Kooij, E.S., Despo, R.W., and Kelly, J.J. (1995) *Appl. Phys. Lett.*, **66**, 2552.

121 Knoll, M. and Ruska, E. (1932) *Z. Physik*, **78**, 318.

122 Siegbahn, K. (1982) *Rev. Mod. Phys.*, **54**, 709.

123 Ardenne, M. (1938) *Z. Physik*, **108**, 553.

124 Binnig, G., Rohrer, H., Gerber, Ch., and Weibel, E. (1982) *Phys. Rev. Lett.*, **49**, 57.

125 Binnig, G., Quate, C.F., and Gerber, Ch. (1986) *Phys. Rev. Lett.*, **56**, 930.

126 Tougaard, S. (1986) *Phys. Rev. B*, **34**, 6779.

127 Shirley, D.A. (1972) *Phys. Rev. B*, **5**, 4709.

128 Neuhold, G., Barman, S.R., Horn, K., Theis, W., Ebert, P., and Urban, K. (1998) *Phys. Rev. B*, **58**, 734.

129 Doniach, S. and Sunjic, M. (1970) *J. Phys. C*, **3**, 285.

130 Bates, D.R., Fundaminsky, A., Leech, J.W., and Massey, H.S.W. (1950) *Phil. Trans. R. Soc. A*, **243**, 93.

131 Lewerenz, H.J. and Dietz, N. (1991) *Appl. Phys. Lett.*, **59**, 1470.

132 Dietz, N. and Lewerenz, H.J. (1993) *Appl. Surf. Sci.*, **69**, 350.

133 Lublow, M. and Lewerenz, H.J. (2008) *Surf. Sci.*, **602**, 1677.

134 Lublow, M. and Lewerenz, H.J. (2009) *ECS Trans.*. **19**, 381.

135 Azzam, R.M.A. and Bashara, N.M. (1989) *Ellipsometry and Polarized Light*, North Holland, Amsterdam.

136 Dietz, N., Fearheiley, M.L., Schroetter, S., and Lewerenz, H.J. (1992) *Mater. Sci. Eng. B*, **14**, 101.

137 Cattarin, S., Guerriero, P., Dietz, N., and Lewerenz, H.J. (1995) *Electrochim. Acta*, **40**, 1041.

138 Dietz, N. and Lewerenz, H.J. (1992) *Appl. Phys. Lett.*, **60**, 2403.

139 Lublow, M. and Lewerenz, H.J. (2005) *Trans. Inst. Met. Finish.*, **83**, 238.

140 Fujiwara, H. (2007) *Spectroscopic Ellipsometry: Principles and Applications*, John Wiley & Sons, Ltd, Chichester.

141 Lublow, M. and Lewerenz, H.J. (2008) *ECS Trans.*, **6**, 10.

142 Lublow, M. and Lewerenz, H.J. (2007) *Surf. Sci.*, **601**, 4227.

143 Lublow, M. and Lewerenz, H.J. (2007) *Surf. Sci.*, **601**, 1693.

144 Yasuda, T. and Aspnes, D.E. (1994) *Appl. Opt.*, **33**, 7435.

145 Humphreys-Owen, S.P.F. (1961) *Proc. Phys. Soc.*, **77**, 949.

146 Ishiguro, K. and Kato, T. (1953) *J. Phys. Soc. Jpn.*, **8**, 77.

147 Bruggeman, D.A.G. (1935) *Ann. Phys.*, **24**, 636.

148 Maxwell Garnett, J.C. (1904) *Phil. Trans. R. Soc. Lond.*, **203**, 385.

149 Neitzert, H.C., Hirsch, W., Kunst, M., and Nell, M.E.A. (1995) *Appl. Opt.*, **34**, 676.

150 Lewerenz, H.J. and Schulte, K.H. (2002) *Electrochim. Acta*, **47**, 2639.

151 Hüfner, S. (2007) *Very High Resolution Photoelectron Spectroscopy*, Springer-Verlag.

152 Bilderback, D.H., Elleaume, P., and Weckert, E. (2005) *J. Phys. B*, **38**, S773.

153 Becker, U., and Shirley, D.A. (eds) (1996) *VUV and Soft X-Ray Photoionization*, Plenum Press, New York.

154 Harada, Y., Taguchi, M., Miyajima, Y., Tokushima, T., Horikawa, Y., Chainani, A., Shiro, Y., Senba, Y., Ohashi, H., Fukuyama, H., and Shin, S. (2009) *J. Phys. Soc. Jpn.*, **78**, 044802.

155 Cerenkov, P.A. (1934) *Doklady Akad. Nauk SSSR*, **2**, 451.

156 Albertano, P., Reale, L., Palladino, L., Reale, A., Cotton, R., Bollanti, S., Di Lazzaro, P., Flora, F., Lisi, N., Nottola, A., Vigli Papadaki, K., Letardi, T., Batani, D., Conti, A., Moret, M., and Grilli, A. (1997) *J. Microsc.*, **187**, 96.

157 Schoenlein, R.W., Chattopadhyay, S., Chong, H.H.W., Glover, T.E., Heimann, P.A., Shank, C.V., Zholents, A.A., and Zolotorev, M.S. (2000) *Science*, **287**, 2237.

158 Lawson, J.D. (1979) *IEEE Trans. Nucl. Sci.*, **NS-26**, 4217.

159 Spence, D.E., Kean, P.N., and Sibbett, W. (1991) *Opt. Lett.*, **16**, 42.

160 Neila, G.R., Carrb, G.L., Gubeli, J.F., Jordana, K., Martinc, M.C., McKinney, W.C., Shinna, M., Tanid, M., Williamsa, G.P., and Zhange, X.-C. (2003) *Nucl. Instrum. Methods Phys. Res. A*, **507**, 537.

161 Nakamura, K.G., Ishii, S., Ishitsu, S., Shiokawa, M., Takahashi, H., Dharmalingam, K., Irisawa, J., Hironaka, Y., Ishioka, K., and Kitajima, M. (2008) *J. Appl. Phys.*, **93**, 61905.

162 Larsson, J., Heimann, P.A., Lindenberg, A.M., Schuck, P.J., Bucksbaum, P.H., Lee, R.W., Padmore, H.A., Wark, J.S., and Falcone, R.W. (1998) *Appl. Phys. A*, **66**, 587.

163 Lewerenz, H.J. (1997) *Chem. Soc. Rev.*, **26**, 239.

164 Hübner, K. (1977) *Phys. Stat. Sol. (a)*, **42**, 501.

165 Thomas, T.D. (1980) *J. Electron Spectrosc. Rel. Phenom.*, **20**, 117.

166 Seah, M.P. (1986) *Surf. Interface Anal.*, **9**, 8598.

167 Rabalais, J.W. (2003) *Principles and Applications of Ion Scattering Spectrometry: Surface Chemical and Structural Analysis*, John Wiley & Sons, Inc., New York.

168 Briggs, D. and Seah, M.P. (1990) *Practical Surface Analysis by Auger and X-Ray Photoelectron Spectroscopy*, John Wiley & Sons, Ltd, Chichester.

169 Lewerenz, H.J., Aggour, M., Murrel, C., Kanis, M., Jungblut, H., Jakubowicz, J., Cox, P.A., Campbell, S.A., Hoffmann, P., and Schmeisser, D. (2003) *J. Electrochem. Soc.*, **150**, E185.

170 Egerton, R.F. (2009) *Rep. Prog. Phys.*, **72**, 16502.

171 Gergely, G. (2002) *Prog. Surf. Sci.*, **71**, 31.

172 Zhong, Q., Inniss, D., Kjoller, K., and Elings, V.B. (1993) *Surf. Sci.*, **290**, L 688.
173 Martin, Y., Williams, C.C., and Wickramasinghe, H.K. (1987) *J. Appl. Phys.*, **61**, 4723.
174 Garcia, R. and San Paulo, A. (1999) *Phys. Rev. B*, **60**, 4961.
175 Derjaguin, B.V., Muller, V.M., and Toporov, P. (1975) *J. Colloid Interface Sci.*, **53**, 314.
176 San Paulo, A. and Garcia, R. (2001) *Phys. Rev. B*, **64**, 193411.
177 Perez, R., Stich, I., Payne, M.C., and Terakura, K. (1998) *Phys. Rev. B*, **58**, 10835.
178 Kolb, D.M. (2002) *Surf. Sci.*, **500**, 722.
179 Engelmann, G.E. and Kolb, D.M. (2003) *Electrochim. Acta*, **48**, 2897.
180 Tersoff, J. and Hamann, D. (1983) *Phys. Rev. Lett.*, **50**, 1998; Tersoff, J. and Hamann, D. (1985) *Phys. Rev. B*, **31**, 805.
181 Lewerenz, H.J. (1993) *J. Electroanal. Chem.*, **356**, 121.
182 Itaya, K. (1998) *Prog. Surf. Sci.*, **58**, 121.
183 Lublow, M. and Lewerenz, H.J. unpublished work.
184 Levy-Clement, C., Lagoubi, A., and Tomkiewicz, M. (1994) *J. Electrochem. Soc.*, **141**, 958.
185 Canham, L.T. (1991) *Nature*, **353**, 335.
186 Lehmann, V. and Föll, H. (1990) *J. Electrochem. Soc.*, **137**, 653.
187 Lewerenz, H.J., Jungblut, H., and Rauscher, S. (2000) *Electrochim. Acta*, **45**, 4615.
188 Jakubowicz, J., Lewerenz, H.J. and Jungblut, H. (2004) *Electrochem. Commun.*, **6**, 1243.
189 Lewerenz, H.J., Jakubowicz, J., and Jungblut, H. (2004) *Electrochem. Commun.*, **6**, 838.
190 Yazyev, O.V. and Pasquarello, A. (2006) *Phys. Rev. Lett.*, **96**, 157601.
191 Peter, L.M., Li, J., Peat, R., Lewerenz, H.J., and Stumper, J. (1990) *Electrochim. Acta*, **35**, 1657.
192 Parr, R.G. and Yang, W. (1989) *Density Functional Theory of Atoms and Molecules*, Oxford University Press, New York.
193 Gerischer, H., Allongue, P., and Costa Kieling, V. (1993) *Ber. Bunsenges. Phys. Chem.*, **97**, 753.
194 Lehmann, V. and Gösele, U. (1991) *Appl. Phys. Lett.*, **58**, 856.
195 Dmol (1996) *A Density Functional Theory Program with the Insight Molecular Modeling Package*, MSI, San Diego.
196 Perfetti, P., Quaresima, C., Coluzza, C., Fortunato, C., and Margaritondo, G. (1996) *Phys. Rev. Lett.*, **57**, 2065.
197 Guittet, M.J., Crocombette, J.P., and Gautier-Soyer, M. (2001) *Phys. Rev. B*, **63**, 125117.
198 Pehlke, E., and Scheffler, M. (1993) *Phys. Rev. Lett.*, **71**, 2338.
199 Peter, L.M., Borazio, A.M., Lewerenz, H.J., and Stumper, J. (1990) *J. Electroanal. Chem.*, **290**, 229.
200 Turner, D.R. (1960) *J. Electrochem. Soc.*, **107**, 810.
201 Hirata, H., Suwazono, S., and Tanigawa, H. (1988) *Sens. Actuators*, **13**, 63.
202 Offereins, H.L., Sandmaier, H., Maruszcyk, K., Kühl, K., and Plettner, A. (1992) *Sens. Mater.*, **3**, 127.
203 Parette, L., Racine, G.-A., de Rooij, N.F., and Bornand, E. (1991) *Sens. Actuators*, **27**, 597.
204 Allongue, P., Costa-Kieling, V., and Gerischer, H. (1993) *J. Electrochem. Soc.*, **140**, 1018.
205 Palik, E.D., Bermudez, V.M., and Glemboki, O.J. (1985) *J. Electrochem. Soc.*, **132**, 871.
206 Baum, T. and Schiffrin, D.J. (1997) *J. Electroanal. Chem.*, **436**, 239.
207 Glembocki, O.J., Stahlbush, R.E., and Tomkiewicz, M. (1985) *J. Electrochem. Soc.*, **132**, 145.
208 Minks, B.P., Oskam, P., Vanmaekelbergh, D., and Kelly, J.J. (1989) *J. Electroanal. Chem.*, **273**, 119–131.
209 Frank, F.C. (1958) On the kinematic theory of crystal growth and dissolution processes, in *Growth and Perfection of Crystals* (eds D.H. Doremus, B.W. Roberts, and D. Turnbull), John Wiley & Sons, Inc., New York.
210 Krysko, M., Franssen, G., Suski, T., Albrecht, M., Lucznik, B., Grzegory, I.,

Krukowski, S., Czerniecki, R., Grzanka, S., Makarowa, I., Leszczynski, M., and Perlin, P. (2007) *Appl. Phys. Lett.*, **91**, 211904.
211 Garcia, S.P., Bao, H., and Hines, M.A. (2004) *J. Phys. Chem. B*, **108**, 6062.
212 Garcia, S.P., Bao, H., and Hines, M.A. (2004) *Phys. Rev. Lett.*, **93**, 166102.
213 Skorupska, K., Lublow, M., Kanis, M., Jungblut, H., and Lewerenz, H.J. (2005) *Appl. Phys. Lett.*, **87**, 262102.
214 Skorupska, K., Pettenkofer, Ch., Sadewasser, S., Streicher, F., Haiss, W., Lewerenz, H.J., *Phys. Stat. Sol.(b)*, (2010) DOI: 10.1002/pssb.201046454
215 Siebentritt, S., Sadewasser, S., Wimmer, M., Leendertz, C., Elisenbarth, T., and Lux-Steiner, M.C. (2006) *Phys. Rev. Lett.*, **97**, 146601.
216 Skorupska, K., Lewerenz, H.J., and Vo-Dinh, T. (2009) *Phys. Scr.*, **79**, 065801.
217 Chazalviel, J.N., Ozanam, F., Etman, M., Paolucci, F., Peter, L.M., and Stumper, J. (1992) *J. Electroanal. Chem.*, **327**, 343.
218 Carstensen, J., Prange, R., and Föll, H. (1999) *J. Electrochem. Soc.*, **146**, 1134.
219 Parkhutik, V., Rayon, E., Pastor, E., Matveeva, E., Sasano, J., and Ogata, Y. (2005) *Phys. Stat. Sol. A*, **202**, 1586.
220 Grzanna, J., Jungblut, H., and Lewerenz, H.J. (2000) *J. Electroanal. Chem*, **486**, 181.
221 Grzanna, J., Jungblut, H., and Lewerenz, H.J. (2000) *J. Electroanal. Chem*, **486**, 190.
222 Li, Y.-J., Oslonovitch, J., Mazouz, N., Plenge, F., Krischer, K., and Ertl, G. (2001) *Science*, **291**, 2395.
223 Miethe, I., García-Morales, V., and Krischer, K. (2009) *Phys. Rev. Lett.*, **102**, 194101.
224 Rappich, J. and Lewerenz, H.J. (1996) *Electrochim. Acta*, **43**, 675.
225 Lewerenz, H.J. (1997) *J. Phys. Chem B* **101**, 2421.
226 Schmeisser, D. (2004) Private communication.
227 Lewerenz, H.J. and Aggour, M. (1993) *J. Electroanal. Chem.*, **351**, 159.
228 Grzanna, J., Jungblut, H., and Lewerenz, H.J. (2007) *Phys. Stat. Sol. A*, **204**, 1245.
229 Christophersen, M., Langa, S., Carstensen, J., Tiginyanu, I.M., and Föll, H. (2003) *Phys. Stat. Sol. A*, **197**, 197.
230 Aggour, M., Giersig, M., and Lewerenz, H.J. (1995) *J. Electroanal. Chem.*, **383**, 67.
231 Feller, W. (1971) *An Introduction to Probability Theory and Its Applications* vol. II, John Wiley & Sons, Inc., New York.
232 Feynman, R.P. (1985) *The Strange Theory of Light and Matter*, Princeton University Press, Princeton, NJ.
233 Feynman, R.P. and Hibbs, A.R. (1965) *Quantum Mechanics and Path Integrals*, McGraw-Hill, New York.
234 Lee, M.L. and Fitzgerald, E.A. (2003) *J. Appl. Phys.*, **94**, 2590.
235 Lublow, M. and Lewerenz, H.J. (2009) *Electrochim. Acta*, **55**, 340.
236 Rauscher, S., Dittrich, TH., Aggour, M., Rappich, J., Flietner, H., and Lewerenz, H.J. (1995) *Appl. Phys. Lett.*, **66**, 3018.
237 Bachmann, K.J., Lewerenz, H.J., and Menezes, S. (1983) US Patent 4388382.
238 Wagner, S., Shay, J.L., Bachmann, K.J., and Buehler, E. (1975) *Appl. Phys. Lett.*, **26**, 229.
239 Li, C.H., Sun, Y., Law, D.C., Visbeck, S.B., and Hicks, R.F. (2003) *Phys. Rev. B*, **68**, 085320.
240 Heller, A., Aharon-Shalom, E., Bonner, W.A., and Miller, B. (1982) *J. Am. Chem. Soc.*, **104**, 6942.
241 Hannappel, T., Töben, L., Visbeck, S., Crawack, H.J., Pettenkofer, C., and Willig, F. (2000) *Surf. Sci.*, **470**, L1.
242 Sander, M., Lewerenz, H.J., Jägermann, W., and Schmeisser, D. (1987) *Ber. Bunsenges. Phys. Chem.*, **91**, 416.
243 Lewerenz, H.J., Aspnes, D.E., Miller, B., Malm, D.L., and Heller, A. (1982) *J. Am. Chem. Soc.*, **104**, 3325.
244 Lewerenz, H.J., Jungblut, H., and Rauscher, S. (2000) *Electrochim. Acta*, **45**, 4615.
245 Schulte, K. (2000) Dissertation, Free University Berlin.
246 Freeland, B.H., Habeeb, J.J., and Tuck, D.G. (1977) *Can. J. Chem.*, **55**, 1527.

247 Schimper, H.-J., Kollonitsch, Z., Möller, K., Seidel, U., Bloeck, U., Schwarzburg, K., Willig, F., and Hannappel, T. (2006) *J. Cryst. Growth*, **287**, 642.

248 Letzig, T., Schimper, H.-J., Hannappel, T., and Willig, F. (2005) *Phys. Rev. B*, **71**, 033308.

249 Menezes, S., Lewerenz, H.J., and Bachmann, K.J. (1983) *Nature*, **305**, 112; Menezes, S., Lewerenz, H.J., Betz, G., Bachmann, K. J., and Kötz, R. (1984) *J. Electrochem. Soc.*, **131**, 3030.

250 Fischer, C.H., Lewerenz, H.J., Lux-Steiner, M.C., Gudat, W., Karg, F., et al. (2004) *2003 Bessy Highlights* (eds G. André, H. Henneken, and M. Sauerborn), BESSY, Berlin, pp. 15–16.

251 Jaffe, J.E. and Zunger, A. (1983) *Phys. Rev. B*, **28**, 5822.

252 Berenguier, B. and Lewerenz, H.J. (2006) *Electrochem. Commun.*, **8**, 165.

253 Noufi, R. (1982) Proc. 16th IEEE Photovoltaic Specialist sConference, p. 1293.

254 Heller, A. and Vadimsky, R.G. (1981) *Phys. Rev. Lett.*, **46**, 1153.

255 Bansal, A. and Lewis, N.S. (1998) *J. Phys. Chem. B*, **102**, 4058.

256 Miller, B., Heller, A., Menezes, S., and Lewerenz, H.J. (1981) *Discuss. Faraday Soc.*, **70**, 223.

257 Kautek, W., Gerischer, H., and Tributsch, H. (1980) *J. Electrochem. Soc.*, **127**, 2471.

258 Lewerenz, H.J., Tributsch, H., and Spiesser, M. (1985) *J. Electrochem. Soc.*, **132**, 700.

259 Tributsch, H., Lewerenz, H.J., and Spiesser, M. (1986) Patent EP0173641.

260 Ohuchi, F.S., Shimada, T., Parkinson, B.A., Ueno, K., and Koma, A. (1991) *J. Cryst. Growth*, **111**, 1033.

261 Genut, M., Margulis, L., Hodes, G., and Tenne, R. (1992) *Thin Solid Films*, **217**, 91.

262 Ellmer, K. (2001) *Phys. D*, **34**, 3097.

263 Kuang, D., Klein, C., Ito, S., Moser, J.-E., Humphry-Baker, R., Evans, N., Duriaux, F., Grätzel, C., Zakeeruddin, S.M., and Grätzel, M. (2007) *Adv. Mater.*, **19**, 1133.

264 Dloczik, L., Ileperuma, O., Lauermann, I., Peter, L.M., Ponomarev, E.A., Redmond, G., Shaw, N.J., and Uhlendorf, I. (1997) *J. Phys Chem. B.*, **101**, 10281.

265 Menezes, S. (1984) *Appl. Phys. Lett.*, **45**, 148.

266 Dalal, V. (1978) *Design, Technology and Cost of Si Cells for Concentrator Applications, Conference Record. Photovoltaic Specialists Conference*, IEEE, New York.

267 Rappich, J., Jungblut, H., Aggour, M., and Lewerenz, H.J. (1994) *J. Electrochem. Soc.*, **141**, L99.

268 Kötz, R. and Lewerenz, H.J. (1978) *Surf. Sci.*, **78**, L 233.

269 Cooper, J.W. (1962) *Phys. Rev.*, **128**, 681.

270 Hüfner, S., Wertheim, G.K., Buchanan, D.N.E., and West, K.W. (1974) *Phys. Lett. A*, **46**, 420.

271 Hüfner, S. and Wertheim, G.K. (1975) *Phys. Rev. B*, **11**, 678.

272 Campbell, J.P. and Lenahan, P.M. (2002) *Appl. Phys. Lett.*, **80**, 1945.

273 Dueñas, S., Castán, H., García, H., Barbolla, J., Kukli, K., Aarik, J., and Aidla, A. (2004) *Semicond. Sci. Technol.*, **19**, 1141.

274 Bachmann, K.J. and Buehler, E. (1974) *J. Electron. Mater.*, **3**, 279.

275 Schwarz-Schampera, U. and Herzig, P.M. (2002) *Indium: Geology, Mineralogy, and Economics*, Springer, Heidelberg.

276 Demeester, P., Pollentier, I., De Dobbelaere, P., Brys, C., and Van Daele, P. (1993) *Semicond. Sci. Technol.*, **8**, 1124.

277 Seidel, U., Schimper, H.-J., Kollonitsch, Z., Möller, K., Schwarzburg, K., and Hannappel, T. (2007) *J. Cryst. Growth*, **298**, 777.

278 Lewerenz, H.J. and Jungblut, H. (1995) *Photovoltaik. Grundlagen und Anwendungen*, Springer, Berlin.

279 Hsu, L. and Wang, E.Y. (1972) Photovoltaic properties of In2O3/semiconductor heterojunction solar cells, in *Proc. 13th Photovoltaic Specialists Conference*, IEEE, New York, pp. 536–540.

280 Kogelnik, H. (1982) *Prax. Naturwiss. Chem.*, **31**, 353.

281 Förster, Th.v. (1959) *Discuss. Faraday Soc.*, **27**, 7.
282 Xu, Z. and Hu, B. (2008) *Adv. Funct. Mater.*, **18**, 2611.
283 Jang, S., Newton, M.D., and Silbey, R.J. (2004) *Phys. Rev. Lett.*, **92**, 218304.
284 Lezec, H.J., Dionne, J.A., and Atwater, H.A. (2007) *Science*, **316**, 430.
285 Dionne, J., Verhagen, E., Polman, A., and Atwater, H.A. (2008) *Opt. Express*, **16**, 19001.
286 Pillai, S., Catchpole, K.R., Trupke, T., and Green, M.A. (2007) *J. Appl. Phys.*, **101**, 093105.
287 Grundmann, M. and Bimberg, D. (1997) *Phys. Rev. B*, **55**, 9740.
288 Kim, S.J., Kim, W.J., Sahoo, Y., Cartwright, A.N., and Prasad, P.N. (2008) *Appl. Phys. Lett.*, **92**, 031107.
289 Chabr, M. and Williams, D.F. (1977) *Phys. Rev. B*, **16**, 1685.
290 Chen, Z., Concepcion, J.J., Jurss, J.W., and Meyer, T.J. (2009) *J. Am. Chem. Soc.*, **131**, 15580.
291 Hatnean, J.A., Raturi, R., Lefèbre, D.B., Leznoff, G.L., and Johnson, S.A. (2006) *J. Am. Chem. Soc.*, **128**, 14992.
292 Booij-James, I.S., Dube, S.K., Jansen, M.A.K., Edelman, M., and Mattoo, A.K. (2000) *Plant Physiol.*, **124**, 1275.
293 Ishikita, H., Stehlik, D., Golbeck, J.H., and Knappe, E.-W. (2006) *Biophys. J.*, **90**, 1081.

3
Printable Materials and Technologies for Dye-Sensitized Photovoltaic Cells with Flexible Substrates

Tsutomu Miyasaka

3.1
Introduction: Historical Background

The study of dye-sensitized semiconductor electrodes started with the renaissance of photoelectochemistry in the late 1960s when Gerischer and co-workers [1] explored various combinations of organic dyes (sensitizers) and n-type and p-type semiconductors [2]. The 1970s saw rapid advancements in the method of photoelectrochemical energy conversion. In 1972 Fujishima and Honda reported the decomposition of water on a photoelectrode of single-crystal TiO_2 [3], showing the photoelectrochemical cell as a potential model for photosynthetic energy conversion. Before then, Tributsch and Calvin demonstrated for the first time photocurrent generation by a thin film of natural chlorophyll as a sensitizer deposited on n-type semiconductors such as ZnO [4] to mimic the spectral sensitivity of photosynthesis. In the following years it was established, based on the study of Langmuir–Blodgett monolayers [5–7], that a single monolayer of adsorbed dye alone can contribute to photoexcited electron injection, similar to the fact by then established for dye-sensitized silver halide photographic material [8]. The energy conversion efficiency attained by a dye monolayer was too small to be useful for a photovoltaic cell while internal quantum efficiency of photocurrent generation was improved to 30% by regulation of intermolecular distances of dyes [9]. The invention by the Grätzel group in the late 1980s of a mesoscopic semiconductor electrode [10], which had a roughness factor of several thousands, enabled the dye-sensitized cell to join a class of utility-type solar cells. Numerous studies have since been undertaken by focusing on efficiency improvement of the dye-sensitized solar cell (DSSC) mostly with the use of mesoporous TiO_2 as a stable semiconductor and ruthenium bipyridyl complex dyes as sensitizers [11]. After optimization for the dye structure, TiO_2 porosity, thickness, etc., the power conversion efficiency has reached 11% and more [12–14]. Although the top efficiency is still in the range comparable with those of thin amorphous silicon cells, a strong merit of DSSCs over the existing solid-state solar cells has been emphasized on account of their low-cost manufacturing processes without the high vacuum and nanoscale manipulation as required by the solid-state p–n junction cells. Nevertheless, there is a difficulty for the industrializa-

Advances in Electrochemical Science and Engineering. Edited by Richard C. Alkire, Dieter M. Kolb, Jacek Lipkowski, and Philip N. Ross
© 2010 WILEY-VCH Verlag GmbH & Co. KGaA, Weinheim
ISBN: 978-3-527-32859-8

tion of DSSCs unless the issues of efficiency and lifetime (durability) are overcome, both of which are well behind those of the crystalline silicon solar cell.

Proof-of-concept tests of DSSCs have been conducted by installing large integrated modules for windows, walls, and other facade applications. These modules, as well as the cell of top efficiency, employ a glass substrate (SnO_2-coated transparent conductive glass) and have shown thousands hours of life by accelerated tests under exposure to sunlight and high temperatures (60–85 °C) [15–17]; the result seems still not satisfactory for robust roof-top applications. For these reasons, our group has focused on developing a low-cost flexible DSSC fabricated on a plastic flexible substrate by taking advantage of rapid printing technologies. Primarily, the point of this study is that we can minimize the production cost by introducing a roll-to-roll manufacture process that takes the place of the high-temperature sintering process. Secondly, the thin flexible body of the DSSC opens the way to an extended field of applications including, especially, the installation of power sources for consumer electronics, where use of solar power reduces the need for secondary batteries. Although plastic devices generally have shorter lifetimes than glass-based ones, they attract potential users for outdoor long-term installment since the low-cost device can permit the user repeated exchange of the device without the risk of large investment. To work with plastic substrates, it is generally required to prepare printable materials (inks or pastes) that form a solid-state film of high electrochemical activity at processing temperatures below 150 °C. It is also strictly required that high conversion efficiency and stability be ensured. This review will therefore focus on important printable technologies associated with the construction of plastic DSSCs.

3.2
Low-Temperature Coating of Semiconductor Films

For the glass-to-plastic conversion of electrode substrates, low-temperature preparation of an adherent mesoporous semiconductor film is the primary requisite. In addition, special care should also be taken in handling and optimizing of a transparent conductive layer that should exhibit a high conductivity for lateral electron transport on a transformable plastic surface. For the very low surface electric resistance (<15 ohm square^{-1}) required for the electrodes of solar cells, indium-tin-oxide (ITO) is a unique material available for low-temperature vacuum deposition on plastic sheets. ITO films, however, are chemically unstable against exposure to low pH (<4) so that use of acidic compounds for semiconductor treatment, for example $TiCl_4$, cannot be applied. Various methods have been tried for TiO_2 coating on an ITO–plastic film. Pichot *et al.* examined coating of nanocrystalline TiO_2 particles without a sintering treatment to show a low conversion efficiency of 1.2% [18]. Hagfeldt *et al.* demonstrated mechanical compression as a post-coating treatment to show relatively high conversion efficiencies, 3 to 5.5%, depending on the intensity of simulated sunlight [19]. Hydrothermal synthesis of mesoporous TiO_2 has proven to be a useful means of achieving high efficiency

(3.3%) [20]. We have demonstrated a facile method of making a plastic photoelectrode by electrophoretical deposition of TiO_2 films [21]. Post-treatment of films by microwave [22] and UV [23] irradiation has been shown to improve the efficiency by reinforcement of interparticle connections. These methods, however, may not be best suited for rapid electrode fabrication processes. Mechanically, the smooth surface of ITO makes difficult using these methods to achieve strong adhesion of TiO_2 by comparison, for example, with the surface of fluorine-doped tin oxide (FTO) formed on glass, which has a characteristic roughness. For the design of rapid roll-to-roll manufacture, it is desired that all coating processes, including TiO_2 coating and catalyst loading on counter electrode, are completed by simple printing methods using a ink or paste form of electrode materials. A key to preparing such pastes is to achieve strong interparticle connection (necking), and to ensure high adhesion strength of the electrode material with respect to the conductive ITO layer, even in the wet condition when immersed in liquid organic electrolytes.

The mesoscopic TiO_2 paste that works at low temperatures (<150 °C) and simultaneously ensures good photoelectric performance should meet some crucial requirements. Firstly, the content of a binder material, added to the paste for ensuring a sufficient viscosity for printing, should be minimized for it can block carrier transport between particles in a resultant dry film, while the paste should have a sufficiently high viscosity needed for coating work. Secondly, the liquid paste should have good affinity to the surface of a plastic substrate, which is often hydrophobic. Thirdly, any liquid components of the paste should be completely eliminated by evaporation at low temperature (<150 °C). Fourthly, all components of the paste should be inert against the corrosion of the conductive layer (ITO, etc.). We have prepared a viscous suspension of nanocrystalline TiO_2, which is suitable for doctor-blade coating or screen printing on an ITO–plastic sheet [24]. Commercial polyethylene naphthalate (PEN; glass transition temperature of 121 °C) film was chosen as a plastic substrate, on which ITO had been coated by an ionic plating method with a roll-to-roll manufacture system. To reduce the surface tension of the liquid paste, TiO_2 particles were dispersed in a mixed solvent of water and alcohol. Among various kinds of volatile alcohols tested, branched alcohols such as *iso*-propanol and *tert*-butanol proved to be effective to give a good viscosity and coating adhesion in the presence of water and TiO_2 nanoparticles. With *tert*-butanol, the viscosity significantly changes with the ratio of water and alcohol in the range 20–2000 mP s; the water content is adjusted by addition of an aqueous brookite-TiO_2 colloid as an interparticle necking agent. After optimization, viscous coating pastes were obtained in which TiO_2 is the sole solid component. With this binder-free paste, a TiO_2 film is dried at 100–150 °C for water elimination and immediately subjected to dye adsorption. Figure 3.1 shows how this paste forms a mesoporous structure by dehydration condensation reaction which follows the formation of a hydrogen-bonded network of well-dispersed nanocrystalline TiO_2. The best composition of the binder-free paste for printing applications, published elsewhere [24], contains approximately 15% of TiO_2 in a mixed solvent of *tert*-butanol and water (2:1 by volume).

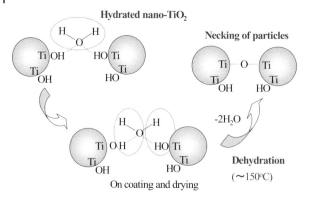

Figure 3.1 Interparticle connection between TiO_2 by dehydration condensation reaction.

3.3
Photoelectric Performance of Plastic Dye-Sensitized Photocells

With doctor-blade or screen-printer methods, the ITO-coated PEN sheet was coated with a 10–15 μm thick TiO_2 film and soaked in a dye solution for sensitization at 40 °C for an hour. A ruthenium complex dye, N719 [21], was employed as the sensitizer. In anti-exfoliation tests, the TiO_2-loaded ITO–PEN sheet showed high stability against bending of the sheet with curvature up to $2\,cm^{-2}$ (a radius of 0.5 cm) and pencil hardness more than H. For cell performance studies, the dye-sensitized ITO–PEN electrode was combined with a platinum-sputtered FTO glass counter electrode to compose a mask-attached test cell with $0.24\,cm^2$ irradiation area. The composition of the liquid electrolyte was varied. Typically, it was LiI (0.4 M), tetrabutylammonium iodide (TBAI) (0.4 M), I_2 (0.04 M), and 4-*tert*-butylpyridine (TBP) (0.3 M) in 3-methoxypropionitrile (MPN). Figure 3.2 summarizes the current–voltage (I–V) characteristic parameters as a function of the average size of TiO_2 nanoparticles measured under incidence of 1 sun intensity [24]. The data were obtained with substantially similar conditions of TiO_2 loading in the range 10–12 g m^{-2} where the TiO_2 layer contained large TiO_2 grains (250 nm) for light scattering enhancement and had a porosity of about 60%. Photocurrent density, J_{sc}, open-circuit photovoltage, V_{oc}, and overall efficiency η tend to exhibit peaks around an average size of 60 nm. This reflects the fact that smaller particles and pore sizes suppress rapid electrolyte diffusion inside the porous TiO_2, while larger particle sizes reduce the surface area for dye adsorption, both leading to a decrease in photocurrent density. The optimized size, 60 nm, is larger than those normally employed for sintered TiO_2 electrodes, namely 20–30 nm. This difference is apparently due to the use of binders in combination with sintering. Firing of a polymer binder and connection of small particles by sintering create new pores which are larger than the initial size while, in the binder-free non-sinter process, pore size distribution is determined by the initial particle size. It is assumed that a larger size distribution is needed for the non-sintered method. However, we found that

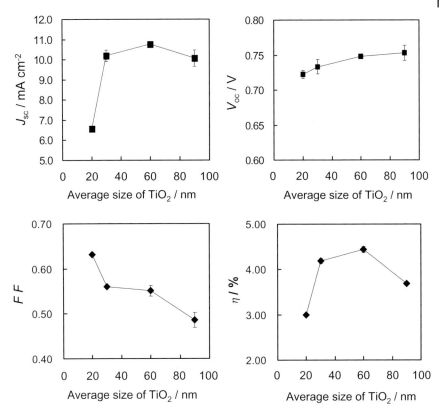

Figure 3.2 Dependence of the *I–V* characteristic parameters measured under 1 sun intensity on the average size of TiO_2 nanocrystalline particles coated on ITO–PEN film from the binder-free paste. (After [24].)

the binder-free TiO_2 film never gives a photocurrent density as high as those of sintered TiO_2 films. This fact indicates that the sintering treatment in the presence of a polymer binder physically reinforces interparticle connection so that the TiO_2 network is improved to have better electron conductivity and a longer electron diffusion length. In effect, incident photon-to-electron conversion quantum efficiency (IPCE) data measured for the dye-sensitized TiO_2-coated ITO–PEN electrode show a maximum value around 65%, which is lower than the best value obtained for sintered TiO_2 glass electrodes (80–90%). Figure 3.3 shows typical examples of a photocurrent density–voltage (*I–V*) characteristic and IPCE action spectrum for an N719-sensitized TiO_2-coated ITO–PEN electrode. This plastic cell yielded a 6.1% conversion efficiency. The IPCE action spectrum shows a sharp drop in the UV region at wavelengths less than 420 nm, which is due to the intrinsic UV absorption of PEN (<390 nm). This filter effectively reduces the overall photocurrent density while it suppresses the photocatalytic activity of TiO_2 that leads to dye decomposition.

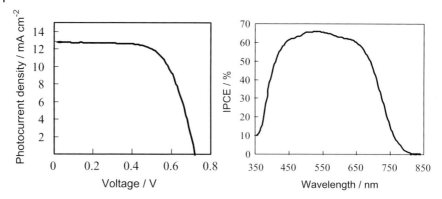

Figure 3.3 Photocurrent density–voltage characteristics under 100 mW cm^{-2} incident intensity of AM1.5 simulated sunlight and IPCE action spectrum for an N719-sensitized TiO$_2$ ITO–PEN electrode. This cell exhibits 6.1% efficiency.

The photovoltaic performance of the plastic cell made by the coating of the specially prepared binder-free paste proved to be good compared to that of cells previously made by our group using the electrophoretic deposition method, in which all TiO$_2$ nanoparticles, free of binders, are deposited in a short time on the ITO surface to form a relatively thick film. These cells made by the electrophoretic method yielded 3.8% efficiency [21]. The adhesion of TiO$_2$ films on ITO, however, is not sufficiently good for practical use, which involves the need to assemble a large integrated module, as will be described below.

It is assumed that the internal electric resistance of the mesoporous TiO$_2$ prepared by the present low-temperature coating method is relatively high compared to that of high-temperature sintered TiO$_2$ films. This is due to imperfections in developing an extended network of interconnected nanoparticles that allows electron diffusion for long distances across particles. Measurements of electron diffusivity (D) and electron lifetime (τ) for an N719-sensitized TiO$_2$ film made from the binder-free paste with a 120 °C treatment demonstrate that electron diffusivity is roughly half that of a sintered film prepared at 450 °C while electron lifetime is comparable. The average electron diffusion length defined as $(D\tau)^{1/2}$ was in the range 13–15 μm depending on the cell structure and external electric conditions (current density and voltage applied), while this value was more than 20 μm for a reference sample of sintered TiO$_2$. This situation affects how the electrode structure is optimized for best photovoltaic performance. For optimization, the relationship between photovoltaic characteristics and TiO$_2$ thickness clearly involves the above issue of electron diffusion. Figure 3.4 shows the dependence of the main photovoltaic parameters on TiO$_2$ loading, where a TiO$_2$ loading of 20 g m^{-2} corresponds to a thickness of 12.5 μm. Figure 3.4 also includes a comparison of data taken for high and low incident intensities, 1 sun and 1/8 sun, where 1 sun is 100 mW cm^{-2}. For photocurrent density, which is normalized for constant incident power, its constant increase with increased TiO$_2$ thickness is no longer sustained

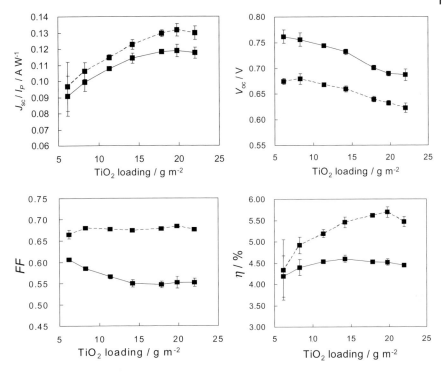

Figure 3.4 Dependences of photocurrent density (J_{sc}), open-circuit voltage (V_{oc}), fill factor (FF), and energy conversion efficiency (η) on the amount of TiO$_2$ loading on an ITO–PEN plastic electrode. Photocurrent density is normalized for constant incident power (I_p) in terms of J_{sc}/I_p (A W^{-1}). Solid and dashed lines were measured for high intensity light (1 sun) and low intensity light (1/8 sun), respectively. TiO$_2$ loading of 20 g m^{-2} corresponds to a thickness of 12.5 μm. (After [24].)

at a thickness around 12 μm (20 g m^{-2}). This affects overall conversion efficiency η which shows a saturation at the same thickness. These facts indicate that carrier transport tends to be inefficient beyond 12 μm, which is also expected from the results of electron diffusion measurements. Another phenomenon to be noted is the large difference in efficiency between high and low light intensities. There is always a loss in photocurrent generation at high intensity, that is, high-density photocurrent generation, as seen in the comparison of intensity-normalized photocurrents. The influence of high incident intensity becomes maximum for electrodes having a thick (highly loaded) TiO$_2$ film, which produces a maximum photocurrent. This difference indicates the existence of a relatively large internal resistance against the flow of photocurrent, which brings about a voltage loss (IR loss) and reduces the fill factor (FF) of the I–V curve. Accordingly, a thick TiO$_2$ film which produces a high density of photocurrent undergoes a large light-intensity hysteresis. This hysteresis is probably more enhanced in the case of binder-free preparation of TiO$_2$ film. It is interesting that post-sintering of the binder-free TiO$_2$ film (prepared at 120 °C) to 500 °C does not cause any essential change in I–V curve and efficiency. In contrast, the conventional sintering method with the use of a binder-mixed TiO$_2$ paste (using polyethylene glycol, etc.) gives an increased photocurrent and IPCE, the latter approaching >80%. The difference in these results with and without the use of binder in the 500 °C sintering indicates that burning of the binder material helps to create new pores, and a mesoporous structure that possesses longer electron diffusion lengths. This result means that the interconnection of TiO$_2$ particles can be completed with the low-temperature treatment of the binder-free film at 120 °C.

In order to improve the photocurrent by increasing the electron diffusion length, mechanical compression of a binder-free film as a post-treatment is known to have a significant effect in efficiency improvement. Recently, an efficiency level close to 8% was obtained by Arakawa and co-workers based on our binder-free TiO$_2$ coating technologies [25].

To conclude concerning the design of a highly efficient cell based on the low-temperature and plastic technologies, the use of a thinner TiO$_2$ film in combination with a dye sensitizer which has a high extinction coefficient is desired to ensure high-density photocurrents. Additionally, because any flexible electrode substrate causes mechanical deformation, a thinner solid film has better stability against physical stresses. In this respect, we are currently assessing different kinds of organic sensitizers such as squalilium dyes and oligothiophene-type dyes, some of which yield high IPCE values comparable to that of N719 on the TiO$_2$-coated ITO–PEN electrode. Apart from organic sensitizers, we have employed inorganic nanocrystalline particles of quantum dot size (2 nm) as visible-light sensitizers for TiO$_2$ as an alternative to organic dyes. Their strong light absorption ability by bandgap excitation is useful for harvesting light with a thinner mesoporous film. A TiO$_2$ electrode sensitized by organo-lead compound perovskite nanocrystals has achieved near 4% efficiency on an FTO electrode [26]. Use of these quantum dot sensitizers is also promising for designing thin-semiconductor-coated plastic electrodes.

3.4
Polymer-Based Counter Electrodes with Printable Materials

For cost reduction of DSSCs, improvement of the preparation of counter electrodes by a vacuum-free printing method becomes important. Typically the surface of a counter electrode substrate is loaded with particles or a thin film of platinum as a catalyst. The role of the catalyst is to minimize the overpotential of the electrochemical reduction of an electron acceptor in the electrolyte. High activity of this cathodic reaction is particularly required for the iodine/triiodide redox system. Although the amount of platinum required is very small, being much less than 10 nm in terms of average thickness [27], preparation processes often employ vacuum sputtering and expensive sources. Only a limited number of materials can replace platinum. One is a group of carbonaceous materials, the other being a group of p-type conductive polymers. The catalytic activities of these alternatives in terms of current density per gram of material, however, are significantly behind that of platinum. Therefore, to realize high cathodic activity, high loadings of these materials are required. Glass-type DSSCs using carbon black-loaded counter electrodes have been studied by Murakami *et al.* [28]. A thick carbon layer of a thickness of 10 μm was shown to give a high efficiency of 9.1% comparable with those of platinum films.

We have studied the performance of poly(3,4-ethylenedioxythiophene) doped with polystyrenesulfonate (PEDOT-PSS) as a printable catalyst for counter electrodes. A coating paste was prepared by mixing PEDOT-PSS with various kinds of metal oxide nanoparticles in order to immobilize a porous catalytic layer on the substrate, and to increase the active surface area. The coating paste comprised an aqueous dispersion of PEDOT-PSS (polymer concentration 1.24%) and metal oxide nanocrystalline particles selected from a group of Al_2O_3, SnO_2, ZnO, and TiO_2. TiO_2 particles (average size 50 nm, specific surface area 30–40 $m^2 g^{-1}$) showed relatively high activity as a cathode catalyst. The PEDOT-PSS/TiO_2 aqueous dispersion was added to an aqueous ITO slurry (15 wt%) at a volume ratio of 1:1 to supply a paste (TiO_2·ITO/PEDOT-PSS) of sufficiently high viscosity for coating. Using an ITO–PEN film as a plastic substrate, doctor-blade coating of the paste was successful and completed by heat-drying at 110 °C for 5 minutes. The mesoporous surface of the counter electrode shows a high roughness factor (>1000) comparable with the TiO_2 photoelectrode.

TiO_2·ITO/PEDOT-PSS films coated on ITO–PEN exhibit a pale blue to gray color and semi-transparency due to the optical characteristics of the nanoparticles. Figure 3.5 shows SEM images of the counter electrode. The thickness of the TiO_2·ITO/PEDOT-PSS film is several micrometers to 10 μm. Figure 3.6 shows an example of a semi-transparent plastic electrode prepared based on this technology. There are several merits of the technology. Firstly, a simple printable process for counter electrode fabrication without vacuum and high temperature is possible. Secondly, a transparent counter electrode (cathode) can be designed for fabrication

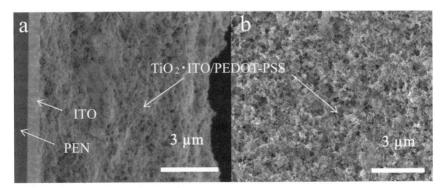

Figure 3.5 SEM images for TiO_2·ITO/PEDOT-PSS nanocomposite film: (a) cross section; (b) surface.

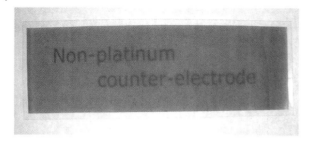

Figure 3.6 Semi-transparent body of the PEDOT-PSS-based plastic counter electrode.

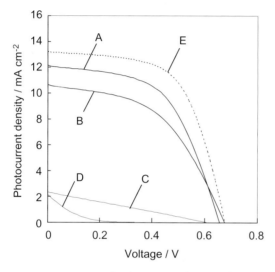

Figure 3.7 I–V curves for dye-sensitized plastic photocells using various counter electrode materials coated on ITO–PEN plastic substrate. A, TiO_2·ITO/PEDOT-PSS film; B, ITO/PEDOT-PSS film; C, PEDOT-PSS film; D, none (ITO–PEN substrate); E, sputtered platinum film on FTO glass as a reference. Electrolyte composition: 0.4 M LiI, 0.4 M TBAI, 0.04 M I_2, 0.3 M NMB in AN/MPN (1:1 by volume). (After [29].)

of a bifacial-type flexible photovoltaic cell in combination with a plastic dye-sensitized electrode (anode). Thirdly, such a method leads to a large cost reduction with use of low-cost materials.

For fabrication of a full plastic DSSC, the plastic counter electrode needs to show high performance in keeping with a high-density photocurrent, with characteristics close to those of a platinum counter electrode. I–V characteristics of plastic DSSCs have been examined. An electrolyte composition of 0.4 M LiI, 0.4 M TBAI, 0.04 M I_2, and 0.3 M N-methylbenzimidazole (NMB) in acetonitrile (AN)/MPN (1:1 by volume) was used and N719 as a dye sensitizer of the TiO_2 photoanode; the I–V curves are compared in Figure 3.7 for PEDOT-PSS-based counter electrodes and a standard platinum sputtered electrode [29].

It is clear from this result that ITO free of catalytic material lacks activity for I_3^- reduction. The non-porous PEDOT-PSS film on ITO–PEN is still insufficient to enhance the I–V performance of the cell yielding a low short-circuit photocurrent density (J_{sc}). Fairly high performance was obtained with ITO/PEDOT-PSS and TiO_2-ITO/PEDOT-PSS films, both having porous surfaces. In effect, J_{sc} of the full-plastic cell using the above PEDOT-PSS-based ITO–PEN counter electrode is still lower by 8% than that of the sputtered platinum counter electrode. The energy conversion efficiency of the PEDOT-PSS-based cell reached 4.38% [29], which approaches the level of a cell using a sputtered platinum-coated FTO glass electrode (5.41%).

It was found that replacement of TiO_2 with other nanoparticles of metal oxides such as Al_2O_3 and NiO equally improves the performance of the non-porous electrodes. Based on this fact, the role of nanoparticles is to create large surface areas for promoting the catalytic reaction activity rather than being involved in improving the surface conductivity. The best performance, however, is obtained with the TiO_2-based material. Photocurrent densities and fill factors of counter electrodes of other metal particles tend to be lower than those of TiO_2. This may be due to the high electrochemical stability of titanium oxides against iodine/iodide compared to nanosized nickel and aluminum oxides. In addition, we consider that the TiO_2 surface interacts better with PEDOT-PSS for forming a polymer-loaded mesoporous structure.

PEDOT derivatives as cathode catalysts can also be immobilized by means of electrodeposition [30]. Poly(3,3-diethyl-3,4-dihydro-2H-thieno-[3,4-b][1,4]dioxepine) (PPDOT-Et2) was employed as an electropolymerizable polymer. Electrolytic polymerization of this polymer on glass and ITO–PEN substrates was carried out under potentiostatic conditions with an applied voltage of 1.1 V vs. Ag/Ag^+ using a 10 mM AN solution of a monomer, 3,4-(2′,2′-diethylpropylene)dioxythiophene, in the presence of $LiClO_4$. The PPDOT-Et2 film showed a wire-like and highly porous structure due to the steric effect of the monomer [31] and resulted in a root-mean-square (RMS) roughness value of about 83.5 nm, which was larger than that of a platinum film of about 23.6 nm over a $5 \times 5 \mu m$ area. The PPDOT-Et2-coated counter electrode is highly transparent compared to the above PEDOT-PSS-coated ones.

The electrochemical activity of the counter electrode can be checked by cyclic voltammetry (CV) with respect to the triiodide/iodide redox couple in a sample electrolyte. Comparison of the peak potentials of reduction and oxidation and current densities thereof with those of various platinum electrodes gives preliminary assessment of the cathode material for use in DSSCs. PPDOT-Et2 conter electrodes, as well as the above-described PEDOT-PSS counter electrodes, demonstrated peak potentials and redox current densities close to those of the platinum electrode. In Figure 3.8, the redox behavior of PPDOT-Et2 in a triiodide/iodide-based electrolyte (1 mM I_2, 10 mM LiI, 0.1 M $LiClO_4$ in AN) is compared with that of the sputtered platinum electrode, where a PPDOT-Et2 film was deposited by a charge capacity of 40 mC cm^{-2} for polymerization. The two anodic and cathodic peaks in the CV curves represent the following two reactions [31]:

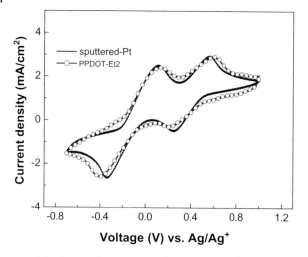

Figure 3.8 Cyclic voltammograms for PPDOT-Et2 film and reference platinum electrode at a scan rate of 100 mV s^{-1}. (After [30].)

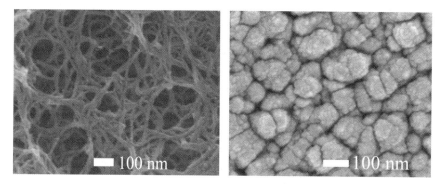

Figure 3.9 SEM images of a PPDOT-Et2 film prepared with deposition capacities of 40 mC cm^{-2} (left) in comparison with a sample of sputtered platinum film (right).

$$3I^- \leftrightarrow I_3^- + 2e^- \quad \text{(for more negative region)}$$

$$2I_3^- \leftrightarrow 3I_2 + 2e^- \quad \text{(for more positive region)}$$

Figure 3.8 reveals that the PPDOT-Et2 film has similar anodic and cathodic peaks and associated charge capacities to the sputtered platinum film and possesses high performance comparable with that of the platinum-based counter electrode. A slightly negative shift of the cathodis peak may indicate the influence of the adsorption of iodide species at the PPDOT-Et2 surface, as reported also in the case of the platinum electrode [32].

Surface morphology of the electrodeposited PPDOT-Et2 film observed by SEM is characterized as a random and highly porous network of polymer fibers, as shown in Figure 3.9. The polymer fiber size tends to depend on the total charge capacity of electrolytic polymerization. For the 40 mC cm^{-2} capacity applied to the sample of Figure 3.9, the fiber thickness is of the order of 10–20 nm.

From further structural characterization of the PPDOT-Et2 films, it is assumed that a dense layer is formed initially, followed by the growth of a porous polymer film with increasing deposited charge capacities from 10 to 80 mC cm^{-2}. Further increase in the deposited charge capacity led to significant aggregations of PPDOT-Et2 that reduced the active surface area. Based on a rotating disk electrode (RDE) voltammetry, the heterogeneous rate constant (k_0) and active surface area of the film are determined from the intercept and slope of Koutecký–Levich plots. It was found that the PPDOT-Et2 films have similar k_0 values of about 5×10^{-5} cm s^{-1}, which is about one order of magnitude lower than that of the platinum electrode. On the other hand, the film with a charge capacity of 40 mC cm^{-2} has the highest active surface area, which is also about one order higher than that of platinum. This explains why the PPDOT-Et2 film prepared with a charge capacity of 40 mC cm^{-2} possesses similar catalytic ability to that of the platinum film.

The above observations relate to FTO glass as the substrate for electrodeposition. ITO-coated plastic substrates can also be used for electrodeposition under the same potentiostatic conditions for polymerization. In this case, however, the homogeneity of the polymerized PPDOT-Et2 film tends to be less than those on FTO substrate. On application to the plastic DSSC, the PPDOT-Et2 counter electrode shows excellent performance in comparison with platinum-sputtered electrodes. Figure 3.10 shows I–V characteristics of plastic DSSCs comprising an N719-sensitized ITO–PEN electrode and PPDOT-Et2-based FTO glass electrodes prepared using different conditions of charging capacities for electrodeposition.

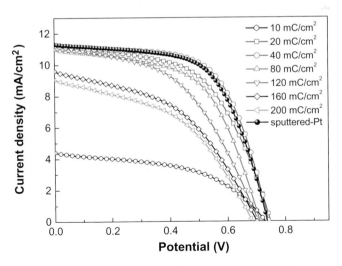

Figure 3.10 I–V curves of plastic DSSCs with use of PPDOT-Et2 counter electrodes prepared with various charging capacities in comparison with platinum counter electrode. Incident intensity: 100 mW cm^{-2} (AM 1.5G). (After [30].)

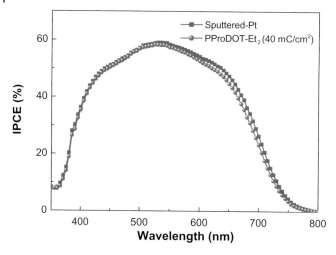

Figure 3.11 IPCE spectra for plastic DSSCs based on PPDOT-Et2 and platinum counter electrodes. (After [30].)

Although the electrolyte composition employed is the same as in Figure 3.7, the two series of counter electrode experiments do not give the same amplitudes of photocurrents due to different conditions of TiO_2 thickness on ITO–PEN.

It is clear from the comparison of I–V curves that there is an optimum condition of charging capacity in regulating film thickness for improving the performance. The DSSC with a $40\,mC\,cm^{-2}$ PPDOT-Et2 counter electrode had the highest short-circuit current density (J_{sc}) and open-circuit voltage (V_{oc}), yielding an I–V curve close to that of the sputtered platinum electrode. The conversion efficiency of the DSSC with the PPDOT-Et2 counter electrode is 5.20%, which is slightly improved over that of the platinum electrode of 5.11% efficiency by the effect of an increased fill factor (0.63). The decrease in J_{sc}, V_{oc}, and fill factor at higher deposited charge capacities is assumed to result from the aggregation of PPDOT-Et2 and the appearance of large pores in the film, as detected by SEM, which reduced the redox rate of I^-/I_3^-. IPCE spectra (Figure 3.11) of the plastic DSSCs based on PPDOT-Et2 ($40\,mC\,cm^{-2}$) and sputtered platinum counter electrodes give the same level of amplitude with a maximum of about 60% at 530 nm. That in the higher wavelength region (550–800 nm) the sputtered platinum-based DSSC has a higher value than PPDOT-Et2 indicates an optical effect of light reflection at the platinum surface which improves photon collection at the working electrode.

Fabrication of counter electrodes by simple coating processes using PEDOT-based printable pastes as described here is an important key for DSSC manufacture with respect to minimum costs and rapid production. Design of semi-transparent counter electrodes also contributes to improved light utilization of the flexible cell and modules.

3.5
Investigation of High-Extinction Sensitizers and Co-adsorbents

Sensitizing dyes with high extinction coefficients are sought after to improve the light-harvesting performance of dye-adsorbed TiO_2 films especially when coating of a thinner TiO_2 is desired in order to avoid cracking on a flexible plastic electrode. The extinction coefficient of the typical ruthenium complex N719 is $1.3 \times 10^4 \, M^{-1} \, cm^{-1}$ at around 520 nm (ethanol solution), which can normally work best in visible-light harvesting with TiO_2 mesopore thickness of more than 15 μm. As described above, the electron diffusion length is relatively short in the TiO_2 film prepared at low temperature on ITO–PEN substrate. With the estimated diffusion length of 12–13 μm, we have to design a plastic electrode loaded with a TiO_2 film of thickness less than 10 μm. Simultaneously, dye sensitizers are required to have high extinction coefficient without losing high efficiency in electron injection.

In collaboration with Ho's group, we have employed an amphiphilic ruthenium sensitizer, SJW-E1, as depicted in Figure 3.12, which has a molar extinction coefficient of $18.7 \times 10^3 \, M^{-1} \, cm^{-1}$ (at 546 nm) [33]. The DSSC based on SJW-E1 yields a high conversion efficiency of 9.02% on glass substrate, exceeding that of a DSSC based on N3 dye (8.42%) under the same conditions [33]. While the method of sensitizing TiO_2 with N719 is well established, use of other kinds of dyes often requires special care for determining optimal conditions for sensitization. One

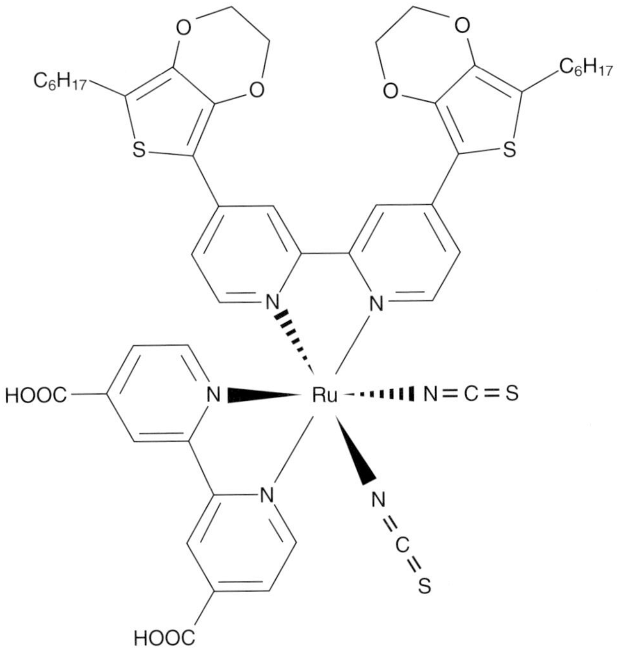

Figure 3.12 Amphiphilic ruthenium sensitizer, SJW-E1.

such requirement is the use of a co-adsorbent in the dye-soaking solution. This is also a well-established method to suppress molecular aggregation of bulky dyes or hydrophobic dyes. In addition to the disaggregation effect, co-adsorbents suppress intermolecular energy transfer of sensitizers that leads to thermal deactivation of excited dyes before electron injection. This energy dissipation matters when fluorescent dyes capable of efficient Förster-type resonance energy transfer (FRET), such as cyanine and merocyanine dyes, are employed for sensitization. The other requirement is the use of a buffer layer (or blocking layer) on a conductive substrate (FTO, ITO, etc.). The role of such a buffer layer is to suppress electron back transfer from the electrode substrate to an ionic acceptor in electrolyte, which is I_3^- in the DSSC. This back transfer is considered to be mediated by dye molecules that directly adsorb on the substrate surface (FTO, ITO, etc.). Such adsorption does not occur for N719. We found, however, that for many other sensitizers, photocurrent density tends to be improved by treating the electrode surface with a thin buffer layer. For SJW-E1, we applied coating of a buffer layer. A dense, non-porous 15 nm thick buffer layer of TiO_x was prepared on the ITO surface by way of reactive sputtering of titanium metal under argon and oxygen atmosphere and subsequent anodic oxidation of titanium in a dilute phosphate buffer solution at a bias potential of 3.0 V. Optically, the transmittance of ITO–PEN with the TiO_x buffer layer decreases by about 10% compared to that without a TiO_x layer (as shown in Figure 3.13).

Low-temperature coating of a mesoporous TiO_2 layer followed this buffer treatment using a binder-free mesoscopic TiO_2 paste. The TiO_2 film was sensitized by soaking in a dye solution for 2 hours at 40 °C. A DSSC fabricated with the SJW-E1-sensitized plastic electrode yielded photocurrent performance which was highly affected by the presence of the TiO_x buffer layer. The basic photocurrent performance shows a TiO_2 thickness dependence that well reflects the higher extinction

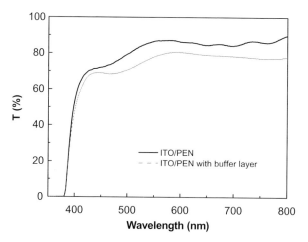

Figure 3.13 The optical transmittances of ITO–PEN substrates with and without sputtered TiO_x buffer layer in the region of 350–800 nm. (After [30].)

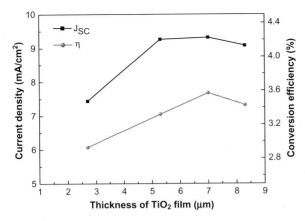

Figure 3.14 Photocurrent density and conversion efficiency of plastic DSSCs based on SJW-E1 dye as a function of the thicknesses of TiO₂ films prepared at low temperature. (After [30].)

coefficient of SJW-E1 compared to N719. Figure 3.14 shows this thickness dependence. The increase of photocurrent density (J_{sc}) with the thickness saturates at around 5 μm, and conversion efficiency, affected by voltage and fill factor (FF), tends to take a maximum at 7 μm. This thickness proved to be less than that normally obtained for the N719 dye, which is around 10 μm. The difference in thickness dependence apparently reflects the 35% larger extinction of SJW-E1 cmpared to N719. Light-harvesting ability of dye molecules plays an important role in the case of non-sintered TiO₂ films on plastic substrates having shorter length for electron diffusion than in sintered TiO₂ films on glass substrates.

It has been reported that FTO and ITO with surface covered by a buffer/blocking layer for preventing the electron back transfer to the electrolyte I^-/I_3^- redox couple can enhance the FF and cell performance [34]. The effect of introducing a TiO$_x$ buffer layer to the SJW-E1-based plastic solar cells was observed from cyclic voltammograms (from −1.3 to 1.2 V vs. Ag/Ag⁺) for the I^-/I_3^- redox reaction with the ITO–PEN electrode. As shown in Figure 3.15, a bare ITO–PEN film causes an oxidation current at potentials higher than 0.2 V, while the oxidation is well suppressed for the TiO$_x$-coated ITO–PEN film. The 15 nm thick TiO$_x$ layer certainly prevents the I^-/I_3^- redox reaction and electron back transfer at the ITO–PEN surface, and blocks the back current, thereby improving cell performance (Figure 3.15).

I–V characteristic curves of the corresponding cells with and without the TiO$_x$ buffer layer are compared in Figure 3.16. With the buffer layer, the FF value of the cell increases effectively from 0.57 to 0.64. In spite of a slight drop occurring in J_{sc} ascribed to the transmittance decrease due to the TiO$_x$ layer, the cell with the buffer layer yields a higher conversion efficiency of 3.84% [34].

In addition to the buffer layer treatment, sensitization with SJW-E1 dye proved to be dependent on the use of co-adsorbent. The use of co-adsorbents in the dye-adsorption process is a well-known strategy for solving the problem of dye aggregation [35]. We investigated the effect of three co-adsorbents, deoxycholic acid (DCA),

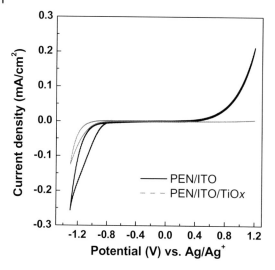

Figure 3.15 CV curves of ITO–PEN electrodes with and without the TiO$_x$ buffer layer measured in AN solution containing 10 mM LiI, 1 mM I$_2$, and 0.1 M LiClO$_4$ with Ag/Ag$^+$ reference electrode and a scan rate of 100 mV s^{-1}. (After [30].)

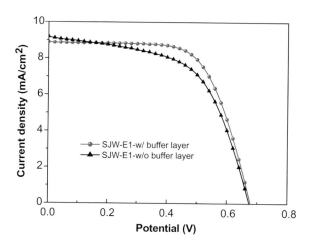

Figure 3.16 Photovoltaic performances of plastic DSSCs based on ITO–PEN substrates with and without TiO$_x$ buffer layer. The thickness of SJW-E1-sensitized mesoporous TiO$_2$ was about 7 μm. (After [30].)

cholic acid (CA), and chenodeoxycholic acid (CDCA). The structures of these cholic acids are shown in Figure 3.17. In the absence of any co-adsorbent, the dye SJW-E1 adsorbed on TiO$_2$ shows an absorption peak which is slightly blue-shifted compared to that of the monomeric state in a dilute dimethylformamide solution as shown in Figure 3.18. This shift is a sign of the H-aggregation behavior. On the contrary, use of DCA as co-adsorbent recovers the monomeric absorption peak for

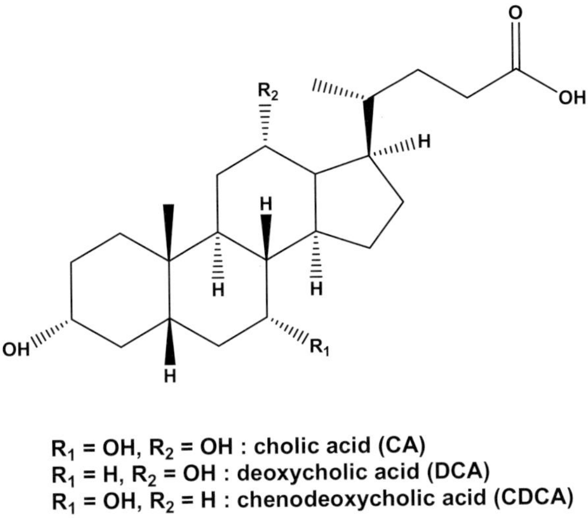

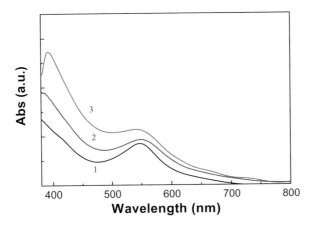

R$_1$ = OH, R$_2$ = OH : cholic acid (CA)
R$_1$ = H, R$_2$ = OH : deoxycholic acid (DCA)
R$_1$ = OH, R$_2$ = H : chenodeoxycholic acid (CDCA)

Figure 3.17 Structures of the three co-adsorbents having cholic acid frameworks.

Figure 3.18 Absorption spectra of SJW-E1 dye with or without DCA on TiO$_2$ films (about 2 μm thick). 1, in dimethylformamide; 2, on TiO$_2$ with 5 mM DCA; 3, on TiO$_2$ without co-adsorbents. (After [30].)

the dye adsorbed on TiO$_2$, indicating that the dye aggregation is reduced. The other co-adsorbents, CA and CDCA, exhibit similar phenomena. Use of these co-adsorbents improves the cell performance in every respect. Table 3.1 summarizes the influence of the co-adsorbents on the *I–V* characteristics of plastic DSSCs, in which the same electrolyte composition as in Figure 3.7 was used. It is noted that all co-adsorbents are capable of improving J_{sc} and V_{oc}. Addition of a significant amount of DCA (10 mM, hundreds times more than the dye concentration) increases V_{oc} by 50 mV to 0.731 V while causing a decrease in photocurrent amplitude. This voltage increase is ascribed to suppression of electron recombination

Table 3.1 The dye loading on TiO_2 electrodes and photovoltaic performances of plastic DSSCs based on SJW-E1 dye with various conditions of co-adsorbents under illumination of 100 mW cm^{-2}.

	Dye loading (mol cm^{-2})	J_{sc} (mA cm^{-2})	V_{oc} (V)	η (%)	FF
—	1.4×10^{-7}	8.88	0.68	3.68	0.61
2 mM DCA	8.7×10^{-8}	9.05	0.700	3.86	0.61
5 mM DCA	9.4×10^{-8}	11.13	0.708	4.88	0.62
10 mM DCA	7.3×10^{-8}	4.22	0.731	2.13	0.69
5 mM CA	9.2×10^{-8}	11.26	0.710	4.96	0.62
5 mM DCDA	8.7×10^{-8}	10.66	0.705	4.28	0.57

between the injected electrons and I_3^- ions on the TiO_2 surface. An optimal concentration of co-adsorbent is found at 5 mM, which gives the best effect of suppressing the quenching of excited energy by aggregated molecules and promotes electron injection to the TiO_2 conduction band. Higher concentration of the co-adsorbent reduces the amount of dye adsorbed on TiO_2, leading to photocurrent reduction. The effect of co-adsorbents on cell performance has been studied by measurement of pulsed-laser-induced current transients: in the presence of co-adsorbents, the dye molecules anchored onto the surface of TiO_2 can increase the electron diffusion coefficient, relative to the case of a bare electrode [36]. This phenomenon must also be a reason for the improvement of J_{sc}. An SJW-E1-sensitized cell prepared by this method achieved a J_{sc} value exceeding 11 mA cm^{-2} and a conversion efficiency close to 5% [35].

Based on the above system assisted by co-adsorbents, other elements of the DSSC structure should be optimized towards fabrication of a cell of best efficiency. Among such elements are TiO_2 thickness and electrolyte composition. Using the SJW-E1 and co-adsorbent sensitization system, we have optimized the electrolyte composition focusing on the concentration of iodine with respect to iodide. It was found that lower concentration of iodine (forming I_3^-) tends to improve the photocurrent density and IPCE value as a result of suppressing back current from TiO_2 to I_3^- in the electrolyte. An optimum condition for the redox couple was found with 0.02 M iodine and 0.4 M LiI + 0.4 M TBAI. After optimization for the thickness of TiO_2 on the TiO_x-coated ITO–PEN, the best plastic cell based on SJW-E1 yielded a photocurrent density of 12.7 mA cm^{-2}, which is 10% higher than the cell containing 0.04 M iodine. V_{oc} could also be slightly improved. The maximum energy conversion efficiency of the cell is 6.31% with a FF of 0.671. In comparison with a standard N719-sensitized plastic DSSC electrode, a higher efficiency was obtained with this dye system, the latter yielding 6.20% for the same conditions of electrolyte and TiO_2 film. This relationship is shown in terms of I–V characteristics and IPCE spectra in Figure 3.19. An enhanced IPCE value of the SJW-E1-sensitized cell over the N719-sensitized one in the long-wavelength region of 700–800 nm reflects the higher extinction coefficient of SJW-E1. The 6.31% efficiency is one of

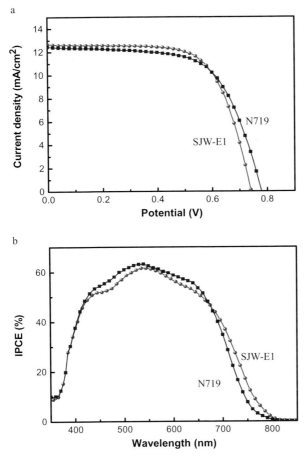

Figure 3.19 The optimal photovoltaic performances of plastic DSSCs based on SJW-E1 and N719. The composition of electrolyte was 0.4 M LiI + 0.4 M TBAI + 0.02 M I_2 + 0.3 M NMB in AN. (a) I–V characteristics and (b) IPCE action spectra.

highest obtained by us with the low-temperature coated TiO_2 and ITO–plastic substrates.

There are other many organic dyes useful for high-efficiency sensitization on TiO_2. Metal-free organic dyes, in particular as represented by polyene and polymethine dyes, have high extinction coefficients as compared to ruthenium complex dyes that involve the metal–ligand charge transfer mechanism for photoexcitation, which thereby exhibit relatively low coefficients. Cyanin and merocyanin dyes, popular for sensitizing photographic materials, possess coefficients of the order of $10^5 \, mol^{-1} \, cm^{-1}$. However, they undergo concentration quenching of excited energy via FRET on semiconductor surfaces and thus require significant dilution of adsorbed dyes with use of co-adsorbents to prevent intermolecular energy transfer. H- and J-aggregates of theses dyes can functions as sensitizers to TiO_2. Some families of coumarin dyes have high coefficients of more than $5 \times 10^4 \, mol^{-1} \, cm^{-1}$, serving as good sensitizers with conversion efficiency more than 8% on TiO_2. Among these organic dyes, indoline-type dyes have recently been extensively studied for their high performance in TiO_2 sensitization with conversion efficiencies close to those of ruthenium complex dyes. Typically, D102, D149, and D205, commercialized by Mitsubishi Paper Mill Ltd, have demonstrated excellent performance on glass-based sintered TiO_2 electrodes yielding IPCEs exceeding 80%. We have investigated the behavior of D149 (Figure 3.20) on non-sintered TiO_2 plastic electrodes.

D149 has a large extinction coefficient in solution of $68\,000 \, mol^{-1} \, cm^{-1}$ at 530 nm [37]. It is known that this dye can be an effective sensitizer without use of co-adsorbents because of less molecular aggregation involved. Following the method described above, a 15 nm thick buffer layer of TiO_2 was coated on an ITO layer by low-temperature processes that comprised vacuum sputtering of titanium metal

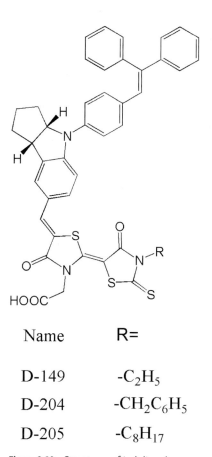

Name	R=
D-149	$-C_2H_5$
D-204	$-CH_2C_6H_5$
D-205	$-C_8H_{17}$

Figure 3.20 Structures of indoline dyes.

and subsequent oxidation to TiO_2. A mesoporous TiO_2 layer was doctor-blade coated on the buffer layer, followed by heat-drying at 130 °C for 5 minutes. Dye adsorption was done by immersing the TiO_2-coated ITO–PEN in an AN/*tert*-butanol (1 : 1 by volume) solution of 0.4 mM D149 under slow shaking at 40 °C to give a dye-sensitized photoanode. Electrolyte composition was 0.4 M LiI, 0.4 M TBAI, 0.04 M I_2, and 0.3 M NMB in a mixture of AN and MPN (1 : 1 by volume). The plastic DSSC made for assessment of photovoltaic performance basically followed the structure and electrolyte composition described above. Optimal conditions for D149 adsorption to the non-sintered TiO_2 film were found to be 1.5 hours of soaking at 40 °C. Use of co-adsorbents such as cholic acid derivatives with a concentration range up to 0.1 M caused no improvement in the photocurrent density. For a reference ITO–PEN substrate without the buffer layer, it was found that D149 adsorbs not only on TiO_2 but also on the ITO surface, indicating strong interaction of the dye with ITO. This caused exfoliation of TiO_2 films after a prolonged adsorption time; for example, the TiO_2 film was completely removed from ITO after 3 hours of adsorption. For the buffer layer-coated ITO–PEN, the dye was totally adsorbed to the TiO_2 layer.

With this high-extinction dye, an important experimental issue is to optimize TiO_2 thickness. Measurements on the thickness dependence of D149-sensitized photocurrent and efficiency showed that the optimal condition for best performance is obtained with a TiO_2 loading of around $10\,g\,m^{-2}$, which corresponds to a thickness of 6 μm. For the same conditions, a reference N719-sensitized electrode gives a maximum photocurrent density and efficiency with a larger TiO_2 loading of more than $15\,g\,m^{-2}$ (>10 μm in thickness). A D149-sensitized plastic DSSC yields J_{sc} close to $8\,mA\,cm^{-2}$ with a thin TiO_2 film. The IPCE action spectrum on an ITO–PEN electrode (Figure 3.21) shows a peak efficiency situated at wavelengths around 600 nm, 50 nm longer than the peak given by N719 sensitization. The

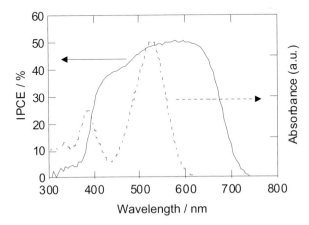

Figure 3.21 IPCE action spectrum of a D149-sensitized plastic DSSC. (After [38].)

longer wavelength edge of the response, however, is 740 nm, which is shorter than that of N719 (800 nm) meaning a poorer light-harvesting performance. The maximum IPCE value of 50% is not high enough by comparison with the N719 dye yielding over 70%. The low IPCE value may indicate that, for TiO_2 prepared at low temperature, electron injection efficiency, that is, internal quantum conversion efficiency, is lower than that achieved by N719. Similar phenomena to this are found with other organic dyes. The dyes that can give a high IPCE value on sintered TiO_2 film, almost comparable with the family of ruthenium complex dyes, do not always give high IPCE values on non-sintered TiO_2 in combination with an ITO conductive layer. This difference is possibly associated with the interaction of the dye with the non-sintered TiO_2 surface having a considerable amount of hydroxyl groups remaining. Electron injection and electron back transfer processes may be affected by this interface and the kind of dye.

D149-sensitized plastic DSSCs are capable of 3.7% efficiency with the use of buffer layer-coated ITO–PEN electrodes [38]. Figure 3.22 compares *I–V* curves in the presence and absence of a TiO_2 buffer layer. Without the buffer layer, the *I–V* curve exhibits a relatively low fill factor and the shunt resistance in an equivalent circuit is relatively low. Higher fill factor and V_{oc} are recovered with the buffer-coated electrode keeping the same level of photocurrent density. Despite a high extinction coefficient, the efficiency obtained with D149 is much less than that obtained with the ruthenium complex dyes due to low IPCE level in addition to the shorter spectral region of the dye (<740 nm). Another indoline-type dye, D205 (see Figure 3.20), was recently examined for plastic DSSCs based on the above

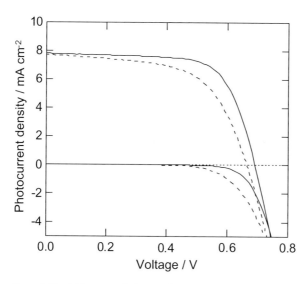

Figure 3.22 *I–V* curves of plastic DSSC using D149 under 1 sun (100 mW cm^{-2}) and dark. Solid lines represent a cell with the TiO_2 buffer layer and dashed lines a cell without the buffer layer. (After [38].)

method without using the co-adsorbent system. With an extinction coefficient of 53 000 at 530 nm, the optical absorption of D205 bound to mesoporous TiO_2 extends to wavelengths of 770 nm. On sintered glass electrodes, a conversion efficiency of 9.5% has been recently reported [39], which is one of highest efficiencies obtained with metal-free organic dye sensitizers. With plastic DSSCs, we have also confirmed that D205 is capable of producing higher J_{sc} than D149 on non-sintered TiO_2. A conversion efficiency exceeding 5% can be achieved with a buffer-coated plastic DSSC with this dye, which is also the highest efficiency obtained with organic dye sensitizers.

Finally, in relation to the light-harvesting function of dyes, use of chlorophyll derivatives in photoelectrochemical cells has long been a subject of intense study in an attempt to simulate the photosynthetic primary processes. Since the first trial in 1971 by Tributsch and Calvin [4], chlorophyll derivatives have been studied as efficient sensitizers for n-type semiconductors because the excited state of chlorophyll is theoretically capable of reducing and oxidizing water similar to TiO_2 [6, 40]. The Grätzel group reported several experimental results using chlorophyll derivatives in mesoscopic TiO_2-based DSSCs [11a,b]. Conversion efficiency then was too small to be accepted as a DSSC that can be compared with those based on the ruthenium complex family. We have attempted to improve the performance of sensitization using chlorin-e6 on mesoporous TiO_2. Similar to natural chlorophylls, chlorin-e6 (Figure 3.23) possesses a high extinction coefficient (58 000 at 660 nm) and spectral sensitivity up to near 700 nm. The photoexcited state of chlorophylls, however, is highly fluorescent with a small Stokes shift and undergoes significant concentration quenching as a result of Förster-type resonance energy transfer. This makes it necessary to add a significant amount of diluents (inert molecules) to the dye system for sensitization, as has been revealed by our

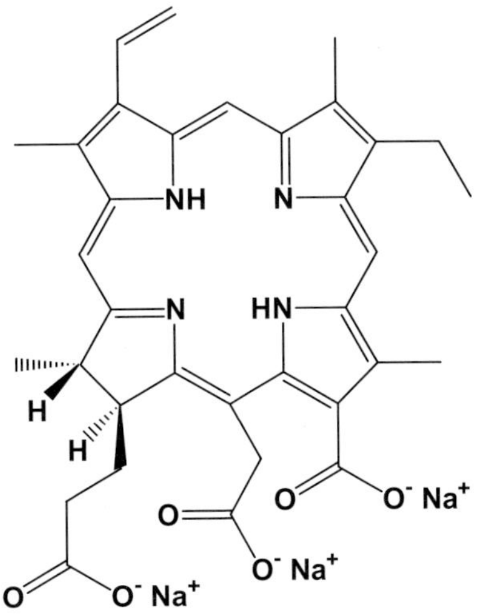

Figure 3.23 Structure of chlorin-e6, a metal-free, water-soluble derivative of chlorophyll.

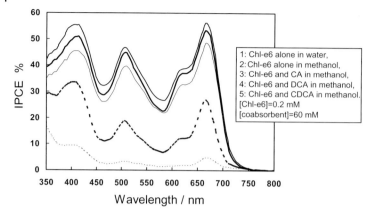

Figure 3.24 IPCE action spectra of chlorin-e6-sensitized photocurrent on TiO$_2$ with various conditions for the kind and concentration of co-adsorbents. Electrode substrate is FTO glass in this experiment.

previous work on SnO$_2$ electrodes [9]. In this respect, cholic acid derivatives were employed as co-adsorbents to the chlorin-e6-sensitized TiO$_2$ electrode. The electrode in this study employed FTO glass substrates. After optimization for co-adsorbent/dye ratio and electrolyte composition, the conversion efficiency of the chlorin-e6-sensitized DSSC reached 4.3% with V_{oc} of 0.61 V [41], which is to our knowledge the highest level with chlorophyll derivatives. The action spectrum of the photocurrent, as shown in Figure 3.24, has a close similarity to the action spectrum of the photosynthesis driven by chlorophylls, exhibiting intense Q and Soret peaks in the blue (400–430 nm) and red (670 nm) regions, respectively. Without the light-harvesting antenna dye system, use of the mesoporous medium enables light collection also in the green region where chlorophylls have a weak absorbance (Figure 3.24).

3.6
Durability Development for Plastic DSSCs

Long-term stability of DSSCs is a key issue for practical applications. Utility-type DSSC modules have been manufactured with the use of glass substrates by many groups such as AISIN [42], Fujikura [17], Sharp [43], Fraunhofer [44], Dyesol, and Solaronix [45], in which TiO$_2$ electrodes are prepared by high-temperature sintering (>450 °C). For these glass-based DSSCs, durability tests in accelerated conditions of high temperature and humidity and under continuous illumination have demonstrated thousands of hours of life. Based on these simulations, DSSCs are assumed to maintain the output power after several years in outdoor conditions. For plastic DSSCs made by the non-sintering process in combination with ITO-coated plastic substrates, cell life tends to be associated with a significant reduction

Table 3.2 Electrolyte compositions of plastic DSSC employed for durability tests.

Electrolyte	I_2 (M)	LiI (M)	TBAI (M)	NMB (M)	Solvent[a]
A	0.04	0.4	0.4	0.3	AN/MPN
B	0.04	0.4	0.4	0.3	GBL/MPN
C	0.04	0.4	0.4	0	GBL/MPN
D	0.04	0.4	0	0.3	GBL/MPN
E	0.04	0	0.4	0.3	GBL/MPN

a) GBL, γ-butyrolactone.

in power production, the cause of which is associated with deterioration of the ITO film and non-sintered TiO_2 film. We have carried out durability tests for plastic DSSCs in an accelerated condition of high temperature under continuous exposure to 1 sun irradiation [46]. A test cell with a somewhat larger active area size of about 6 mm in diameter was first examined. This was made with a gasket film of Himilan 1702 (thickness 25 μm) and platinum-sputtered FTO glass as a sealing material and counter electrode, respectively. A standard N719 was employed in this assessment. Our study focused on the influence of liquid electrolytes by testing various kinds of electrolyte compositions for improvement of cell durability. Table 3.2 gives examples of electrolyte compositions. It is well known that ITO is chemically unstable against both strong acid and alkaline solutions, indicating that the composition of electrolyte directly affects the durability of ITO-based DSSCs. The optimal electrolyte composition for plastic DSSCs seems to differ from that for conventional glass-based DSSCs. Electrolyte composition A has been used as a standard in our plastic cell fabrication, which is capable of yielding high efficiency under intense 1 sun illumination. Electrolyte B employs γ-butyrolactone as a less volatile solvent (boiling point of 203 °C) taking the place of AN. Electrolytes C, D, and E are alternatives where one component is removed from electrolyte A and B. DSSCs made with electrolyte B, C, D, and E were kept in a constant temperature and humidity chamber at 55 °C and 95% relative humidity. In order to check ITO deterioration, the I–V characteristics and AC impedance spectra of sample cells were monitored.

Figure 3.25 shows the change of AC impedance spectra (Nyquist plot) of DSSCs. Intercept on the x-axis represents series resistance (R_s) which includes surface resistance of a substrate. Deterioration of the ITO layer causes an increase of R_s with time, which is the case for the electrolyte compositions C and D. Electrolyte composition E, which does not contain LiI, shows a different trend. The size of the arch with the x-axis intercept decreases and a change of R_s is small. From these results, we deduce that LiI seems to cause a reaction with ITO that leads to deterioration of the cell performance. Durability of plastic TiO_2 electrodes is significantly improved by using LiI-free electrolyte. Figure 3.26 shows the change in I–V characteristics of a plastic DSSC while holding the cell in the dark under

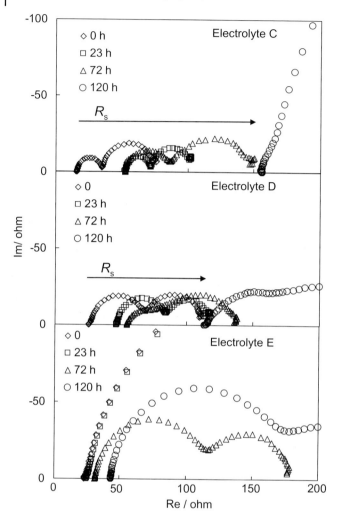

Figure 3.25 Changes in AC impedance spectra for plastic DSSCs with electrolytes C, D, and E in Table 3.2, under the storage condition of 55 °C and 55% relative humidity. (After [46].)

conditions of high temperature, 55 °C, and high humidity, 95%, where a LiI-free electrolyte based on composition E with a solvent of propylenecarbonate was used. Following an initial drop in J_{sn} and FF, the cell maintained its power maximum and initial conversion efficiency (2%) for over 220 hours. For the same conditions, a DSSC using LiI-containing electrolyte (electrolyte A) almost lost photovoltaic activity after 48 hours. These data were obtained from test cells without means of protection against water and oxygen penetration. When water penetrates into plastic films and electrolyte, desorption of sensitizing dyes, especially N719 dye, from the titania surface and destruction of the TiO_2 film can occur due to their hydrophilic interaction. The durability of plastic film-type DSSCs must be

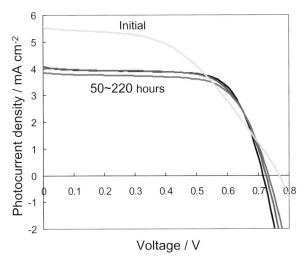

Figure 3.26 Change in I–V characteristics of a plastic DSSC consisting of propylenecarbonate based electrolyte under the storage condition of 55 °C and 95% relative humidity. (After [46].)

significantly improved when the appropriate protection layers are applied, internally or externally, to the plastic substrate.

Durability was also investigated for stability against continuous light exposure at 1 sun (100 mW cm^{-2}) with a solar simulator. This light soaking causes heating of the cell surface up to around 50 °C. Figure 3.27 shows long-term changes in photovoltaic characteristics under continuous irradiation at 1 sun. In Figure 3.28 are compared I–V curves of the cell monitored during the light soaking test. This test cell reached a maximum energy conversion efficiency of 3.7% after an aging of 58 hours and initial drop in J_{sc} and FF. This peak efficiency was reduced to 1.9% (half of the maximum) after 880 hours. The current expectation is that further improvements in electrolyte composition in combination with the use of protection (barrier) layers against water penetration are expected to increase drastically the cell lifetime.

As a second stage of durability testing, we have fabricated large-area plastic DSSCs of various sizes to assess the photovoltaic performance as well as lifetime on exposure to accelerated high-temperature atmospheres. The processes of making these modules, which have a series connection of unit cells to integrate currents and/or voltage, are described in the following section. The durability of these modules depends not only on the chemical stability of materials but also on the mechanical stability of the three-dimensional structure including sealers, electronic wiring, and circuitry. Given that these mechanical issues to ensure durability are overcome, cell lifetimes comparable with those obtained with a small test cell can be achieved. With a large full-plastic cell of 10 × 10 cm in size having a series connection of six cells, for example, we could obtain a lifetime exceeding 200 hours with an accelerated condition.

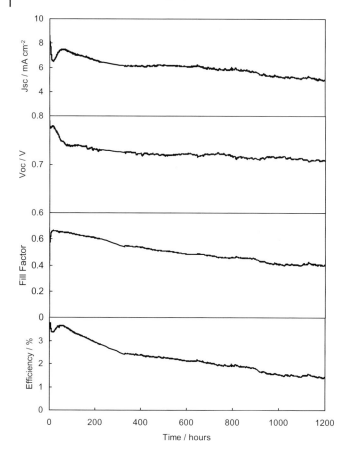

Figure 3.27 Changes of photovoltaic parameters for a plastic DSSC measured under continuous irradiation (100 mW cm^{-2}) at 50 °C. (After [46].)

3.7
Fabrication of Large-Area Plastic DSSC Modules

Basic technologies of low-temperature TiO_2 coating and electrolyte improvements invented in our laboratory have been applied to fabrication of utility-type modules in collaboration with Peccell Technologies, Inc., a Toin University-based venture. Some examples of module developments are introduced in this section.

High-voltage DSSC modules of series DC connection have so far been developed by many groups using FTO glass electrodes [42–45]. In contrast, only a few groups have demonstrated fabrication of DSSC modules of full-plastic configuration. It is now widely accepted that one of major goals of the technologies of plastic electronics is industrialization of a low-weight, flexible, and low-cost device to the consumer electronics market. To this end, for example, a DC voltage of higher than 5 V is usually needed for powering electric devices with secondary batteries such

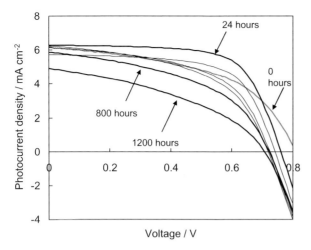

Figure 3.28 Change in I–V characteristics of a plastic DSSC under continuous irradiation (100 mW cm^{-2}) at 50 °C. (After [46].)

as lithium ion-based ones. For the open-circuit voltage of DSSCs in the range 0.6–0.85 V, methods of electrically connecting unit cells become necessary.

A method of DC connection that we have previously tried is that long strip-shaped unit cells with a size of 3.5 cm × 30 cm are connected at the long side with double-stick conductive tape. The number of unit cells connected in this fashion depends on the output voltage required. Unit cells were fabricated on ITO–PEN films. On both edges of the long side of a unit cell, conductive silver grids for collecting current were mounted also using a low-temperature process (<120 °C). A hot-melt type sealer (polymerizable material) which combines two plastic substrates and prevents leaking of electrolyte was specially developed for operation at temperatures around 120 °C. Also, a light-hardening type sealer that works at UV to visible wavelengths was applied to ITO–PEN film substrates of 200 μm thickness. Various electrolyte compositions were investigated for module construction. One of them comprised 0.4 M LiI, 0.04 M I$_2$, 0.4 M TBAI, and 0.3 M *tert*-butylpyridine in a mixture of MPN and γ-butyrolactone (1:1). Figure 3.29 shows a full-plastic DSSC module of 30 × 30 cm with series DC connection of 10 unit cells (weight 60 g, thickness about 450 μm) [46, 47], which employs a highly transparent counter electrode bearing a microscale network of Ti–Pt metal alloy as a cathode catalyst [27]. This titanium-based alloy film shows high resistively against corrosion by iodine. For a proof-of-concept demonstration, the DSSC module was exhibited at the World Exposition (EXPO 2005, Aichi, Japan) by installing it on the green (planted) wall for a month during summertime. The module generates 7.2 V as V_{oc} and 250–300 mA as short-circuit current under exposure to 1 sun. This sample was able to power a commercial mobile phone directly under outdoor sunlight. Placed on green plants over a month, the module showed the ability to supply sunlight to activate photosynthesis with red light passed by the cell body, without reducing the growth rate of the plants.

Figure 3.29 A full-plastic transparent DSSC module with a size of 30 × 30 cm² supplying 7.2 V as open-circuit voltage, made of 10-cell DC series connection. The module was installed on the BioLung (Afforested wall) of EXPO 2005 for outdoor testing.

Flexible, bendable DSSC modules can immediately utilize rounded surfaces and corners onto which thin bodies can be fixed. On such curved surfaces, however, it matters that the incident light intensity varies greatly depending on the position within the module. Figure 3.30 shows time courses of photocurrent output monitored for our plastic DSSC module and a reference a-Si cell module, during a time course of sunny and cloudy hours, and the effect of temperature change on the output of these modules. It is first noted that in cloudy circumstances characterized by weak diffused (scattered) light, the DSSC module shows a much higher photocurrent than the a-Si module. This fact indicates the high efficiency of collecting light of large incident angles (shallow incidence of light). Secondly, while photocurrent rapidly decreases with increasing surface temperature for the a-Si module, it tends to increase for the DSSC module, due to increased ionic diffusion in the electrolyte. Consequently, the DSSC module has good stability against fluctuation of light and change of surrounding temperature. This fact also indicates that integrated electric power by DSSCs can be significantly high and is most effectively utilized by storing the generated energy to a rechargeable battery. In this connection, we have also conducted a study on the construction of a photorechargeable cell, which we named a photocapacitor, based on the dye-sensitized TiO_2 system. The method of storing solar power (visible light energy) by the dye-sensitized photocapacitor is described elsewhere [48].

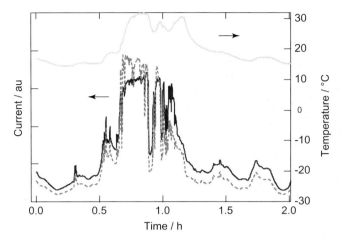

Figure 3.30 Time courses of photocurrents of a plastic DSSC module (solid line) and of an a-Si module (dotted line) placed at the window side when exposed to sunlight. The temperature change at the DSSC module surface, caused by sunshine, is shown at the top.

Durability tests conducted for the full-plastic DSSC modules described above show that long-term stability, in particular at high temperature, is still a significant challenge to overcome. The DSSC module exhibited at EXPO 2005 was exposed to rain, summer-time intense sunlight, and frequent temperature change for 26 days. After this outdoor installment, the module still maintained the 7.2 V maximum voltage but underwent a photocurrent decrease by about 50%. The main cause of deterioration was physical damage occurring at the flexible interface of ITO and TiO_2, and an increased resistance at the ITO layer rather than leakage of electrolyte. Apparently, the physical and chemical stabilities at the ITO–TiO_2 and ITO–electrolyte interfaces are keys to securing long-term durability.

While a series DC connection can be used to produce a high voltage, parallel connection of unit cells gives an alternative way for power production by way of using an electronic circuit (charge pump) for amplifying voltage. A large-area plastic module (Figure 3.31) was manufactured for such applications. The module outputs a DC voltage of 0.72 V being equivalent to a single cell. It can supply a large photocurrent by exposure to sunlight without a large loss in fill factor (IR loss). High current capability of this type of cell is supported by metal grid patterns as current collectors attached on the ITO plastic substrate by printing technology. Using this cell, we could confirm a powerful advantage of DSSCs over other types of solid-state photovoltaic devices, in particular crystalline silicon-based devices. An advantage is the ability of DSSCs to respond to and utilize low-intensity light and weak diffused light. Photovoltage generated by DSSCs is fairly stable in amplitude against the change of light intensity and can be maintained even for exposure to weak indoor illumination, which is normally two orders of magnitude lower than that of sunlight. Stable voltage output of DSSCs is especially useful for powering or charging batteries of electronic equipment such as computers and mobile

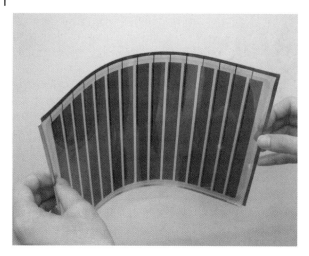

Figure 3.31 A plastic DSSC module with parallel connection of 15 cells, capable of supplying 0.72 V and a large current depending on light intensity.

Figure 3.32 Full-plastic flexible DSSC module exhibited at PVEXPO 2008 (Tokyo), producing more than 110 V under indoor illumination.

phones. It is noted that a visible-light-sensitized DSSC has high conversion efficiency, twice that corresponding to the sunlight spectrum, to light sources such as fluorescent light.

Based on the above concept, a large integrated plastic module of 1×2 m in size and 0.5 mm in thickness was constructed by combining 96 sheets of a flexible sub-module having a standard size of 10×10 cm. This module, as shown in Figure 3.32, was exhibited for proof-of-concept testing at PVEXPO 2008 in Tokyo. Being flexible and lightweight (0.8 kg m^{-2}), the bifacial-type module was hung from the ceiling of an exhibition booth, which was exposed to indoor light with intensity of 1000–1500 lx (1.5–2.3 W m^2) supplied by visible illumination systems. The module generated a stable DC voltage of more than 110 V. In the same circumstances, crystalline silicon photovoltaic cells tested as references were unable to generate normal voltage more than 0.3 V per unit cell. Those devices with the relatively low internal resistance of doped silicon crystals do not maintain a high fill factor or open-circuit voltage for low-density photocurrents under weak light. Large-area plastic DSSC modules also showed excellent performance in utilizing (absorbing) light incident to the module surface with shallow angles (high incident angles). An example of experimental data for the light angle dependence of photocurrent is shown in Figure 3.33, in which the target light source is sunlight. It is clear from this result that a large drop in the output power only occurs at incidence angles larger than 60° for the DSSC, at which significant power reduction occurs for the polycrystalline silicon cell. This difference is simply due to low refractive

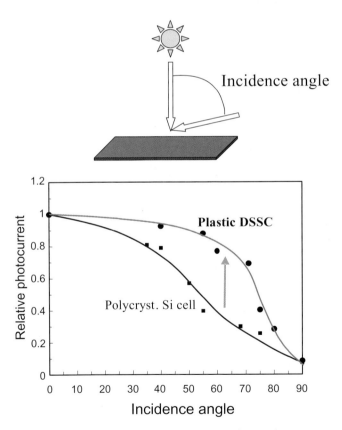

Figure 3.33 Comparison of photocurrent outputs from a plastic DSSC module and a crystalline silicon module as a function of incidence angle of sunlight. Amplitude is normalized as unity at the right incidence (zero angle).

index and low optical reflectivity of dye-coated mesoporous layers compared to polycrystalline silicon. Accordingly, we expect that DSSC modules, especially those with plastic substrates having lower refractive index, will exhibit superb performance in photovoltaic conversion in indoor circumstances or in the outdoor condition of cloudy and rainy days.

3.8
Concluding Remarks

The goal of developing printable materials and technologies is not only to realize high-speed and low-cost manufacturing processes, but also to meet various demands of users who need freely designed and fine patterns of electrodes. The combination of plastic substrates and screen printing is one of the best suited means for this purpose. Despite relatively low IPCEs (60–70%), efficient power generation of DSSCs under indirect diffused light becomes a major advantage for flexible modules, which can be positioned on non-flat places being exposed to light of various incident angles. Intrinsic properties of plastic materials such as thermal expansion, water absorption, and gas penetration, however, make it difficult to ensure long-term stability of DSSCs under high temperature and humidity. These issues must be solved by using a composite plastic film and/or a gas barrier coating as additional components to the commercial substrate. Fortunately, because DSSCs are much more stable against water and oxygen contaminations than the solid-state organic electronic devices like organic light-emitting diodes, low-cost barrier materials and films are expected to be applied to DSSCs, especially those for use in consumer electronics. For roll-to-roll printing processes, solidification and quasi-solidification of liquid electrolyte layers, which have been much studied for glass-type cells, are successfully applied to the cell construction. Printing of viscous pastes as electrolyte layers can further simplify DSSC manufacture without the need of a liquid injection process. To this end we have developed a carbon and ionic liquid-based high-viscosity paste as a quasi-solid electrolyte, which proved to work at high efficiency with a small amount of iodine as oxidant [49, 50]. These methods will contribute to further improve the cost performance of DSSCs towards the innovation of green photovoltaic devices.

References

1 Tributsch, H. and Gerischer, H. (1969) *Ber. Bunsenges. Phys. Chem.*, **73**, 851 and references therein.
2 Memming, R. (2001) *Semiconductor Electrochemistry*, Wiley-VCH Verlag GmbH, Weiheim.
3 Fujishima, A. and Honda, K. (1972) *Nature*, **238**, 37.
4 Tributsch, H. and Calvin, M. (1971) *Photochem. Photobiol.*, **14**, 95.
5 Kuhn, H., Möbius, D., and Bücher, H. (1972) *Physical Methods in Chemistry*, vol. 1 (eds A. Weissberger and B.W. Rossiter), John Wiley & Sons, Inc., New York, p. 577.

6 Miyasaka, T., Watanabe, T., Fujishima, A., and Honda, K. (1978) *J. Am. Chem. Soc.*, **100**, 6657.
7 Miyasaka, T., Watanabe, T., Fujishima, A., and Honda, K. (1980) *Photochem. Photobiol.*, **32**, 217.
8 James, T.H. (1977) *The Theory of the Photographic Process*, 4th edn, Macmillan, New York; Tani, T. (1995) *Photographic Sensitivity: Theory and Mechanisms*, Oxford University Press.
9 Miyasaka, T., Watanabe, T., Fujishima, A., and Honda, K. (1979) *Nature*, **277**, 638.
10 O'Regan, B. and Grätzel, M. (1991) *Nature*, **353**, 737.
11 (a) Nazeeruddin, Md.K., Pechy, P., and Grätzel, M. (1997) *Chem. Commun.*, 1705; (b) Wang, P., Zakeeruddin, S.M., Comte, P., Charvet, R., Humphry-Baker, R., and Grätzel, M. (2003) *J. Phys. Chem. B.*, **107**, 14336; (c) Chen, C.-Y., Wu, S.-J., Wu, C.-G., Chen, J.-G., and Ho, K.-C. (2006) *Angew. Chem. Int. Ed.*, **45**, 5822; (d) Jiang, K.-J., Masaki, N., Xia, J.-B., Noda, S., and Yanagida, S. (2006) *Chem. Commun.*, 2460; (e) Wang, P., Klein, C., Humphry-Baker, R., Zakeeruddin, S.M., and Grätzel, M. (2005) *J. Am. Chem. Soc.*, **127**, 808.
12 Grätzel, M. (2005) *Chem. Lett.*, **34**, 8; Nazeeruddin, M.K., De Angelis, F., Fantacci, S., Selloni, A., Viscardi, G., Liska, P., Ito, S., Besho, T., and Grätzel, M. (2005) *J. Am. Chem. Soc.*, **127**, 16835.
13 Chiba, Y., Islam, A., Watanabe, Y., Komiya, R., Koide, N., and Han, L. (2006) *J. Appl. Phys.*, **45**, L638.
14 Chen, C.-Y., Wang, M., Li, J.-Y., Pootrakulchote, N., Alibabaei, L., Ngoc-le, C., Decoppet, J.D., Tsai, J.-H., Grätzel, C., Wu, C.-G., Zakeeruddin, S.M., and Grätzel, M. (2009) *ACS Nano*, **3**, 3103.
15 Wang, P., Zakeeruddin, S.M., Moser, J.E., Nazeeruddin, M.K., Sekiguchi, T., and Grätzel, M. (2003) *Nat. Mater.*, **2**, 402.
16 Kato, N., Takeda, Y., Higuchi, K., Takeuchi, A., Sudo, E., Tanaka, H., Motohiro, T., Sano, T., and Toyoda, T. (2009) *Solar Energy Mater. Solar Cells*, **93**, 893.
17 Matsui, H., Okada, K., Kitamura, T., and Tanabe, N. (2009) *Solar Energy Mater. Solar Cells*, **93**, 1110.
18 Pichot, F., Pitts, J.R., and Gregg, B.A. (2000) *Langmuir*, **16**, 5626.
19 Lindström, H., Hormberg, A., Magnusson, E., Malmqvist, L., and Hagfeldt, A. (2001) *J. Photochem. Photobiol. A*, **145**, 107; Boschloo, G., Lindström, H., Magnusson, E., Hormberg, A., and Hagfeldt, A. (2002) *J. Photochem. Photobiol. A*, **148**, 11.
20 Zhang, D., Yoshida, T., and Minoura, H. (2003) *Adv. Mater.*, **15**, 814.
21 Miyasaka, T., and Kijitori, Y. (2004) *J. Electrochem. Soc.*, **151**, A1767.
22 Miyasaka, T., Kijitori, Y., Murakami, T.N., Kimura, M., and Uegusa, S. (2002) *Chem. Lett.*, **31**, 1250.
23 Murakami, T.N., Kijitori, Y., Kawashima, N., and Miyasaka, T. (2004) *J. Photochem. Photobiol. A*, **164**, 187.
24 Miyasaka, T., Ikegami, M., and Kijitori, Y. (2007) *J. Electrochem. Soc.*, **154**, A455.
25 Yamaguchi, T., Tobe, N., Matsumoto, D., Nagai, T., and Arakawa, H. (2010) *Solar Energy Mater Solar Cells*, **94**, 812.
26 Kojima, A., Teshima, K., Shirai, Y., and Miyasaka, T. (2009) *J. Am. Chem. Soc.*, **131**, 6050.
27 Ikegami, M., Teshima, K., Miyoshi, K., Miyasaka, T., Wei, T.C., Wan, C.C., and Wang, Y.Y. (2007) *Appl. Phys. Lett.*, **90**, 153122 and references therein.
28 Murakami, T.N., Itoh, S., Wang, Q., Nazeeruddin, M.K., Bessho, T., Cesar, I., Liska, P., Humphry-Baker, R., Comte, P., Pechy, P., and Grätzel, M. (2006) *J. Electrochem. Soc.*, **153**, A2255.
29 Muto, T., Ikegami, M., and Miyasaka, T. (2010) *Electrochem. Soc.*, **157**, B1195.
30 Lee, K.-M., Hsu, C.-Y., Chen, P.-Y., Ikegami, M., Miyasaka, T., and Ho, K.-C. (2009) *Phys. Chem. Chem. Phys.*, **18**, 3375.
31 Gaupp, C.L., Welsh, D.M., and Reynolds, J.R. (2002) *Macromol. Rapid Commun.*, **23**, 885.
32 Popov, A.I., and Geske, D.H. (1958) *J. Am. Chem. Soc.*, **80**, 1340.
33 Chen, C.Y., Wu, S.J., Li, J.Y., Wu, C.G., Chen, J.G., and Ho, K.C. (2007) *Adv. Mater.*, **19**, 3888.

34 Cameron, P.J. and Peter, L.M. (2003) *J. Phys. Chem. B*, **107**, 14394; Fabregat-Santiago, F., Bisquert, J., Palomares, E., Otero, L., Kuang, D., Zakeeruddin, S.M., and Grätzel, M. (2007) *J. Phys. Chem. C*, **111**, 6550; Wu, J., Lana, Z., Wang, D., Haoa, S., Lin, J., Huang, Y., Yin, S., and Sato, T. (2006) *Electrochim. Acta*, **51**, 4243.

35 Hara, K., Sato, T., Katoh, R., Furube, A., Yoshihara, T., Murai M., Kurashige, M., Ito, S., Shinpo, A., Suga, S., and Arakawa, H. (2005) .*Adv. Funct. Mater.*, **15**, 246; Horiuchi, T., Miura, H., Sumioka, K., and Uchida, S. (2004) *J. Am. Chem. Soc.*, **126**, 3014; Wang, P., Zakeeruddin, S.M., Humphry-Baker, R., Moser, J.-E., and Grätzel, M. (2003) *Adv. Mater.*, **15**, 2101.

36 Nakade, S., Saito, Y., Kubo, W., Kanzaki, T., Kitamura, T., Wada, Y., and Yanagida, S. (2003) *Electrochem. Commun.*, **5**, 804.

37 Horiuchi, T., Miura, H., Sumioka, K., and Uchida, S. (2004) *J. Am. Chem. Soc.*, **126**, 12218; Ito, S., Zakeeruddin, S.M., Humphry-Baker, R., Liska, P., Charvet, R., Comte, P., Nazeeruddin, M.K., Péchy, P., Takata, M., Miura, H., Uchida, S., and Grätzel, M. (2006) *Adv. Mater.*, **18**, 1202.

38 Miyoshi, K., Numao, M., Ikegami, M., and Miyasaka, T. (2008) *Electrochemistry*, **76**, 158.

39 Kuang, D., Uchida, S., Humphry-Baker, R., Zakeeruddin, S.M., and Grätzel, M. (2008) *Angew. Chem. Int. Ed.*, **47**, 1923.

40 Miyasaka, T. and Honda, K. (1981) *ACS Symp. Ser.*, **146**, 231.

41 Ikegami, M., Ozeki, M., Kijitori, Y., and Miyasaka, T. (2008) *Electrochemistry*, **76**, 140.

42 Toyoda, T., Sano, T., Nakajima, J., Doi, S., Fukumoto, S., Ito, A., Tohyama, T., Yoshida, M., Kanagawa, T., Motohiro, T., Shiga, T., Higuchi, K., Tanaka, K., Takeda, Y., Fukano, T., Katoh, N., Takeichi, A., Takechi, K., and Shiozawa, M. (2004) *J. Photochem. Photobiol. A*, **164**, 203.

43 Fukui, A., Fuke, N., Komiya, R., Koide, N., Yamanaka, R., Katayama, H., and Han, L. (2009) *Appl. Phys. Express*, **2**, 082202; Han, L., Fukui, A., Chiba, Y., Islam, A., Komiya, R., Fuke, N., Koide, N., Yamanaka, R., and Shimizu, M. (2009) *Appl. Phys. Lett.*, **94**, 013305.

44 Hinsch, A., Brandt, H., Veurman, W., Hemming, S., Nittel, M., Würfel, U., Putyra, P., Lang-Koetz, C., Stabe, M., Beuker, S., and Fichter, K. (2009) *Solar Energy Mater. Solar Cells*, **93**, 820.

45 DSSC manufactures (2010), for examples, Dyesol Ltd., http://www.dyesol.com/; Solaronix SA, http://www.solaronix.com/ (accessed on July 25th, 2010).

46 Ikegami, M., Suzuki, J., Teshima, K., Kawaraya, M., and Miyasaka, T. (2009) *Solar Energy Mater. Solar Cells*, **93**, 836.

47 Miyasaka, T., Kijitori, Y., and Ikegami, M. (2007) *Electrochemistry*, **75**, 2.

48 Miyasaka, T., and Murakami, T.N. (2004) *Appl. Phys. Lett.*, **85**, 3932; Murakami, T.N., Kawashima, N., and Miyasaka, T. (2005) *Chem. Commun.*, 3346.

49 Ikeda, N., Teshima, K., and Miyasaka, T. (2006) *Chem. Commun.*, 1733.

50 Ikeda, N., and Miyasaka, T. (2007) *Chem. Lett.*, **36**, 466.

4
Electrodeposited Porous ZnO Sensitized by Organic Dyes – Promising Materials for Dye-Sensitized Solar Cells with Potential Application in Large-Scale Photovoltaics

Derck Schlettwein, Tsukasa Yoshida, and Daniel Lincot

4.1
Introduction

The supply of a fast growing world population and an even faster growing world economy with energy has become one of the major problems in the world not only because of the depletion of fossil fuel resources but also because of the corresponding threats of climate change caused by pollution of the atmosphere by carbon dioxide and other greenhouse gases [1, 2]. In view of limited resources of nuclear fuels and the still unsolved questions of nuclear waste storage, the use of renewable energy sources has become a major pathway to tackle the problems. Aside from the established use of hydroelectric and wind power, an increasing use of solar radiation represents an almost mandatory path to solve the problem: the world annual energy demand of 2006 (5×10^{20} J [3]), for example, corresponds to only about 1/8000 of the net annual incoming solar radiation of 4×10^{24} J on the earth. Using this enormous amount of incoming energy at least to a small extent would yield a considerable contribution to the solution of one of the major problems in the world. The two main ways to transform solar radiation to flexibly usable forms are presently seen in photovoltaic converters [4] or in solarthermal [5] converters of solar radiation to electrical power. Both ways are presently merging towards economically feasible technologies with large solar parks developing [6–8]. As another alternative technology, a direct conversion of solar heat to electricity realized in thermoelectric converters [9] is also discussed with good options to close the technological gap between photovoltaic and classic solarthermal conversion. All present technologies, however, need to decrease the cost of the produced electrical energy in order to lead to their widespread use and hence considerable contributions to solve the human energy problem. Photovoltaics presently represents the largest installed power, since photovoltaic cells can easily be used in differently sized modules and parks, and since no moving parts are needed, represent a very promising technology with potentially low maintenance requirements.

Photovoltaic cells are typically categorized in terms of development generations. The first generation of cells consist of semiconductor wafers (~0.3 mm of Ge, Si,

Advances in Electrochemical Science and Engineering. Edited by Richard C. Alkire, Dieter M. Kolb, Jacek Lipkowski, and Philip N. Ross
© 2010 WILEY-VCH Verlag GmbH & Co. KGaA, Weinheim
ISBN: 978-3-527-32859-8

GaAs) typically cut from single crystals of the material [10] or, more recently, also from multicrystalline blocks of Si [11, 12]. As a second generation, cells were developed based on thin films deposited by sputtering or vapor deposition (physical or chemical) or, more recently, by depositions from solution of the active semiconducting layers on an independent substrate. Cells of this type consist of amorphous or microcrystalline Si, CdTe, or differently composed chalcopyrites based on $CuInSe_2$ or CuInS including also Ga and Al in different concentrations. A third generation of cells are typically summarized as those that tackle the classic Shockley–Queisser limit which clearly stated from theory that the maximum possible efficiency of any cell concept that utilizes non-concentrated solar radiation and one absorber material with one bandgap and hence one absorption edge cannot exceed 31% because of thermal losses for energies larger than the absorption edge of the semiconductor and transmission losses for energy lower than the absorption edge [2, 10]. Aside from optical up-conversion of the radiation otherwise not absorbed or optical down-conversion of the radiation that otherwise would be transformed to heat, which are economically barely feasible, many concepts are followed that utilize multiple absorbers or materials with multiple optical transitions and separate electrical harvest [10]. Tandem cells using two absorber layers of different bandgaps use this concept, and increasing the number of absorber layers of different bandgaps can further increase the maximum possible efficiency [10]. In a single semiconductor material, quantum confinement is needed to realize multiple efficient transitions [10].

Organic dye and pigment molecules represent a class of materials by means of which clearly enhanced optical absorption coefficients can be realized caused by increased oscillator strengths in aromatic systems. Therefore the film thickness needed for efficiently absorbing cells can be further reduced beyond the level already reached for the classic thin-film technologies discussed above [2]. Nevertheless the loss of generated carriers in initially tested bilayer cells was too large to realize efficient cells since only a very narrow interfacial zone of two materials was active in either Schottky-type or p/n-type cells [13, 14]. Two very successful approaches [15] have developed recently: (i) the use of multiple ultrathin films of high structural control using doping of layers to decrease the series resistance [16, 17] and (ii) the use of bulk heterojunctions consisting of a conductive pathway of a crystalline organic acceptor with large interfacial contact area to a matrix of a donor conductive polymer, prepared from a mixed solution of both constituents [18]. Also, hybrid approaches of layered structures with mixed interlayers [19, 20] or interpenetrating interfaces [21] were successfully studied. An alternative way to make use of the strong absorption of organic dye molecules is realized in dye-sensitized solar cells (DSSCs) [22, 23] consisting of a porous wide-bandgap semiconductor as electron-conducting phase, the organic dye as absorber, and a hole-conducting contact phase, typically an iodine-containing electrolyte, which are discussed in this chapter (Figure 4.1). Their development although presently focused on TiO_2 as the wide-bandgap component indeed started following fundamental work of dye-sensitized photocurrents at ZnO single crystals [24] in which only dyes adsorbed at the surface of ZnO worked as sensitizers [25]. Sintered ZnO with a larger surface area already led to improved performance of about 2% under weak monochromatic illumination [26]. A DSSC based on electrodeposited porous yet crystalline ZnO which combines the benefits of both these historic approaches recently reached 5.6% efficiency under AM 1.5 white light conditions [27]. The three different classes of cells based on organic dye or pigment molecules as absorbers are often referred to as a new generation of photovoltaic cells because of new conversion principles, new ways of preparation, an extended variety of substrate materials, and hence a largely changed cost structure of prospective technologies based on these cells.

Photoelectrochemical cell concepts based on light absorption in a semiconductor and charge transfer to a redox-active electrolyte (Figure 4.1a) have been studied extensively as an alternative to the cells based on semiconductor junctions [23, 28]. Such junctions could be useful in particular to generate chemical fuels and hence provide easy storage, although corrosion reactions have severely limited the use of such concepts in technical cells. As an important example TiO_2 has proven to be a very stable semiconductor even under illumination with very high oxidative power of holes in the valence band and reductive power of electrons in the conduction band because of its large bandgap. Electrodes of TiO_2 can therefore be used in a number of reactions, mainly useful in water cleaning and the decay of toxic pollutants. The large bandgap of TiO_2, however, requires ultraviolet light to electronically excite it, which is still possible to drive some of these reactions but which hinders an efficient use of such junctions in the conversion of solar radiation into

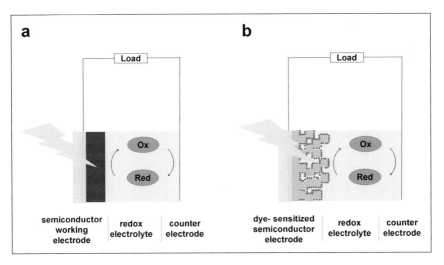

Figure 4.1 Working principle of regenerative photovoltaic cells based on photoelectrochemical reactions (a) using an optically absorbing bulk semiconductor and (b) using a sensitized nanoporous wide-bandgap semiconductor ("dye-sensitized solar cell").

forms usable by households or industry. Sensitization by strong absorbers is therefore needed to efficiently utilize such junctions for the conversion of solar light (Figure 4.1b). Sensitization consists of light absorption in the sensitizer, electron transfer to the conduction band of TiO_2, recovery of the neutral sensitizer by oxidation of a redox electrolyte, and subsequent reduction of the redox electrolyte at a counterelectrode.

In order to provide an efficient injection into the conduction band of the semiconductor, a monolayer of sensitizer is needed on the semiconductor surface. A monolayer on a flat surface, however, would not absorb sufficiently, and therefore nanostructured porous materials are needed as wide-bandgap semiconductors. A very well-established approach to realize such a matrix is the preparation of nanocrystalline TiO_2 films based on sol–gel techniques followed by annealing to provide sufficiently conducting thin films by sintering together the grain boundaries of the initially formed TiO_2 nanoparticles, even protected against agglomeration by surface modifications in the starting solution. These processes have been widely optimized and efficient cells are reproducibly prepared that provide conversion efficiencies η of up to 10.4% in stable cells when contacted by the classic acetonitrile-based iodine/iodide-containing electrolyte [23]. Since this arrangement still poses some problems in module engineering and long-term stability, alternative arrangements are sought that replace the redox electrolyte by organic hole conductors as contact phases, and in such cells η values of 0.5% [29], 1.1% [30], 1.8% [31] or 4% [32] in strongly differing cell concepts and at strongly differing level of development were reached. When looking at these last examples, the close relationship to the concept of the organic bulk heterojunction [18] (see above) becomes obvious in the large interfacial area of two interpenetrating conductive solid networks and the future development of such cells will be accompanied by intense cross-fertilization among these concepts.

Another optimization direction in the present research and development of DSSCs consists of an extension of the preparation conditions toward lower temperatures. Two reasons are seen for this strategy: (i) high processing temperatures generally mean high energy consumption and therefore high costs and (ii) low processing temperatures allow the use of flexible, low-weight, and low-cost substrates like polymers, textiles, or perhaps even paper. To avoid expensive glass substrates is already a good motivation but the option to prepare solar cells in completely new geometries, to keep mechanical flexibility even after cell preparation and during their use would allow the exploration of new application fields for solar cells [33–36]. In this context recently prepared DSSCs based on ZnO as semiconductor and prepared from aqueous solution by electrodeposition represent a very attractive new way to prepare efficient new solar cells, which may open new fields of application [27] and which represent attractive candidates for very large-scale (and hence low-cost) applications of photovoltaics since they are based on easily scalable preparation techniques and on abundant sources of materials. In this chapter the underlying principles, basics of the strategy, and recent success will be discussed to provide an overview and motivate the conclusions and outlook.

4.2
Electrodeposition – A Well-Established Technology

Electroplating rather noble metals like Pt, Au, Ag, and Cu but also less noble ones like Zn from aqueous solutions is one of the classic methods of metal coating on conductive preferably metal surfaces dating presumably back as far as 2000 years into the Parthian period [37] in arts and manufacturing. The process was optimized empirically with respect to a very well-defined surface finishing long before a scientific basis of these procedures was established by controlled potential or controlled current techniques and by the use of appropriate additives like brighteners or electrocatalysts [38–41]. More recently the electrodeposition of metals has seen a remarkable renaissance because it was very successfully used in the recent generations of microelectronic processors to contact individual layers and contact sites by electrodeposited Cu rather than by the hitherto established vapor-deposited Al contacts [42–44]. This success was reached by adjustments in the electrodeposition technology that allowed the filling of even very narrow holes and trenches. Thereby contact could be made through small vias by use of appropriate electrocatalytic and blocking additives which allowed such superconformal growth of Cu. The detailed mechanistic description of electrodeposition on an atomic level is therefore still the subject of very active research in modern electrochemistry. Excellent books [45, 46] and reviews [42–44] are available and it is certainly beyond the scope of this chapter to even try to give an overview.

Less established than the electrodeposition of metals but nevertheless known for quite some time are processes to electrodeposit compound semiconductors [47–49]. Caused by the basic need of charge transport across the growing film during electrodeposition, such depositions are of interest that use a cathodic reaction to electrodeposit n-type materials or an anodic reaction to electrodeposit p-type materials because an applied negative potential favors the presence of donor sites and an applied positive potential favors the formation of acceptor sites in the electrodeposited compound semiconductors. Examples of electrodeposited n-type conductors are CdSe, ZnO, and $CuInSe_2$ among a large number of other oxides, sulfides, and chalcopyrites [49]. CdTe, CuSCN, and CuI represent examples of electrodeposited p-type semiconductors [49]. Depending on additives and coupled chemical reactions, compact films, porous structures, or assemblies of individually crystallized nanoparticles can be formed [50–52]. Highly ordered crystal growth was shown in the epitaxial deposition of single-crystalline thin films, when suitable single-crystal substrates were chosen [53–57]. Because of the high crystallinity of chemical bath-deposited CdS and its epitaxial growth onto the surface of $CuInSe_2$ [58], the use of such chemically grown CdS as a buffer layer for $Cu(In,Ga)Se_2$ thin-film solar cells significantly improved their conversion efficiency [59, 60]. Light absorbers in thin-film photovoltaic cells such as CdTe and $Cu(In,Ga)Se_2$ were also prepared by electrodeposition and achieved promising efficiencies [61–63]. Because cost reduction and large-scale production have to be achieved in order to successfully implement photovoltaics on a level needed to significantly contribute to a sustainable supply of energy to a growing human population, electrodeposition processes of compound semiconductors can become large-scale key technologies.

4.3
Electrodeposition of ZnO Thin Films

The attractive way of film preparation by electrodeposition can be used to realize an economic technology for DSSCs. In this context it is important to note that ZnO can serve as an alternative to TiO$_2$ in DSSCs and that ZnO can be prepared by electrodeposition. DSSCs can be realized based on porous ZnO as a wide-bandgap semiconductor [27, 64–68]. Since typically an annealing step is not needed to obtain well-crystallized electrodeposited ZnO the use of new substrate materials becomes possible, leading to a number of new options in the applications of solar cells. The use of mechanically flexible, lightweight, and low-cost substrate electrodes is of special interest to increase the range of application for photovoltaics and/or to reduce costs and thereby increase the overall efficiency of the technology. ZnO as a material is further especially interesting since Al-doped ZnO (AZO) typically prepared by magnetron sputtering can serve as a transparent conductive oxide to replace indium tin oxide (ITO) which has become very costly because of a limited supply of In [69]. Since very well-controlled homoepitaxial growth of ZnO can be obtained not only by chemical vapor deposition (CVD) [70] but also by aqueous chemical growth [71] or directly by electrodeposition [72], the combination of a compact AZO back electrode with electrodeposited porous ZnO layers as sensitized electrodes represents a very promising combination of materials with the prospect of minimized interface defects despite low production costs [72].

Different precursor routes exist to cathodically electrodeposit ZnO from aqueous electrolytes with either O$_2$ [73, 74], NO$_3^-$ [75–77], or H$_2$O$_2$ [78, 79] as oxygen precursors in solution. In general, the reactions can be best described as electrochemically induced precipitation reactions rather than simple Faradaic reactions to convert soluble, typically ionic, species into solids as is the case for the electrodeposition of metallic films. The respective precursor is reduced at the cathode accompanied by a pH increase:

$$\frac{1}{2}O_2 + H_2O + 2e^- \rightarrow 2\,OH^-$$

$$NO_3^- + H_2O + 2e^- \rightarrow NO_2^- + 2OH^-$$

$$H_2O_2 + 2e^- \rightarrow 2OH^-$$

This increase in pH leads to a situation in which the solubility product of Zn(OH)$_2$ or ZnO is exceeded and a supersaturation situation is reached [49]. In the presence of appropriate crystallization seeds ZnO crystals nucleate and grow under very well-controlled conditions close to equilibrium at a growth rate approximately five times faster on the {0001} plane compared to the {10-10} or {01-10} planes [80]. As a consequence, precipitation on the cathode is observed since here the concen-

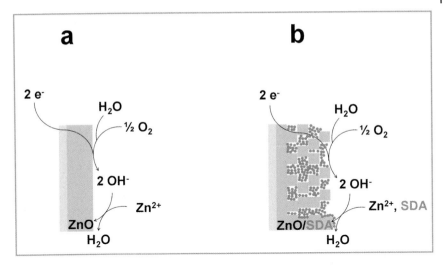

Figure 4.2 Electrodeposition of (a) bulk ZnO induced by the reduction of oxygen in the presence of an aqueous zinc salt solution and (b) porous ZnO in the presence of the above solution and in addition a structure-directing agent to bind to the growing ZnO surface, terminate the growth locally, and fill the pores.

tration of OH⁻ is highest and since typically crystallization seeds exist on the electrode to allow facile growth of the films (Figure 4.2a). The conditions are further chosen such that crystalline ZnO rather than a hydroxide, hydroxychloride, mixtures, or amorphous ZnO is formed. An increased temperature of 70–80 °C is a key parameter in this respect. The difference in growth rates on the different crystal faces leads to a pronounced texture of the films. Adsorbates can be further used to either block or catalyze the crystal growth on a given face and thereby tune the observed texture of films.

4.4 Sensitization of ZnO

To use ZnO as an electrode material in DSSCs, porous rather than bulk material is needed. A pathway in parallel to that established for TiO_2 consists of the preparation of ZnO nanoparticles, film formation, and subsequent conditioning [64, 81]. As in the case of TiO_2, high temperatures have to be applied in order to decompose the capping layers typically needed to stabilize the nanoparticles and to establish good electrical contact among the particles by sintering. Using such a strategy, cells were obtained with remarkable 4.7% efficiency [81]. As an alternative, a nanoparticulate powder was ground and attached as an electrode onto a conductive foil by a high-pressure method, reaching a similar conversion efficiency of about 5% [82, 83]. It turned out that nanoparticulate ZnO could compete well with TiO_2 in a number of semiconductor characteristics and even showed a higher average lifetime of electrons in the conduction band, promising for further developments [64].

As a redox electrolyte, although Co complexes were also suggested as viable alternatives [84], solutions of the I_2/I_3^- redox couple in either organic solvents or ionic liquids or mixtures despite the difficulties they pose to efficient sealing of the cells still provide the best available contact phases for dye-sensitized electrodes. I_2/I_3^- provides a strongly attenuated back reaction relative to the forward reaction leading to a low degree of back transfer of injected electrons from the conduction band of the semiconductor to the electrolyte [85–90]. A Grotthuß-like transport mechanism for this couple leads to a low series resistance in completed cells [91–93].

ZnO provides an increased surface reactivity when compared, for example, with TiO_2 which can be nicely utilized to establish the equilibrium conditions needed for a highly ordered growth during the electrodeposition reactions (Figure 4.2). On the other hand, however, this increased surface reactivity also leads to an increased threat of corrosion reactions. ZnO is less stable under acidic as well as alkaline conditions compared with TiO_2, the established wide-bandgap semiconductor used in a number of dye-sensitized solar cell studies and developments. In these developments the path has been paved for a number of sensitizers, surface blocking agents, or other adsorbates for their use in DSSCs. Because of the increased chemical sensitivity of ZnO, however, these molecules and conditions are not always suitable for ZnO, and can lead to corrosion reactions and therefore not function as they do on TiO_2 [94]. Therefore alternative molecules and conditions have to be established for ZnO to make best use of ZnO as a semiconductor in DSSCs and take full advantage of electrodeposition as a preparation route.

4.5
Alternative Sensitizer Molecules

Since the use of ZnO as active semiconductor electrodes asks for the design of specifically suitable sensitizer molecules, an opportunity is given to either adapt the Ru complexes that had been optimized for TiO_2-based DSSCs to the specific needs of ZnO, or search for an optimum sensitizer from a broader choice of molecules. The goal is to find an even better sensitizer and good prospects to base it upon abundant sources rather than the limited supply of Ru which presently leads to high costs of sensitizers. The working principle of a DSSC is shown in Figure 4.3 to discuss the photoelectrochemical requirements of a suitable sensitizer.

To provide a stable attachment of a sensitizer to a wide-bandgap oxide semiconductor and to allow for a fast electron injection from the excited state of a sensitizer molecule to the conduction band of the oxide, covalent links are preferable as opposed to simple coordination or van der Waals bonds [22]. On the other hand, very acidic sensitizer molecules favor corrosion of ZnO and thereby lead to inefficient charge transfer at the interface and to a low cell stability [94]. Carboxylic or

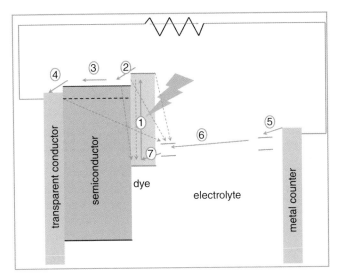

Figure 4.3 Schematic diagram of the DSSC working principle with the individual steps of light absorption (1), electron transfer to the conduction band of the semiconductor (2), electron diffusion in the semiconductor (3), collection of electrons at the back electrode (4), electron transfer to the oxidized species in the redox electrolyte at the counterelectrode (5), charge transport in the electrolyte (6), and regeneration of the neutral dye molecule by the reduced species of the redox electrolyte (7) for which the rate constants have to be optimized. Also shown are the major loss mechanisms (dotted lines) which have to be suppressed to reach a high efficiency of the solar cell.

sulfonic acid side groups at the sensitizer molecules are suitable as a chemical anchor to the presumably hydroxide-terminated oxide surface during growth from an aqueous electrolyte but the overall sensitizer acidity has to be kept in a range so as not to corrode the ZnO surface.

A further fundamental prerequisite is a high extinction coefficient in the visible range of the solar spectrum. The absorption spectrum has to be chosen such that either a broad band is provided by a given sensitizer or more than one sensitizer has to be used to efficiently cover the solar spectrum (panchromatic sensitization) [95–97].

Last but not least, the electronic energy levels of the sensitizer have to fit to both the semiconductor and the contact electrolyte, to allow facile and efficient electron transfer. In the excited state, the electron to be transferred should be best positioned in a level just slightly above the conduction band edge of ZnO to still provide sufficient driving force for its injection but not sacrifice too much energy upon its transfer (Figure 4.3). In the thereby oxidized sensitizer molecule, the electronic hole to be filled by transfer from the redox electrolyte should be in an energy level just below the redox potential of the redox electrolyte for the same reasons. If these criteria are missed, a large portion of the available open-circuit photovoltage is lost. Since a number of factors will influence the line-up of energy levels on a chemically reactive surface by interactions with the sensitizer itself, adsorbed water, blocking agents or constituents of the redox electrolyte calculated values, photoelectron spectroscopy data, or redox potentials can be used as a guide. But in most cases the preparation of the electrode and a photoelectrochemical characterization accompanied by spectroscopic measurements is the best way to fully understand the sensitizer/surface interactions and to judge about the applicability of a given sensitizer [98–101]. The recombination reaction by charge transfer of an electron back from the semiconductor conduction band to the sensitizer molecule was found to be strongly suppressed by an increased tunneling distance between semiconductor and the electronic system of the sensitizer providing another optimization parameter in the length of electronically insulating linker groups between sensitizer core and chemical anchor group [102].

4.5.1
Porphyrins and Phthalocyanines as Alternative Metal Complexes

Aside from an attractive variety of Ru complexes as sensitizers with different complexing groups, different degree of electronic asymmetry, and different anchoring groups to bind to an oxide surface [22, 103–106], mainly porphyrins [107–113] and phthalocyanines (Pc) [96, 114–123] have been investigated as alternative metal complex sensitizers in DSSCs (Figure 4.4). Whereas initially these alternative metal complexes often showed poor performance in DSSCs they have recently been become quite competitive to the established Ru complexes. The Ru complexes show an advantageous broad absorption spectrum in the visible range of the solar spectrum and have been widely optimized in their ligand structure to provide fast injection of an electron from the excited state of the sensitizer to the semiconductor conduction band and widely suppressed recombination of such

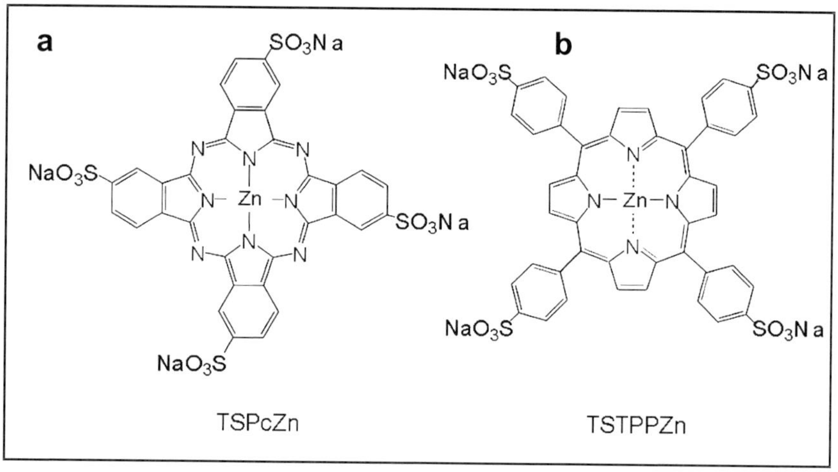

Figure 4.4 Structural formulae (zinc complexes) of (a) tetrasulfonated phthalocyanine (TSPc; different isomers present) and (b) tetrasulfonated tetraphenylporphyrin (TSTPP).

electrons to the oxidized sensitizer. Porphyrins and, in particular, phthalocyanines on the other hand provide record molar extinction coefficients of up to $\varepsilon_{(680\,nm)} = 1.6 \times 10^5\,l\,mol^{-1}\,cm^{-1}$ [124, 125] in an attractive wavelength range at about 650 nm [115], about 11 times that of the typically used Ru sensitizers at their absorption maximum around 540 nm [126] at, however, significantly decreased widths of the absorption bands. In concepts that use more than one sensitizer these dyes therefore can play a significant role. By such increased absorption the overall thickness of the porous sensitized semiconductor film can be decreased at still sufficient overall absorbance which offers good chances to increase the conversion efficiency of DSSCs. This is because of a reduced series resistance for charge transport through the porous semiconductor network and a reduced probability of back electron transfer from the semiconductor to the electrolyte [127]. Porphyrins and phthalocyanines like the Ru complexes have to be optimized with respect to the appropriate chemical anchors and linker groups in order to reach a fast injection of an electron from the excited state of the sensitizer to the conduction band of the semiconductor at widely suppressed back reaction to the oxidized sensitizer molecule [98, 99, 101, 102, 121]. The molecules also have to be adjusted mainly by appropriate choice of the central metal group for a suppression of the back electron transfer reaction to the redox electrolyte [119].

4.5.1.1 Frontier Orbital Positions

Knowledge of the highest occupied molecular orbital (HOMO) and lowest unoccupied molecular orbital (LUMO) positions of a given molecule on a semiconductor electrode is crucial to evaluate and predict feasibility of charge injection, back transfer, as well as catalysis of recombination (Figure 4.3). Experimental methods to elucidate these positions aim at ionization of the sample and analysis of the energetic change involved. Two different methods are frequently used: either redox electrochemistry (Figure 4.5) or ultraviolet photoelectron spectroscopy (UPS)

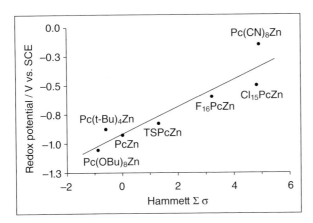

Figure 4.5 Redox potential for the first reduction of zinc complexes of phthalocyanines dissolved in DMF and Hammett $\Sigma\sigma$ values determined for a variety of differently substituted ligands. Reprinted with permission from [160].

and inverse photoemission spectroscopy (IPES). In UPS, pure materials are analyzed by measuring the kinetic energy of photoelectrons from occupied orbitals following ionization by irradiation of the sample molecule in the vacuum UV (e.g., He_I 21.2 eV) or soft X- ray (e.g., synchrotron) regimes. Molecules can either be in the gas phase or in a thin film on a conductive substrate. The choice of central metals, axial atoms at the metal, and substitutions in the ligand showed clear consequences for the frontier orbital positions in thin films [128–130]. A clear stabilization of the HOMO position by up to about 1 eV was caused by electron-withdrawing substituents. Compared to this strong influence of the ligand, the influence of the central metal and axial atoms was rather weak (≤ 0.2 eV), of the same order as differences in the solid-state structure would cause [131]. Gas-phase data were found at about 1–1.3 eV higher binding energy than the solid-state data. This apparent destabilization of the solid is caused by stabilization of the intermittently produced ions by a polarization energy of the solid well consistent with values of typically 1–1.5 eV reported earlier for aromatic materials [132–134]. UPS is further well suited to analyze details of the energetic alignment of molecules in semiconductor contacts [22, 135–138]. An analysis of interfacial energy level lineup of phthalocyanine thin films with organic or inorganic semiconductors (Figure 4.3) as opposed to energy levels in individual molecules [128, 132–134, 139] can help to understand the effect of sensitizer aggregation in dye-sensitized semiconductor electrodes, beyond a decreased electron injection probability into the conduction band of the semiconductor because of competing radiationless decay in aggregates. Aside from the establishment of a space charge in such aggregates, an interface reaction will typically occur and lead to surface states, a surface dipole, and hence a shift of the vacuum level across the junction. Adsorbed traces of water, solvents, ions from the electrolyte, and blocking agents will further contribute to such relative shifts of energy levels across the photoelectrochemically active interface. Complementary to UPS, IPES by irradiation of the sample with an electron beam and spectral analysis of emitted photons allows a direct analysis of unoccupied states, but is rather limited in its applicability due to frequent radiation damage of the samples. Electrochemical redox potentials on the other hand reflect free enthalpies of ionization of the molecules either in a solvent environment or as a thin film, then accompanied by intercalation of charge-balancing counterions. For a number of phthalocyanines, only the redox potential of reduction (characteristic for unoccupied levels) or oxidation (characteristic for occupied levels) can be obtained in common solvents because of a limited stability of the solvents towards reduction or oxidation. To allow a discussion about both HOMO and LUMO despite these experimental limits, optical absorption, emission, or electron energy loss spectroscopy (EELS) were used to estimate the frontier orbital gap. A rather constant gap of about 1.6 eV was determined for most phthalocyanines [128, 140–143].

4.5.1.1.1 Influence of Ligand Substitutions

It is well known that ligands strongly influence the electrochemical properties of phthalocyanines [143]. Electron-withdrawing groups lead to a positive shift of the redox potentials and electron-donating groups to a negative shift. These effects are most clearly seen in solution and for complexes with a central metal like Cu or Zn which have a closed d-shell and hence are widely redox-inactive. The equilibrium potentials of octa-*n*-butoxyphthalocyaninatozinc ((OBu)$_8$PcZn), for example, with electron-donating alkoxy groups were shifted negatively by about 200 mV compared to the unsubstituted phthalocyaninatozinc (PcZn) for the first oxidation as well as for the first two reductions [144]. The opposite direction was found for octacyanophthalocyanines, (CN)$_8$Pc, where the equilibrium potentials for the first four reductions of the complexes of Cu and Zn were shifted positively by about 700 mV, again compared to the unsubstituted Pc [145, 146]. To understand the role of phthalocyanine aggregates in DSSCs, the redox properties of thin films can be studied and interactions with intercalating counterions have to be considered, a significant difference compared with the study of their photoelectrochemical activity. Films for electrochemical studies were either drop-coated or vapor-deposited on conductive back electrodes. The unsubstituted phthalocyanines PcFe, PcCo, PcNi, PcCu, and PcZn undergo stoichiometric oxidation and re-reduction in aqueous electrolytes of neutral pH, however, under a loss of charge in subsequent cycles [147], which could be avoided by limiting the potential range and by a proper choice of counteranions [148, 149]. Films of (CN)$_8$PcH$_2$, (CN)$_8$PcCu, and (CN)$_8$PcZn were reversibly reduced and re-oxidized in acidic aqueous media under conditions where unsubstituted Pc could not be reduced, pointing towards a positive shift of the redox potential as also observed in solution [150]. Redox reactions at the π-system of Pc and related macrocycles are accompanied by characteristic changes in the optical spectra due to changes in the population of electronic levels [151]. This electrochromic behavior further stimulated interest in thin-film electrochemistry of these materials in view of device applications as smart windows or color displays, especially as the absorption coefficients of Pc and therefore the observed spectral changes are relatively large.

Redox reactions of thin films require charge balancing by counterions from the contacting electrolyte solution. Therefore the rate-limiting step for the overall reaction can be either the diffusion of electrons (electron hopping) in the electrode material characterized by their diffusion constant D_e, or the diffusion of counterions in the film characterized by their diffusion constant D_i [152–156]. A low mobility of ions in the films will lead to irreversible redox processes as in the case of unsubstituted Pc [147–149]. Less kinetic hindrance was obtained for films of substituted Pc that could be reduced in aqueous acidic electrolytes. Proton intercalation turned out to be rapid enough to ensure reversibility of the redox process. Consequently, large spectral changes could be observed for films of (CN)$_8$Pc [150], octacyano-tetrapyrazino-tetraazaporphyrines (TPz(CN)$_8$TAP) [157], tetrapyrazino-tetraazaporphyrines (TPzTAP) [157, 158], tetrapyrido-tetraazaporphyrines (TPyTAP) [159], and hexadecafluorophthalocyaninatozinc (F$_{16}$PcZn) [160, 161]. The observed shifts in redox potential can be compared to shifts in photoelectron spectra [128, 161] but also to chemical reactivity (Figure 4.5) as reflected in Hammett coefficients [128, 162, 163]. A value of −0.6 V vs. SCE was determined for the redox potential of the first reduction to the radical anion of F$_{16}$PcZn$^{·−}$ in

N,N'-dimethylformamide (DMF) and −0.9 V vs. SCE for the second reduction to $F_{16}PcZn^{2-}$. Both potentials were shifted about 0.4 V positive when compared to the unsubstituted PcZn caused by stabilization of the π-system due to the electron-withdrawing fluorine atoms in the ligand [161]. The electrode kinetics were studied in detail by cyclic voltammetry under variation of the intercalating ionic species, film thickness, and sweep rate. Charge uptake and the dependence of peak current densities on the square root of the sweep rate showed that the reaction rate is limited by the diffusion of intercalating cations and a diffusion constant D_i for K^+ in $F_{16}PcZn$ in the range 1.6×10^{-12} to $8.0 \times 10^{-12} \, cm^2 s^{-1}$ was calculated and confirmed by potential step experiments. Such rather slow diffusion of counterions was caused by strong intermolecular interaction in the film. A fast and rather complete reduction of films could only be achieved for thin films and slow sweep rates [160, 161]. Those reactions which allow the use of protons as intercalating ions therefore have clear technical benefits over those requiring slower counterions.

4.5.1.1.2 Influence of the Central Metal Group

Phthalocyanine complexes of central metals with open d-shells (e.g., Mn, Fe, Co) showed additional reduction and oxidation reactions due to changes in the oxidation state of the metal [143, 164, 165]. A number of the catalytic properties observed in homogeneous solutions of these complexes were preserved when adsorbed to an electrode surface leading to rich electrocatalytical properties, for example, in the oxidation or reduction of organohalides but also other reactants like phenols, thiols, O_2, CO_2, NO_2^-, or SO_3^{2-} to be kept in mind when these molecules are discussed as sensitizers [119, 166–171]. In a study on the influence of the ligand on the redox properties of thin films of, for example, Co complexes [172], unsubstituted phthalocyaninatocobalt(II) (PcCo) and the cobalt complexes of the substituted ligands octabutoxyphthalocyanine ($(OBu)_8PcCo$), octacyanophthalocyanine ($(CN)_8PcCo$), phthalocyaninetetracarboxylic acid ($(COOH)_4PcCo$), and tetrapyridotetraazaporphyrine (TPyTAPCo) were adsorbed on the basal plane of pyrolytic graphite (BPG) from solutions in organic solvents to form ultrathin films. The films were studied by cyclic voltammetry in the presence of an inert aqueous electrolyte. The observed peak separations, the widths at half height, and the charge in each cycle indicated complete reduction and re-oxidation of an adsorbed monolayer in the cases of $(CN)_8PcCo$ and TPyTAPCo. An incomplete reduction found for $(OBu)_8PcCo$ and PcCo was caused by formation of films thicker than a monolayer and hence hindered transport of electrons and/or ions. The films underwent reduction and re-oxidation in the ligand inner π-system (L/L^-) as well as in the metal center Co(II)/Co(I). The potentials were shifted according to the stabilization or destabilization of the frontier orbitals by the respective substituents in the ligand. Protons from the electrolyte interacted with the films during reduction and were desorbed during re-oxidation which led to the expected pH dependence (0.059 mV per pH unit in the case of a reversible reaction involving one proton for each electron) of the redox potential. Different slopes, however, were observed in the dependence of the redox potentials of the metal center and the ligands on the pH of the electrolyte in different pH ranges. The acid/base properties of the ligands determined a threshold pH above which the potential of the metal-centered first reduction was no longer dependent on pH, indicating that protons could no longer be stabilized in the film. The ligand-centered second reduction, however, led to interaction with protons over the whole accessible pH range. The basicity of the outer aza nitrogen atoms in the ligand ring was decreased by electron-withdrawing substituents as seen in the threshold pH up to which protons were the only cations to compensate the electronic charge to provide electroneutrality. Only under acidic conditions and in the first reduction was the electronic charge balanced by a stoichiometric number of protons. Under more alkaline conditions other cations from the electrolyte participate in this reaction. In the second reduction, protons compensated the charge over the whole pH range, but the increased slope pointed towards a number larger than stoichiometric indicating the additional interaction with anions [172].

In summary, the position of the frontier electron energy levels (HOMO and LUMO) of porphyrins and phthalocyanines can be finely tuned by the appropriate combination of central metal and substituted ligand as detected in photoelectron spectroscopy or in the redox potentials. A rich redox chemistry in interplay with a number of reactants and counterions has been established. A systematic consideration of the electrochemical and photoelectrochemical characteristics of porphyrins and phthalocyanines as individual molecules in solution, as molecules adsorbed at surfaces, and as molecular thin films serving to mimic the characteristics of molecular aggregates is of much relevance to the choosing or designing of optimized porphyrin or phthalocyanine sensitizers for DSSCs.

4.5.1.2 Photosensitization by Porphyrins and Phthalocyanines

Following absorption of a photon in a molecule, this molecule in general is a stronger reducing agent due to an electron in a state of higher energy (LUMO in the case of the Q band of phthalocyanines), but also a stronger oxidizing agent due to the electron vacancy in a state of lower electron energy (HOMO in the case of the Q band of phthalocyanines) when compared with the molecule in the ground state [173, 174]. Therefore an increased driving force for both reduction and oxidation reactions is expected. Photosensitization of a semiconductor electrode as discussed in this chapter therefore is only one of the many chemical reactions that can be driven by molecular centers in an electronically excited state. Since the reactions are then becoming dependent on illumination, the role of the molecular center is generally referred to as "photosensitizer". The support of the chemical reaction can consist of an increased thermodynamic driving force stemming from an increased gradient of the (electro)chemical potential as in the presently discussed examples, or in a facilitation of reaction kinetics through an increased reaction rate by photocatalysis. Porphyrins and phthalocyanines are of special interest as photosensitizers because they absorb with high extinction coefficients in the visible range, they can have quite extended lifetimes of the excited state [124, 125, 173, 174], and the phthalocyanines in particular have proven to be very stable against chemical oxidation or reduction, also under illumination [175].

Through an influence of different central metal groups and substituents on the electronic structure of the macrocyclic ligand in porphyrins and phthalocyanines, the optical absorption spectra are influenced by these groups, but the principal characteristics of a π–π^* transition in the ligand are widely preserved [125, 176]. The subsequent steps of fluorescence emission, internal conversion, intersystem crossing, and phosphorescence emission (Figure 4.6), are very strongly influenced, in particular by the choice of the central metal group [173, 174, 177] with spin–orbit coupling as the main mechanism [178]. In particular for the Mg, Al, Si, and Zn complexes long lifetimes of the first excited singlet state of the order of 5 ns and of the first excited triplet state of up to 2 ms were observed [177, 178]. Consequently a large number of reactions were reported that were facilitated by porphyrins or phthalocyanines as photosensitizers in solution, mainly based upon charge transfer, radical formation, or energy transfer from the first excited triplet state because its lifetime allows for a large probability of interaction with a reactant [173, 174, 177, 179]. Aggregation of porphyrins and phthalocyanines [180] significantly enhanced the rate of internal conversion and thereby led to decreased lifetimes of the first excited singlet state, decreased probability of intersystem crossing, and decreased lifetimes of the first excited triplet state with the consequence of decreased fluorescence and phosphorescence yield as well as decreased photosensitization efficiency [173, 174, 177, 178]. Despite this decreased quantum efficiency in the solid state, however, porphyrins and phthalocyanines are (and this is of particular interest in the context of this chapter) also well known as active photosensitizers when adsorbed to an electrode surface [166]. This is of particular interest since the photostability of the already quite stable individual molecules is significantly altered in the solid state [175]. Depending on the central metal, axial atoms, ligand substitutions, and solvent environment chosen, either photooxidative or photoreductive activity was observed [166]. These reactions are also used in a number of applications, mainly in chemical sensing of such different reactants as phenols, organohalides, pesticides, thiols, SO_2, sulfur oxoanions, CO_2, CO, NO_3^-, NO_2^-, cyanides, thiocyanades, hydrazine, and hydroxylamine [166]. This list of examples shows the great potential of porphyrins and phthalocyanines as photosensitizers, also in the context of DSSCs, but it also shows that catalytic and photocatalytic processes that may interfere in a negative way with the desired direction of net electron transfer from the electrolyte to the sensitizer to the wide-bandgap semiconductor have to be kept in mind and under control.

4.5.1.2.1 Photoelectrochemistry at Phthalocyanine Thin Films

In view of a quite pronounced tendency of phthalocyanines towards dimer and aggregate formation, the photoelectrochemical characteristics of such phthalocyanine molecular aggregates have to be considered when phthalocyanine-sensitized semiconductors are studied. Experiments at phthalocyanine thin films can serve as a good model system of such reactions [181]. Following light absorption in a solid molecular thin film, the film should be both more oxidizing and more reducing than the film in the dark as was also seen for molecules in solution (see preceding paragraph). For thin films, however, partial oxidation or reduction of the

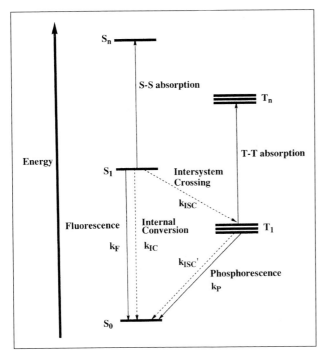

Figure 4.6 Diagram of excited states (ground state S_0, first excited singlet state S_1, lowest triplet state T_1, and higher triplet states T_n) and transition processes between these sates for a phthalocyanine with a closed-shell central metal. Reprinted with permission from [178].

molecular thin films has to be considered. Molecules that are easily oxidized, for example, will constitute a p-type semiconducting electrode rather than an intrinsic one if the films were exposed to ambient conditions. The opposite holds for molecular thin films that are easily reduced which can easily lead to n-type conduction. At such doped molecular semiconductor electrodes, the relative changes in the concentrations of excess electrons (LUMO) and electron vacancies (HOMO) will occur unsymmetrically, leading to preferentially cathodic photocurrents at p-type electrodes and anodic photocurrents at n-type electrodes [182], a rule well established in semiconductor photoelectrochemistry [183, 184].

In an early comparison of the catalytic activity of thin films of PcH_2, PcCu or PcFe in the dark and under illumination [185], it was found that the photocatalytic activity was barely correlated to the catalytic activity in the dark and this difference was discussed as arising from a photoconductivity effect in the bulk of the phthalocyanine films. A photocurrent and also photovoltage observed at H_2Pc [186] was discussed as being due to a photovoltaic effect at the phthalocyanine film. Phthalocyanine thin films used in early photogalvanic cells [187] were described as p-type semiconducting electrodes that established a region of space charge at the Pc/electrolyte interface. Nevertheless small anodic photocurrents were also observed and explained by light-induced changes in the concentration of both types of charge carriers in a weakly doped semiconductor [188, 189] or by a weak sensitization effect at the back electrode in the case of SnO_2 substrates [190, 191]. This reaction at the SnO_2 back contact was characterized in detail by the analysis of experiments at varied illumination conditions and a thickness of the photoactive layer on the electrolyte side of 35 nm was established, a narrow zone but significantly increased above direct molecular interaction [192], in agreement with potential-dependent impedance measurements at electrochemical contacts with films of PcH_2, PcNi, and PcZn [193]. It was noticed that the role of acceptor molecules in the volume of the films is crucial for the observed characteristics and that oxygen from air in most cases caused the observed p-doping and hence cathodic photocurrents. Also attempted was a more controlled doping by reactions with halogen molecules like iodine, or organic molecules like *ortho*-chloranil, tetracyanoquinodimethane, tetrafluorenone, etc. This work quickly led to the need for a detailed investigation of the electrical properties in the solid state which have been reviewed in great detail earlier [194] and also recently [195].

The photoelectrochemical characteristics of thin films of phthalocyanines with higher-valent central metals carrying additional electron-withdrawing axial ligands like PcAl(OH), PcAl(Cl), PcGa(F), PcGa(Cl), PcIn(Cl), PcTi(O), and PcV(O) were also investigated in great detail [196–204]. For this group of materials, although all films were vapor-deposited, large differences in film characteristics were observed with either anodic or cathodic photocurrents dominating the photoelectrochemical behavior. The role of oxygen as a dopant leading to the typically observed p-type characteristics of phthalocyanines could be clearly elucidated in these studies [197, 202, 203]. An optimized orientation of crystals led to larger exciton diffusion lengths and charge carrier mobilities as shown in an increased photoelectrochemical efficiency [204]. A detailed analytical investigation of different ways of chemical treatment of PcAl(Cl) as compared to PcGa(Cl) and PcIn(Cl) revealed the hydrolysis of the Al–Cl bond as the key step towards the formation of more efficient electrode materials [196].

Chemical substitution at the phthalocyanine ligand with electron-withdrawing substituents led to a dominance of anodic photocurrents as opposed to that of cathodic photocurrents observed for the unsubstituted materials. This was shown for $(CN)_8PcZn$ [140, 205], TPyTAPZn [140, 206, 207], and TPzTAPZn [140, 206]. Based on these results, n-type conductivity was assigned to these materials of molecules with electron-withdrawing substituents based on the argument outlined above [129, 140, 205–209]. Electrodes of $F_{16}PcZn$ showed photocurrents of almost identical significance in both photocurrent directions [209, 210]. Since films of $F_{16}PcZn$ investigated directly following the preparation still showed n-type characteristics, the photoelectrochemical behavior was explained by compensation of n-dopants by air. The assignments were supported by studies of the electrical properties of the materials [129, 130, 207, 211–213] and in solid heterojunctions [128, 214, 215].

Phthalocyanines show two distinct absorption bands in the visible range, namely the Q band around 650–700 nm and the B or Soret bands around 300–350 nm (Figure 4.7). These bands correspond to a transition from the HOMO to the LUMO

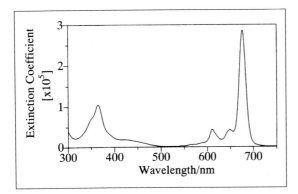

Figure 4.7 UV–visible absorbance spectrum of a solubilized PcZn in an organic solvent. Clearly seen are the range of the B or Soret band at 300–400 nm and the range of the Q band at 600–700 nm. Also seen is the vibrational fine structure of the Q band at 600–650 nm. Reprinted with permission from [178].

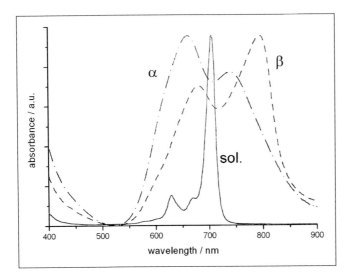

Figure 4.8 Optical absorbance spectra of PcZn thin films in the two most easily obtained crystal structures α and β as indicated in the plots compared to a solution spectrum of PcZn in DMF solution. Clearly seen is the prominent splitting of the Q band in the solid state caused by resonance coupling of the molecular electronic systems dependent on the packing in the crystals.

in the case of the Q band and to a transition from the second highest occupied molecular orbital (SHOMO) to the LUMO in the case of the B band (Figure 4.9c) [125]. In the solid state these optical transitions of Pc molecules are preserved, just slightly shifted and broadened considerably (Figure 4.8) [125]. Most of the light of a white light source that leads to the observed photocurrents will be absorbed in

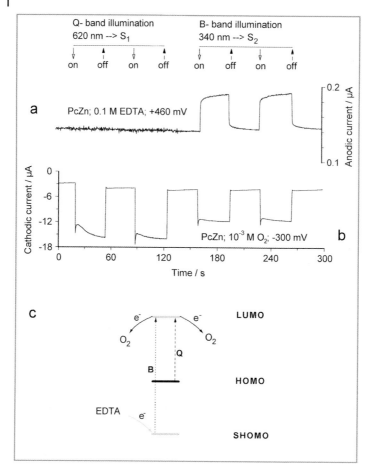

Figure 4.9 Current at 100 nm vapor-deposited thin films of PcZn on ITO (1 cm²) measured upon illumination either in the B band (3 × 10¹⁵ photons cm⁻² s⁻¹) or in the Q band (7 × 10¹⁵ photons cm⁻² s⁻¹) during potentiostatic polarization in contact with aqueous electrolytes with either (a) 0.1 M EDTA at +460 mV vs. SCE or (b) 10⁻³ M O₂ at −300 mV vs. SCE. (c) Summary of observed photocurrents upon changing occupation of the frontier energy levels by the different wavelengths. Reprinted with permission from [209].

the Q bands. Excitation in the B band of PcZn (about 340 nm), however, led to a strongly contrasting behavior to the situation described above for Q band illumination (Figure 4.9). Caused by the separate character of the resulting excited electronic states, phthalocyanine electrodes showed the specific characteristics of a switchable photocurrent direction (Figures 4.9a and 4.9b) and showed the possibility of electron transfer from a state higher than the first excited state [209], a concept well suited to realize a third-generation concept (see Section 4.1) based on molecular materials since such effects are not easily observed in classic semiconductors.

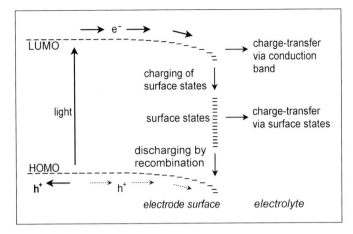

Figure 4.10 Schematic representation of different observed reactions involved in the light-induced charge transfer at phthalocyanine surfaces with adsorbed reactant species.

In studies of electrochemical photocurrents obtained at molecular semiconductor thin films (PcZn, $(CN)_8$PcZn, TPyTAPZn) and their dependence on the concentration of the reactant in the electrolyte (O_2, ethylthiolate (RS$^-$), hydroquinone/benzoquinone (HQ/BQ), $Fe(CN)_6^{3-/4-}$), a saturation behavior of the photocurrents was found arising from reactant adsorption prior to charge transfer [205, 216, 217]. Often functions according to Langmuir's adsorption isotherm based on the presence of only one kind of independent adsorption sites and a maximum coverage of a monolayer of reactant were found to give a reasonable fit to the observed concentration dependence of the photocurrent data. Reaction kinetics were found that support the formation of chargeable surface states (adsorbates) at the molecular semiconductor surface (Figure 4.10) [182, 210, 214].

4.5.1.2.2 Sensitization of Semiconductor Electrodes by Porphyrins and Phthalocyanines

The investigation of the photoelectrochemical properties of phthalocyanine thin films discussed in the preceding paragraph originated from studies in which phthalocyanine films had been deposited on inorganic oxide electrodes to investigate their role as sensitizers and protecting layers against photocorrosion. Anodic sensitization could be achieved by vapor-deposited PcCu and by surface-bound tetrasulfonated copper phthalocyanine on n-SnO_2 conducting glass [218] and also by vapor-deposited thin films of PcH$_2$ on single-crystal electrodes of n-TiO_2, n-$SrTiO_3$, n-WO_3, n-ZnO, n-CdS, n-CdSe, n-Si, n-GaP, or slides of SnO_2 conducting glass [219]. The sensitization capability of the phthalocyanine layers could clearly be shown by anodic photocurrents that followed the absorption spectrum in their spectral dependence but the quantum efficiency of the sensitization was generally low and photocorrosion of the semiconductor substrate electrodes (evident for those materials of smaller band gap) could not be suppressed sufficiently. The same conclusion was drawn from an extended study in which also films of PcMg, PcZn, PcAl(Cl), PcTi(O), PcCo, and PcFe on single-crystalline n-TiO$_2$ or n-WO$_3$ were investigated [220]. In more recent studies it was found that although surface defects on TiO$_2$ quenched the photocurrent from light absorption in the oxide electrode, the (sensitized) photocurrents following light absorption in the phthalocyanine were left widely unchanged [221] and details of the energy level alignment were discussed. It was also shown that the film morphology of PcTi(O) or PcV(O) played a crucial role. Closed layers of the materials led to the occurrence of cathodic photocurrents originating from the phthalocyanine as active semiconductor material as discussed in the preceding paragraph, whereas islands of the materials led to sensitized anodic photocurrents [222].

The photoelectrochemical properties of PcIn(Cl) thin films were also investigated on single-crystalline layered semiconductor electrodes of SnS$_2$ or MoS$_2$ [223, 224]. These surfaces turned out to provide a suitable substrate to deposit highly ordered epitaxial phthalocyanine films by organic molecular beam epitaxy and also to allow the study of such films as sensitizers for n-type semiconductors. Very narrow absorption spectra and also photocurrent action spectra could be obtained speaking for a well-defined crystallization of PcIn(Cl) on n-SnS$_2$ and injection of electrons from the excited state of the dyes in the crystallites into SnS$_2$. A quite constant quantum efficiency of about 10% was found when such highly ordered films were compared with polycrystalline films indicating a direct injection of charge from molecules adjacent to SnS$_2$ since otherwise more pronounced differences should have been observed among the different crystalline modifications [224]. An increased quantum efficiency of up to 44% was obtained for films of phthalocyanines carrying eight dodecylamide or dodecylester functions dropcoated from solutions on SnS$_2$ [225]. These films showed liquid-crystalline properties and small chromophore interaction. The high quantum efficiency was reached for films of sub-monolayer coverage and decreased for thicker films pointing toward efficient injection again only from molecules close to the interface with SnS$_2$. Consequently the incident photon to current conversion efficiency (IPCE) for sub-monolayer or thicker films (up to 30 equivalent monolayers) was found quite constant at only 0.8%.

Higher efficiencies of phthalocyanine-modified oxide semiconductor electrodes could be achieved if nanoparticulate oxide films were used since the electrode surface and hence the interface area of the phthalocyanine with the oxide could be increased significantly. In these cases a monolayer of phthalocyanine on the nanoparticles provided sufficient light absorption. A high quantum efficiency can be maintained if aggregation of the phthalocyanines can be suppressed. Vapor deposition of the phthalocyanines therefore turned out to be not appropriate. The chromophores were typically adsorbed from solutions and preferably of phthalocyanines that carried a chemical substituent that provided good anchoring to the oxide surface. Tetrasulfonated metal-free phthalocyanine and the complexes of Zn, Ga(OH), Co, In(OH), and Ti(O) were adsorbed to nanocrystalline films of TiO$_2$ prepared by the sol–gel technique and their photoelectrochemical activity was measured under monochromatic illumination [226]. Among these, the zinc complex showed the highest IPCE at a sensitization quantum yield of still only 5.7%. Also investigated was the Al complex of tricarboxymonoamidephthalocyanine adsorbed on commercial TiO_2 particles and photocatalytic oxidations of organic molecules like phenols or hydroquinone could be achieved [227]. A composite electrode of TiO_2/phthalocyanine could also be prepared by spray pyrolysis of a mixed solution of titanium oxyacetylacetonate and the Cu complex of a tetrasulfonated phthalocyanine with a sensitized photocurrent obtained in the phthalocyanine Q band absorption range. However, the photocurrents reached only a few $\mu A\,cm^{-2}$ under illumination with $100\,mW\,cm^{-2}$ of white light [228]. Also, attempts have been made to prepare very complex materials of TiO_2 particles modified by quantum-sized CdS on the surface and then further modified by Ga(OH), Zn, In(OH), or V(O) complexes of tetrasulfonated phthalocyanines. Efficient sensitization by monomers of the Ga(OH) complex was claimed [229] at an IPCE of up to 10%, but photocorrosion was a severe problem for such electrodes as reported earlier. Significantly more efficient electrodes sensitized by phthalocyanines have been prepared by use of an Ru complex of octamethylphthalocyanine with two additional axial ligands of pyridine-biscarboxylic acid to bind to the oxide surface grafted to nanoparticulate TiO_2 [114]. An IPCE of 60% and photocurrents of $10\,mA\,cm^{-2}$ were reached under illumination with white light under AM 1.5 conditions. In a study using a number of differently carboxylated or sulfonated phthalocyanine complexes of Zn or Al(OH), similar values were reached with the Zn complexes of tetracarboxyphthalocyanine or tetrasulfophthalocyanine. These showed the best performance within this group of materials reaching an IPCE of 30–45% and a conversion efficiency of about 1% under AM 1.5 ($100\,mW\,cm^{-2}$) conditions [115]. Recently the conversion efficiency of phthalocyanine sensitizers for TiO_2 could be increased to 3.5% at an IPCE of 80% by use of a zinc phthalocyanine with a carboxylic acid group as anchor at one of the four benzo groups of the ligand and a tertiary butyl group attached to each of the other four benzo groups [121]. Among electrodes sensitized by phthalocyanines these are very promising values, but it should be kept in mind that around 10% efficiency is typically reached by dyes of the Ru trisbipyridyl class when adsorbed at TiO_2. Also, zinc tetraphenylporphyrins have been widely optimized by additional substituents in the phenyl group of the ligand and by the choice of an additional carboxylate anchoring group [110–112]. A zinc tetraphenylporphyrin with a styrylbenzoic acid anchor led to a conversion efficiency of 4.2% under AM 1.5 conditions and at an IPCE in the Soret band maximum of 80% [110]. The same anchor group, but with four xylyl groups instead of the phenyl groups in the ligand led to an increase of 4.8% conversion efficiency at 75% maximum IPCE. A zinc tetraphenylporphyrin with a cyanoacrylic acid anchor provided a conversion efficiency of 5.6% [111]. Recently, a zinc tetraphenylporphyrin with a toluyl group instead of the phenyl group and a malonic acid anchor group was reported as having reached a conversion efficiency of 7.1% [112]. It was found that the substituted phenyl groups acted as electron donor and the conjugated anchor groups as electron acceptor and that this push–pull arrangement led to such excellent values compared to other sensitizers of the porphyrin and phthalocyanine type with respect to the efficiency and also when compared to the Ru-sensitized cells with respect to cost, availability, and environmental benignity [112].

4.5.2
Purely Organic Dyes

Aside from porphyrins and phthalocyanines, classic organic dyes without any metal center are also suitable candidates for efficient sensitizers in DSSCs. A number of molecules from different classes of dyes have been studied adsorbed to nanoparticulate TiO_2 as established semiconductor material. Results have been summarized in two recent reviews [127, 230]. Xanthene dyes, and particularly coumarin-based dyes have been studied, reaching incident photon-to-current efficiencies of IPCE = 70–80% in their absorption maximum around 470 nm [231]. By covalent attachment of methine bridges to a substituted coumarin moiety, an IPCE of 84% at 540 nm and a conversion efficiency of $\eta = 6.7\%$ could be reached [232], extended to $\eta = 7.4\%$ by use of an oligothiophene bridge to the carboxylate anchor group [233]. Oligoene systems with substituted dialkylaniline or diarylaniline instead of coumarin moieties have also been studied and reached $\eta = 6.6$–6.8% [234, 235] at IPCE > 80% [236]. Dialkylaniline substituents were also used to introduce asymmetry into squaraine dyes leading to increased efficiency compared to the symmetric analogs and even providing injection from aggregated sensitizers [237]. Indoline dyes have been successfully optimized as sensitizers for TiO_2 [238] with efficiencies claimed of up to $\eta = 8.0\%$, at, however, unrealistically high photocurrents of 18.5 mA cm^{-2} at IPCE < 90%. Thienylfluorene dyes with arylaniline donor and cyanoacrylic acid acceptor and anchor substituents in conjugation led to sensitized photocurrents at $\eta = 5.5\%$ [239], and hemicyanine dyes also performed at IPCE = 74% and $\eta = 5.2\%$ [240], indicating the great potential of a variety of organic dyes as sensitizers for DSSCs when appropriately substituted with electronic donor and acceptor as well as chemical binding groups.

4.6
Electrodeposition of Hybrid ZnO/Organic Thin Films

Motivated by the great potential of electrodeposition as a technology (Section 4.2) and for the preparation of active ZnO semiconductor electrodes in particular (Section 4.3) and in view of the applicability of ZnO as sensitized photoelectrode (Section 4.4), electrodeposition of hybrid ZnO/dye materials was developed as an alternative to prepare dye-sensitized electrodes, to make optimum use of ZnO as a material, and to avoid the high processing temperatures otherwise used that are disadvantageous in view of a free choice of substrate and optimized energy efficiency in cell preparation. The need of a chemical anchor group (carboxylates, sulfonates, etc.) at the sensitizers (Section 4.5) leading to water solubility in a number of cases [177] allowed the addition of the prospective sensitizers to the deposition bath of ZnO, allowing them to adsorb to the growing ZnO film and thereby produce hybrid

4.6 Electrodeposition of Hybrid ZnO/Organic Thin Films | 245

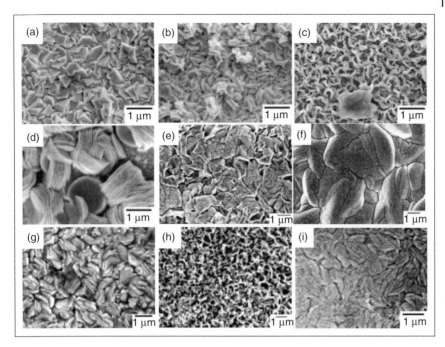

Figure 4.11 Scanning electron microscopy analysis of ZnO/dye hybrid thin films electrodeposited at −0.9 V and 70 °C for 60 min from a 0.1 M Zn(NO$_3$)$_2$ aqueous solution: (a) pure ZnO; (b) ZnO/TSPcZn hybrid; (c) ZnO/TSPcAl hybrid; (d) ZnO/TSPcSi hybrid; (e) ZnO/EY hybrid; (f) ZnO/PB hybrid; (g) ZnO/TB hybrid; (h) ZnO/N3 hybrid; (i) ZnO/R5P hybrid. Reprinted with permission from [241].

materials of ZnO with the prospective sensitizers. This work has been thoroughly reviewed [241], also more recently [27]. In the presence of water-soluble dyes, porous but still crystalline ZnO semiconductor thin films were deposited (Figure 4.2b) that contained the dye molecules in their pores leading to organic/inorganic hybrid materials [242]. Such materials were successfully grown in the presence of tetrasulfonated phthalocyanines (TSPcMt) [243–246], xanthene dyes such as eosin Y (EY) or phloxine B (PB) [247–250], tetrabromophenol blue (TB) [251, 252], riboflavin phosphonate (R5P) [253], and cis-dithiocyanato-bis(4,4'-dicarboxylic acid-2,2'-bipyridine)ruthenium(II) (N3) [254] as shown in Figure 4.11. This one-step preparation with the dye in the nitrate-based bath for the ZnO electrodeposition led to simultaneous self-assembly of ZnO and the adsorbed dye and thereby yielded homogeneously colored crystalline ZnO films with porous morphology. The dyes showed good compatibility with the aqueous deposition conditions and a suitable range of redox potentials to allow proper alignment of the energy levels in the contact to ZnO. Adsorption of dye molecules onto the growing surface of ZnO strongly affected the crystal growth of the ZnO and led to a significantly higher surface area when compared to pure ZnO films formed without addition of the dyes, opening up a new synthetic route to photoactive materials for dye-sensitized semiconductor electrodes. Strongly differing crystal sizes, morphologies, textures of films, and porosities of the crystalline ZnO, as well as different degrees of aggregation of the dye molecules could be obtained dependent upon the adsorbed dye molecules and proper choice of the deposition conditions (Figure 4.11) [241, 244, 247]. For TSPc, for example (Figures 4.11b–d), the choice of the central metal group allowed the growth of films of quite remarkably different morphology and relative orientation of the ZnO nanocrystals [244]. The formation of ester-like bonds through the sulfonic acid groups of a TSPcMt molecule to the surface of ZnO was assumed to play a decisive role in this assembly, since a formation of such bonds was also seen in TSPcCo/TiO_2 composites [255]. The adsorption of TSPcMt preferentially occurred onto the a/b-plane of ZnO crystals, since crystal growth predominantly along the a-axis and b-axis was observed. The overall growth direction is defined vertical to the electrode. Therefore those ZnO crystallites with their a-axis and b-axis perpendicular to the substrate grew faster and represented the preferential crystal orientation of the final film as detected by X-ray diffraction [244]. Differences in the stability of TSPcMt adsorption to different crystallographic faces of ZnO led to a significant anisotropy of the crystal growth.

The dye molecules in the hybrid materials not only adsorbed to ZnO but also formed well-defined intermolecular structures which were studied by UV–visible absorption spectroscopy. The interaction of the chromophores in condensed dye assemblies led to specific changes of the electronic structure [125, 177, 180, 181] discussed in detail for the electrodeposited ZnO/TSPcZn hybrid thin films [244, 245]. Films electrodeposited at −0.7 (slow growth) or −0.9 V vs. SCE (faster growth) showed characteristic differences. Both films were blue and less scattering than the pure ZnO, but the film deposited at −0.7 V obviously contained a higher amount of dye and showed formation of π-stacking aggregates of TSPcZn by a split spectrum in the Q band range as opposed to the film deposited at −0.9 V which was pale blue and showed only the peak expected for monomeric TSPcZn [256]. Washing the ZnO/TSPcZn hybrid thin film deposited at −0.7 V with a surfactant solution led to partial desorption of TSPcZn molecules from the composite films and converted their absorption spectra to those very similar to those of the hybrid thin film deposited at −0.9 V [244]. Formation of dye multilayers due to dye/dye interaction in the films deposited at −0.7 V was thereby indicated.

Photocurrent spectra, time-resolved photocurrent measurements, and intensity-modulated photocurrent spectroscopy (IMPS) revealed sensitized photocurrents at these electrodes in contact with an organic I_2/I_3^- electrolyte. In line with typical homogeneous photosensitization efficiencies (Section 4.5.1.2) it was found that increased aggregation led to decreased electrode efficiencies [245, 246, 257]. A considerably higher quantum efficiency of monomeric dye when compared with aggregated dye was thereby shown (Figure 4.12). The achieved IPCE values were generally low. TSPcSi(OH)$_2$ showed the highest quantum efficiency of the studied compounds, a factor of about 3 higher compared with monomeric TSPcZn and a factor of about 17 higher than TSPcAl(OH). More efficient electron injection from the excited state of the dye to the conduction band of ZnO was explained by a more suitable relative position of the electron energy levels combined with a strong

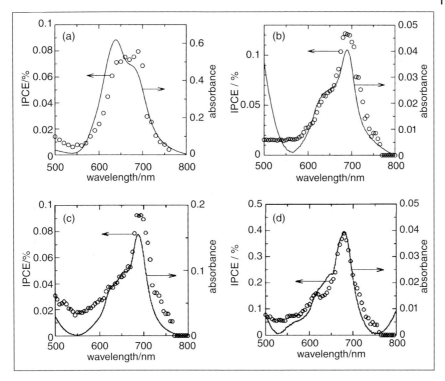

Figure 4.12 Spectral dependence of the IPCE (circles) in the oxidation of iodide at −0.2 V vs. Ag/AgNO$_3$ and optical absorption spectra (solid curves) of TSPc/ZnO hybrid materials grown from a 0.1 M Zn(NO$_3$)$_2$ aqueous solution maintained at 70 °C: (a) TSPcZn/ZnO grown at −0.7 V vs. SCE; (b) TSPcZn/ZnO grown at −0.9 V vs. SCE; (c) TSPcAl(OH)/ZnO grown at −0.9 V vs. SCE; (d) TSPcSi(OH)$_2$/ZnO grown at −0.9 V vs. SCE. Reprinted with permission from [257].

chemical interaction of TSPcSi(OH)$_2$ with ZnO [246, 257]. Aggregates of TSPcMt, however, showed a more facile regeneration of the neutral dye, i.e., faster transfer of electrons from I$_3^-$ to the aggregates compared to the monomers [245, 257]. In these experiments it was also found that TSPcSi(OH)$_2$ provided faster and more efficient electron injection into ZnO compared with TSPcZn or TSPcAl(OH) [245, 246, 257].

The interplay of different dye molecules present during the electrodeposition reactions of dye-modified ZnO was studied for zinc complexes of tetrasulfonated phthalocyanine (TSPcZn, Figure 4.4a) and tetraphenylporphyrin (TSTPPZn) in which case both dyes (Figure 4.4b) were simultaneously adsorbed to form a hybrid material with ZnO [258]. The typical absorption bands for both dyes were detected. Films of TSPcZn/ZnO consisted of larger particulate domains when compared with TSTPPZn/ZnO or (TSTPPZn + TSPcZn)/ZnO. The presence of the porphyrin stabilized the phthalocyanine on the ZnO since a greater amount of TSPcZn was adsorbed. In the photosensitization of ZnO, TSPcZn and TSTPPZn worked in parallel using light in their respective absorption range [259]. The presence of TSPcZn led to a threefold increase of the photoelectrochemical quantum efficiency of TSTPPZn [259]. An increased rate constant of electron transfer to oxidized TSPcZn (regeneration reaction) suppressed the competing recombination via TSTPPZn and TSPcZn took over a role as charge mediator in the regeneration reaction of oxidized TSTPPZn. Nevertheless only poor IPCE in the 0.1% range could be reached for these electrodes with aggregates of TSPcZn and an overall small surface area of the hybrid material.

Hybrid electrodes of ZnO with EY deposited in one step from such nitrate-based solutions already showed higher efficiency than those with TSPcZn, TSTPPZn, or other dyes mentioned above [247, 249, 257]. The current decreased throughout the potential range upon addition of EY during electrodeposition of ZnO and highly porous, large grains of ZnO were formed by specific hindrance of crystal growth at the predominant face of ZnO for the adsorption of EY [241]. A systematic broadening of the nucleation peak in the chronoamperogram occurred: it was shifted to later times and a decrease of the steady-state current was observed upon increased concentrations of EY. Furthermore, when the dye concentration was higher than 100 µM, no film was deposited, confirming the hindrance of growth by adsorbed EY [249]. The films were colorless following deposition but gradually turned to red when kept in air afterwards, indicative of an electrochemical reduction of EY during film deposition and subsequent re-oxidation. EY was present in these films as aggregates in a remarkably increased concentration but nevertheless led to similar photosensitization quantum yields as the most efficient TSPc in the lowest concentration. Photocurrents a factor of about 100 higher in the $1\,\text{mA}\,\text{cm}^{-2}$ range under white light illumination were thereby obtained and facile electron transfer from EY to ZnO and from I_3^- to oxidized EY was concluded from time-resolved photocurrent measurements [247, 249, 257]. An almost rectangular photocurrent response was obtained with almost no overshoot above the stationary value, indicating only a small amount of surface charging. When the illuminating beam was shut, a small cathodic current was observed, suggesting a slightly hindered electron transfer from I_3^- to the oxidized EY. From IMPS and intensity-modulated photovoltage spectroscopy (IMVS) it was concluded that no electric field existed in the porous electrodes that would change with the applied potential and thereby would lead to changes in the concentration of free electrons at the surface. The observed transient time showed a fast diffusion of electrons in the porous single-crystalline ZnO. Values of the electron lifetime of about two seconds were determined. Compared with previous data, it was concluded that the lifetime of the free electrons in the porous single crystal of ZnO electrodeposited in the presence of EY was considerably larger than that in nanoparticulate ZnO [260] and comparable to TiO_2 typically used in the most efficient DSSCs [85]. Also from this analysis, the thin films of porous single crystals of ZnO as obtained by electrodeposition of ZnO under structural control by EY can be considered promising electrode materials [249].

The role of EY as a structure-directing agent (SDA) in the electrodeposition of ZnO could be developed to an even greater extent (Figure 4.2b) when the deposition was performed from deposition solutions with O_2 as oxygen precursor for ZnO rather than NO_3^- [27, 261, 262]. Also hybrid materials of TSPc with ZnO were deposited from O_2-based solutions as in the example of TSPcNi leading to a range of materials stretching from monomers of TSPcNi embedded into ZnO crystals to amorphous networks of TSPcNi with corresponding contrasting optoelectronic properties [263]. The EY/ZnO hybrid materials consisted of well-adhering hybrid thin films with a high content of EY and excellent crystallinity. The reaction allowed such precise control that even epitaxial growth could be achieved on the (0001) face of GaN [264] as well as ZnO [72] substrates. The adsorption of EY during film growth was shown to have a strong influence on the lattice constant of the growing highly porous sponge-like crystals by an expansion of the ZnO lattice by 3.6% in the c-direction [72]. Hybrid EY/ZnO thin films could be obtained over a wide potential range, even at potentials more positive than the reduction of the dye. If the electrodeposition was carried out at potentials where EY was reduced, a significantly larger amount was incorporated into the films caused by an increased stability by formation of a complex between the reduced dye and Zn^{2+} [249]. In contrast to the situation during depositions from nitrate-based solutions, the addition of EY accelerated the reduction of O_2 and the subsequent film growth [261]. Addition of EY even at a concentration as low as 1 mM largely enhanced the cathodic current. With increasing additions of EY, the current systematically increased. The increase of current was caused by electrocatalysis of EY in the O_2 reduction to different extent, however, when different electrode potentials were discussed [27] and led to an increased rate of ZnO deposition with a Faradaic efficiency of 100% [265]. Quite different mechanisms of electrocatalysis for the reduction of O_2 were found relevant for EY in its neutral or reduced state [27]. EY/ZnO electrodes prepared by these means from O_2-saturated solutions showed photoelectrochemical sensitization characteristics towards ZnO in standard iodine-containing electrolytes of 2.3 mA cm^{-2} under illumination by 200 mW cm^{-2}. But with this performance their efficiency was only slightly higher than for those from the nitrate deposition solutions despite the clearly improved electrode structure combining high crystallinity and high surface area, also leading to good accessibility of the sensitizer in the electrode and fast electrode kinetics as seen in time-resolved photocurrent measurements [261]. If Coumarin 343 (C343) was used as SDA instead of EY during electrodeposition, significantly different textures of the growing porous crystalline ZnO were observed [27]. Whereas ZnO was oriented with its c-axis parallel to the substrate surface normal in the case of EY/ZnO, it was oriented with its c-axis perpendicular to it in the case of C343/ZnO.

4.7
Porous Crystalline Networks of ZnO as Starting Material for Dye-Sensitized Solar Cells

The good accessibility of EY incorporated in the porous ZnO crystalline network detected in the photoelectrochemical characteristics was used to desorb the dye

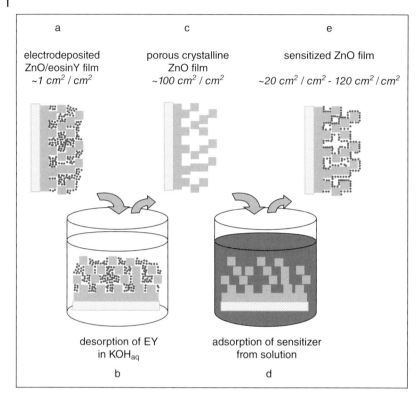

Figure 4.13 Optimized utilization of (a) electrodeposited ZnO/eosinY by (b) desorption of eosinY leaving the (c) pure porous ZnO matrix to which the (d) sensitizer is adsorbed from solution to yield the (e) optimized sensitized ZnO with accessible inner surface.

and dissolve it in aqueous KOH solution (Figures 4.13a–c). This reaction was originally performed to determine the amount of EY in the films and to provide a lower estimate of the pore volume within the ZnO films. A complete desorption of the dye was observed for hybrid films deposited at potentials at which EY was present in reduced form. The amount of dye desorbed from the films corresponded to a volume of 40% of the total volume of the hybrid film [265]. The porous crystalline structure was preserved after extraction of EY [27, 247, 249]. These electrodeposited films therefore were found to consist of well-defined ZnO crystals stretching over a few micrometers in size with a large internal surface area because of the formation of nanopores within the grains, originally created by incorporated EY during electrodeposition but then empty following its desorption [27].

This porous crystalline network of ZnO was used as a matrix to adsorb a variety of sensitizer molecules (Figures 4.13d–e). Electrodes prepared this way performed at significantly higher efficiency than the electrodes sensitized by dyes in the as-deposited films [266–269], and also reached the presently achieved record efficiency for cells based on electrodeposited ZnO [27] sensitized by a well-adapted

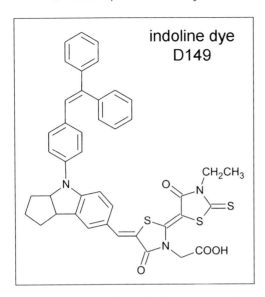

Figure 4.14 Structural formula of 5-[[4-[4-(2,2-diphenylethenyl)phenyl]-1,2,3,3a,4,8b-hexahydrocyclopent[b]indol-7-yl]methylene]-2-(3-ethyl-4-oxo-2-thioxo-5-thiazolidinylidene)-4-oxo-3-thiazolidineacetic acid (D149).

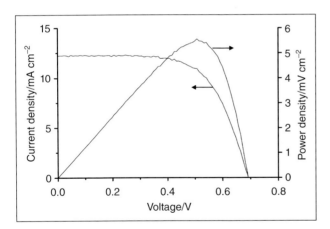

Figure 4.15 Current–voltage curve (left axis) and achievable power density (right axis) of a state-of-the-art photoelectrochemical cell consisting of electrodeposited porous ZnO sensitized by D149 in contact to a standard iodine-based electrolyte. Reprinted with permission from [27].

indoline sensitizer [270], that is, reaching IPCE > 80% close to the absorption maximum. The best performing indoline dye (D149, Figure 4.14) [270] achieved a short-circuit photocurrent of $I_{SC} = 12.23\,\mathrm{mA\,cm^{-2}}$, an open-circuit voltage of $V_{OC} = 691\,\mathrm{mV}$, a fill factor of FF = 0.658, and an overall conversion efficiency of $\eta = 5.56\%$ (Figure 4.15) [27]. Electrodes based on ZnO deposited in the presence of EY combine a high internal surface area for sufficient sensitizer adsorption securing a good light-harvesting efficiency with rapid electron transport in the crystalline ZnO films proven by time-resolved photocurrent and photovoltage measurements as well as IMPS and IMVS experiments [268, 269, 271, 272] leading to the observed collection efficiency close to unity [27]. These electrodes offer good prospects to study other sensitizers of the organic dye or metal complex type [266, 270, 273] to investigate their suitability for ZnO as semiconductor material. This work has been thoroughly reviewed recently [27].

4.8
Adaptation of Electrodeposition Towards Specific Demands of Alternative Substrate Materials

Active semiconductor structures consisting of electrodeposited porous yet crystalline ZnO sensitized by organic dyes were prepared from precursor solutions under moderate temperature and using techniques that are easily scalable to large-area applications. In most of the cited studies, a transparent conductive oxide (TCO) on glass was used as substrate electrode because of good availability and convenient working conditions in preparation and analysis. Glass, however, is a rather costly material, heavy, mechanically fragile, and therefore not flexible. Further, glass can economically be produced as flat sheets, but the variety of shapes and structures is quite limited. Electrodeposition, on the other hand, does not require the high processing temperatures that glass is often applied for, and can be performed for a variety of shapes and structures. The technology of electrodeposition and subsequent sensitization as well as the dye-sensitized cells as assemblies are perfectly compatible with glass as substrate material, but electrodepoosition does not fully utilize the potential of glass as a substrate material and, vice versa, glass as a substrate material does not allow to fully benefit from the advantages of electrodeposition as a truly three-dimensional coating technique. New applications can therefore be created and the potential of electrodeposition fully utilized if adapted to alternative substrates.

4.8.1
Plastic Solar Cells

Although plastic foils are limited in their thermal stability, and hence the preparation of conductive oxide films on plastic substrates is still a challenge, TCO on plastic foils have now become available [274–278]. In the present context it may be of special relevance that also ZnO-based coatings on plastic are available [277, 278], presumably good substrates for the electrodeposition of ZnO. Demonstrator DSSCs based on electrodeposited ZnO on plastic substrates [27] have been presented by Gunze Ltd, Japan, as part of a wearable solar charger for mobile electronics (Figures 4.16a and 4.16b) and by Sekisui Jushi Corporation, Japan, within concepts of illuminated road signs whose batteries could be charged by these cells

4.8 Adaptation of Electrodeposition Towards Specific Demands of Alternative Substrate Materials

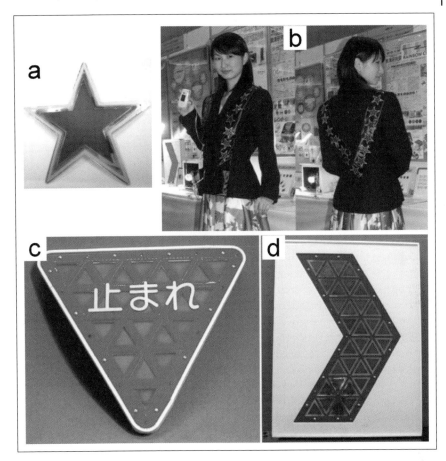

Figure 4.16 Demonstrator objects of photoelectrochemical cells based on porous ZnO (see text) on plastic substrates: (a) star-shaped individual cell; (b) cells mounted on a wearable belt and connected to charge the battery of a mobile phone; (c) triangular cells mounted into a stop-sign; (d) triangular cells mounted into a directional arrow road sign. Reprinted with permission from [27].

(Figures 4.16c and 4.16d). A conversion efficiency of $\eta = 1.4\%$ was achieved in the demonstrator but was driven to 4% for miniature cells. These demonstrators clearly showed the feasibility of such plastic DSSCs and large-area applications will largely depend on the availability and quality of these or other conductively coated transparent plastic foils.

4.8.2
Textile-Based Solar Cells

To allow direct human access to mobile electronics, their integration into textiles is very attractive because of the ruggedness and flexibility of textiles and hence increased comfort of use. In particular, sensors that monitor body functions or the environmental conditions of mobile work places (firefighters, etc.) are needed in direct contact to the user, preferably integrated into textiles. Sensors and electronic components have been miniaturized and have become very modest in their demand of energy. The ideal energy supply, however, would be independent of the electric grid and would not demand any replacement of parts. Such independent supply could be provided by photovoltaic cells if the user will be occasionally exposed to light.

Textiles have become attractive electrode materials. Metalized threads and threads of fine steel wires show remarkable conductivity and mechanical flexibility. Good integration of conducting pathways into textiles could thereby be achieved using spatially well-controlled positioning of individual threads in textiles by traditional textile technologies, for example, weaving or stitching [279–282]. Textile-based electronic and sensor devices are therefore sought that realize their function on textile threads to be integrated into textile structures with their extreme flexibility and ruggedness and maintain the attractive functionality of textiles and wearing comfort also for such functionalized textiles. Photovoltaic cells by nature of their function need to cover a considerable area and hence the presence of classical photovoltaic devices considerably interferes with the textile characteristics of a garment. Foil-based cells are preferred to bulk devices in this respect (Figure 4.16) but even they represent a clear alteration of the wearing comfort. Therefore textile-based photovoltaic cells, directly realized on threads and garments, will represent a big step forward. Photovoltaic active junctions are therefore sought that can be prepared at temperatures low enough to be compatible with textile structures and that are compatible with electrode distances in the 100 µm range, a typical distance of threads in ultrafine woven textiles [279, 280].

DSSCs represent a technology which can routinely use electrode distances in the cells already of about 50 µm [283]. Following further optimization such cells should be compatible with the electrode distance that can be realized by the distance of threads in textiles. Recently DSSCs based on porous TiO_2 prepared on wires of stainless steel [284–286] or titanium [287] have been reported and their function could be proven. TiO_2, however, required annealing temperatures far beyond the stability range of typical textile substrates. ZnO as discussed above (Section 4.3) can be prepared as a porous semiconductor for DSSCs in low-temperature processes well compatible with textile electrodes (<150 °C). Electrodeposition occurs to all exposed conductive sides of a three-dimensional object and is technically well compatible with typical textile finishing technologies. It therefore represents a straightforward strategy to prepare, study, and optimize sensitized ZnO on textile threads and on thin wires that are compatible with the manufacturing of textiles to realize textile-based photovoltaic cells. Aside from the motivation to provide a grid-independent energy supply for textile-based electronics, the development of textile-based photovoltaics may turn out to be an attractive strategy towards mechanically very flexible and rugged solar cells.

As a conductive textile electrode, Shieldex®, an Ag-coated polyamide yarn produced by Statex Produktions & Vertriebs GmbH (Bremen, Germany) represented

4.8 Adaptation of Electrodeposition Towards Specific Demands of Alternative Substrate Materials

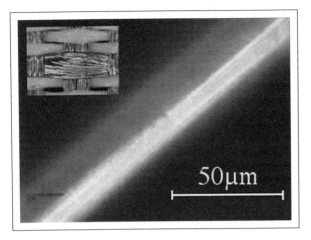

Figure 4.17 Optical microscopy image of an ELITEX® thread and a waver structure thereof (inset). Reprinted with permission from [279].

a reasonable textile starting point. Electroplating of additional Ag [279, 280] led to highly conductive ELITEX® filaments, yarns, and knitted fabrics (Figure 4.17). ELITEX® filaments (typically 27 µm in diameter), threads (consisting of 34 filaments), or knitted fabrics of either threads or filaments were connected as a working electrode but sealed between laminating foil as mechanical support and electrical insulation to electrodeposit highly porous ZnO films at 70 °C from oxygen-saturated aqueous zinc salt solutions containing EY as a structure directing agent (Figure 4.18b) [288]. The current density during deposition was increased compared with planar electrodes by enhanced diffusion at the filaments operating as cylindrical microelectrodes (Figure 4.18a). The film morphology was strongly influenced by geometrical constraints within the threads and by the hydrodynamic flow rate in the deposition solution. EY was desorbed from the electrodes and N535, an established Ru sensitizer, was adsorbed (Figure 4.18c). The photocurrent observed for such textile electrodes increased to a peak value of 1.3 mA cm^{-2} in the beginning of a white light pulse (AM 1.5) indicating facile electron injection from the photoexcited sensitizer (Figure 4.18d) but then significantly decreased accompanied by corrosion of the Ag electrode in the iodine-containing electrolyte [288].

Considerably improved morphology and uniformity were achieved by the use of a nitrate-based deposition bath for EY/ZnO under pulsed potentiostatic control [289]. These conditions turned out advantageous to control the conditions of mass flow and to suppress parasitic currents which otherwise severely interfered with the desired film growth. Homogeneous porous films of ZnO were formed from which EY could be desorbed in a method adapted from the one developed for films from O_2-based solutions (Figure 4.19). The textile-supported ZnO electrodes were then sensitized by D149, one of the best indoline sensitizers for ZnO [27] but still showed negligible photovoltaic parameters of $\eta < 0.1\%$, still caused by electron back transfer from the Ag electrode and by its corrosion [289].

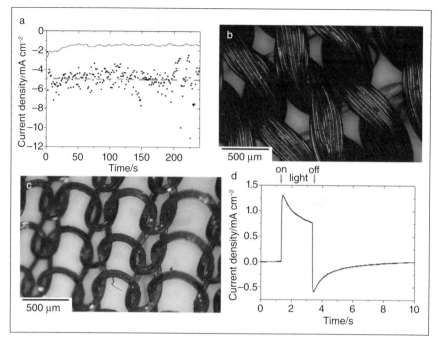

Figure 4.18 Results of electrodeposition of sensitized ZnO on ELITEX® conductive textiles. (a) Current density during electrodeposition at 70 °C from a resting oxygen-saturated aqueous 0.1 M $ZnCl_2$ solution with 50 µM eosinY (squares averaged by the dashed line) compared to the current density observed at a planar substrate rotating at 500 rpm under otherwise identical conditions (solid line). (b) Optical microscopy of a knitted fabric of an ELITEX® bundle covered by ZnO/eosinY. (c) Knitted fabric of an individual ELITEX® filament following desorption of eosin Y and adsorption of N3 as sensitizer. (d) Photocurrent observed at the N3-sensitized electrode (see (c)) in an iodine-containing standard organic electrolyte. Reprinted with permission from [288].

Significant progress in the field towards efficiencies at least in the region of 1% was achieved when textile-compatible threads and filaments were used with a protective passivation layer (Figure 4.20) which provided decreased rates of electron back transfer and corrosion [290]. Suitable conditions for electrodeposition on such surfaces are being developed and their optimization is being undertaken.

4.9
State of the Art and Outlook

Electrodeposition not only of metals but also of compound semiconductors has reached a level of quality and control that large-scale applications have become

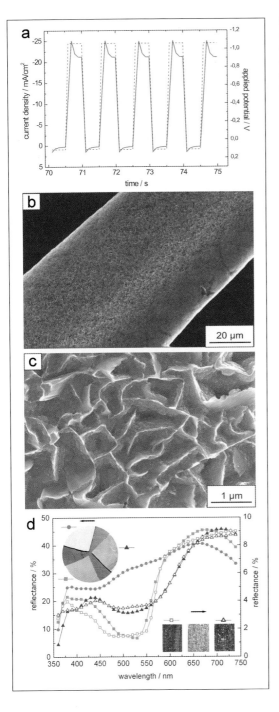

Figure 4.19 Results following pulsed electrodeposition of ZnO/eosin Y on ELITEX®. (a) Current density (solid curve) during controlled potential (dashed curve) polarization pulses of individual filaments at 70 °C immersed into 0.1 M aqueous $Zn(NO_3)_2$ with 50 µM eosin Y as structure-directing agent, (b, c) Scanning electron micrographs showing the homogeneous coverage of the textile electrodes with porous composite material. (d) Reflectance spectra of electrodeposited ZnO/eosin Y (open squares) and D149-sensitized ZnO (open triangles) on ELITEX® (right axis) and of ZnO/eosin Y (filled squares), ZnO following desorption of eosin Y (circles), and D149-sensitized ZnO (filled triangles) on Ag-coated FTO (left axis). In the inset, photographs of as-deposited ZnO/eosin Y, films after desorption of eosin Y, and films after adsorption of D149 are shown on Ag-coated FTO (upper left) and on ELITEX® (lower right) to provide a visual impression. Reprinted with permission from [289].

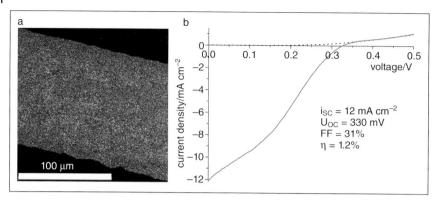

Figure 4.20 Results following pulsed electrodeposition of ZnO/eosin Y on a 70 μm diameter tungsten wire and subsequent sensitization by D149. (a) Confocal laser microscopy image following a pulsed electrodeposition from aqueous 0.1 M Zn(NO$_3$)$_2$ and 50 μM eosin Y at 70 °C under 0.5 s pulsed potential alternating between −1.26 V and −0.86 V vs. Ag/AgCl for 500 s, measured following desorption of eosin Y and adsorption of D149. (b) Current–voltage characteristics in the dark (dashed curve) and under illumination by 100 mW cm^{-2} close to AM 1.5 conditions by a xenon arc lamp equipped with a water filter (solid curve) of the D149-sensitized ZnO on the tungsten wire, measured as working electrode in (1 : 4 by volume) acetonitrile and ethylene carbonate with 0.5 M tetra-n-butylammonium iodide with a Pt wire as counterelectrode and an Ag/AgNO$_3$ reference electrode. 0 V in the plot refers to the rest potential of the electrode in the dark, as illuminated cell area the product of diameter and length of the electrode (0.018 cm^2) was chosen. Reprinted with permission from [290].

achievable. ZnO is a very attractive candidate for large-area applications of semiconductors because of only small toxicity problems and abundance of the materials. ZnO therefore has good potential to contribute to photovoltaics on the very large scale with a perspective to significantly contribute to a regenerative energy supply of a growing world population. The crystallographic and morphologic control of electrodeposited ZnO has increased tremendously over the last decade and now devices have been proposed on the prototype level which reach energy conversion efficiencies of about 5.6% under AM 1.5 conditions. Given the fast progress thus achieved and given the fact that no reason in principle can be seen why ZnO should not reach the 11% at which cells with TiO$_2$ can presently work, good prospects are seen for ZnO electrodes. With a considerably decreased input of energy needed to realize the cells in all-low-temperature processes, the economic feasibility seems promising. ZnO is presently investigated on a large scale also as a transparent conductive oxide and these two technologies might favorably merge in the production of DSSCs for the very large-scale photovoltaic market. Aside from this main line, electrodeposition of ZnO provides such unique features that a number of niche applications can be developed which may further assist to promote the field.

References

1. United Nations Environment Programme (UNEP) (2006) *Energy, Finance and Climate Change, Our Planet*, vol. 16, United Nations Environment Programme (UNEP).
2. Würfel, P. (2009) *Physics of Solar Cells*, 2nd edn, Wiley-VCH Verlag GmbH, Weinheim.
3. International Energy Agency (2008) *Key World Energy Statistics*, International Energy Agency, Paris.
4. Ginley, D., Green, M.A., and Collins, R. (2008) Solar energy conversion towards 1 terawatt. *MRS Bull.*, **33** (4), 355–364.
5. Mehos, M. (2008) Another pathway to large-scale power generation: concentrating solar power. *MRS Bull.*, **33** (4), 364–366.
6. http://www.pvresources.com/en/top50pv.php (accessed 31 August 2009).
7. http://www.energy.ca.gov/siting/solar/index.html (accessed 31 August 2009).
8. http://www.desertec.org/en/concept/ (accessed 31 August 2009).
9. Tritt, T.M., Böttner, H., and Chen, L. (2008) Thermoelectrics: direct solar thermal energy conversion. *MRS Bull.*, **33** (4), 366–368.
10. Nelson, J. (2003) *The Physics of Solar Cells*, Imperial College Press, London.
11. Green, M.A., Emery, K., King, D.L., Igari, S., and Warta, W. (2003) Solar cell efficiency tables (Version 22). *Prog. Photovolt Res. Appl.*, **11**, 347–352.
12. http://www.qcells.de/en/products_services/solar_cells/index.html (accessed 31 August 2009).
13. Wöhrle, D. and Meissner, D. (1991) Organic solar cells. *Adv. Mater.*, **3** (3), 129–138.
14. Wöhrle, D., Kreienhoop, L., and Schlettwein, D. (1996) Phthalocyanines and related macrocycles in organic photovoltaic junctions, in *Phthalocyanines: Properties and Applications*, vol. 4 (eds C.C. Leznoff and A.B. Lever), Wiley-VCH Verlag GmbH, Weinheim, pp. 219–284.
15. Shaheen, S.E., Ginley, D.S., and Jabbour, G.E. (2005) Organic-based photovoltaics: toward low-cost power generation. *MRS Bull.*, **30** (1), 10–15.
16. Pfeiffer, M., Beyer, A., Plönnings, B., Fritz, T., Leo, K., Hiller, S., Schlettwein, D., and Wöhrle, D. (2000) Controlled doping of organic dyes: basic mechanisms and implications for their use in organic photovoltaic cells. *Solar Energy Mater. Solar Cells*, **63**, 83–99.
17. Forrest, S.R. (2005) The limits to organic photovoltaic cell efficiency. *MRS Bull.*, **30** (1), 28–32.
18. Günes, S., Neugebauer, H., and Sariciftci, N.S. (2007) Conjugated polymer-based organic solar cells. *Chem. Rev.*, **107** (4), 1324–1338.
19. Männig, B., Drechsel, J., Gebeyehu, D., Simon, P., Kozlowski, F., Werner, A., Li, F., Grundmann, S., Sonntag, S., Koch, M., Leo, K., Pfeiffer, M., Hoppe, H., Meissner, D., Sariciftci, N.S., Riedel, I., Dyakonov, V., and Parisi, J. (2004) Organic p-i-n solar cells. *Appl. Phys. A*, **79**, 1–14.
20. Rand, B.P., Xue, J., Uchida, S., and Forrest, S.R. (2005) Mixed donor-acceptor molecular heterojunctions for photovoltaic applications. *J. Appl. Phys.*, **98**, 124902.
21. Hong, Z.R., Männig, B., Lessmann, R., Pfeiffer, M., and Leo, K. (2007) Improved efficiency of zinc phthalocyanine/C60 based photovoltaic cells via nanoscale interface modification. *Appl. Phys. Lett.*, **90**, 203505.
22. Hagfeldt, A. and Grätzel, M. (2000) Molecular photovoltaics. *Acc. Chem. Res.*, **33** (5), 269–277.
23. Grätzel, M. (2001) Photoelectrochemical cells. *Nature*, **414**, 338–344.
24. Gerischer, H. and Tributsch, H. (1968) Elektrochemische Untersuchungen zur spektralen Sensibilisierung von ZnO. Einkristallen. *Ber. Bunsenges. Physik. Chem.*, **72**, 437–445.

25 Gerischer, H. (1972) Electrochemical techniques for the study of photosensitization. *Photochem. Photobiol.*, **16**, 243–260.

26 Tsubomura, H., Matsumura, M., Nomura, Y., and Amemiya, T. (1976) Dye sensitized zinc oxide: aqueous electrolyte: platinum photocell. *Nature*, **261**, 402–403.

27 Yoshida, T., Zhang, J., Komatsu, D., Sawatani, S., Minoura, H., Pauporté, T., Lincot, D., Oekermann, T., Peter, L., Schlettwein, D., Tada, H., Wöhrle, D., Funabiki, K., Matsui, M., Miura, H., and Yanagi, H. (2009) Electrodeposition of inorganic/organic hybrid thin films. *Adv. Funct. Mater.*, **19**, 17–43.

28 Memming, R. (2001) *Semiconductor Electrochemistry*, Wiley-VCH Verlag GmbH, Weinheim.

29 Coakley, K.M., Liu, Y., Goh, C., and McGehee, M.D. (2005) Ordered organic–inorganic bulk heterojunction photovoltaic cells. *MRS Bull.*, **30** (1), 37–40.

30 Beek, W.J.E., Slooff, L.H., Wienk, M.M., Kroon, J.M., and Janssen, R.A.J. (2005) Hybrid solar cells using a ZnO precursor and a conjugated polymer. *Adv. Funct. Mater.*, **15**, 1703–1707.

31 Milliron, D.J., Gur, I., and Alivisatos, A.P. (2005) Hybrid organic–nanocrystal solar cells. *MRS Bull.*, **30** (1), 41–44.

32 Grätzel, M. (2005) Dye-sensitized solid-state heterojunction solar cells. *MRS Bull.*, **30** (1), 23–27.

33 Pichot, F., Pitts, J.R., and Gregg, B.A. (2000) Low-temperature sintering of TiO_2 colloids: application to flexible dye-sensitized solar cells. *Langmuir*, **16** (13), 5626–5630.

34 Rowell, M.W., Topinka, M.A., McGehee, M.D., Prall, H.-J., Dennler, G., Sariciftci, N.S., Hu, L., and Gruner, G. (2006) Organic solar cells with carbon nanotube network electrodes. *Appl. Phys. Lett.*, **88**, 233506.

35 Blom, P.W.M., Mihailetchi, V.D., Koster, L.J.A., and Markov, D.E. (2007) Device physics of polymer : fullerene bulk heterojunction solar cells. *Adv. Mater.*, **19** (12), 1551–1566.

36 Peet, J., Kim, J.Y., Coates, N.E., Ma, W.L., Moses, D., Heeger, A.J., and Bazan, G.C. (2007) Efficiency enhancement in low-bandgap polymer solar cells by processing with alkane dithiols. *Nat. Mater.*, **6** (7), 497–500.

37 http://www.iranchamber.com/history/articles/parthian_battery.php (accessed 1 September 2009).

38 Despic, A.R. and Popov, K.I. (1971) The effect of pulsating potential on the morphology of metal deposits obtained by mass-transport controlled electrodeposition. *J. Appl. Electrochem.*, **1**, 275–278.

39 Ibl, N. (1980) Some theoretical aspects of pulse electrolysis. *Surf. Technol.*, **10**, 81–104.

40 Datta, M. and Landolt, D. (1985) Experimental investigation of mass transport in pulse plating. *Surf. Technol.*, **25**, 97–110.

41 Schmidt, W.U., Alkire, R.C., and Gewirth, A.A. (1996) Mechanic study of copper deposition onto gold surfaces by scaling and spectral analysis of in situ atomic force microscopic images. *J. Electrochem. Soc.*, **143** (10), 3122–3132.

42 Datta, M. (2003) Electrochemical processing technologies in chip fabrication: challenges and opportunities. *Electrochim. Acta*, **48**, 2975–2985.

43 Drews, T.O., Webb, E.G., Ma, D.L., Alameda, J., Braatz, R.D., and Alkire, R.C. (2004) Coupled Mesoscale-continuum simulations of copper electrodeposition in a trench. *AIChE J.*, **50** (1), 226–240.

44 Moffat, T.P., Wheeler, D., Edelstein, M.D., and Josell, D. (2005) Superconformal film growth: mechanism and quantification. *IBM J. Res. Dev.*, **49** (1), 19–36.

45 Bockris, J.O.M. and Khan, S.U.M. (1993) *Surface Electrochemistry*, Plenum Press, New York/London.

46 Milchev, A. (2002) *Electrocrystallization: Fundamentals of Nucleation and Growth*, Kluwer Academic, Boston/Dordrecht/London.

47 Gregory, B.W. and Stickney, J.L. (1991) Electrochemical atomic layer epitaxy (ECALE). *J. Electroanal. Chem.*, **300** (1–2), 543–561.

48 Therese, G.H.A. and Kamath, P.V. (2000) Electrochemical synthesis of metal oxides and hydroxides. *Chem. Mater.*, **12** (5), 1195–1204.

49 Lincot, D. (2005) Electrodeposition of semiconductors. *Thin Solid Films*, **487**, 40–48.

50 Penner, R.M. (2001) Hybrid electrochemical/chemical synthesis of semiconductor nanocrystals on graphite, in *Electrochemistry of Nanomaterials* (ed. G. Hodes), Wiley-VCH Verlag GmbH, Weinheim, pp. 1–24.

51 Hodes, G. and Rubinstein, I. (2001) Electrodeposition of semiconductor quantum dot films, in *Electrochemistry of Nanomaterials* (ed. G. Hodes), Wiley-VCH Verlag GmbH, Weinheim, pp. 25–65.

52 Switzer, J.A. (2001) Electrodeposition of superlattices and multilayers, in *Electrochemistry of Nanomaterials* (ed. G. Hodes), Wiley-VCH Verlag GmbH, Weinheim, pp. 67–101.

53 Froment, M., Claude Bernard, M., Cortes, R., Mokili, B., and Lincot, D. (1995) Study of CdS epitaxial films chemically deposited from aqueous solutions on InP single crystals. *J. Electrochem. Soc.*, **142**, 2642–2649.

54 Froment, M. and Lincot, D. (1995) Phase-formation processes in solution at the atomic level: metal chalcogenide semiconductors. *Electrochim. Acta*, **40**, 1293–1303.

55 Hodes, G., Golan, Y., Behar, D., Zhang, Y., Alperson, B., and Rubinstein, I. (1998) Electrodeposited quantum dots: size control by semiconductor-substrate lattice mismatch, in *Nanoparticles and Nanostructured Films* (ed. J.H. Fendler), Wiley-VCH Verlag GmbH, Weinheim, pp. 1–22.

56 Switzer, J.A. (1998) Electrodeposition of superlattices and nanocomposites, in *Nanoparticles and Nanostructured Films* (ed. J.H. Fendler), Wiley-VCH Verlag GmbH, Weinheim, pp. 53–70.

57 Lincot, D., Froment, M., and Cachet, H. (1999) Chemical deposition of chalcogenide thin films from solution, in *Advances in Electrochemical Science and Engineering*, vol. 6 (eds R.C. Alkire and D. Kolb), Wiley-VCH Verlag GmbH, Weinheim, pp. 165–235.

58 Furlong, M.J., Froment, M., Bernard, M.C., Cortès, R., Tiwari, A.N., Krejci, M., Zogg, H., and Lincot, D. (1998) Aqueous solution epitaxy of CdS layers on CuInSe. *J. Cryst. Growth*, **193**, 114–122.

59 Stolt, L., Hedström, J., Kessler, J., Ruckh, M., Velthaus, K.-O., and Schock, H.W. (1993) ZnO/CdS/CuInSe$_2$ thin-film solar-cells with improved performance. *Appl. Phys. Lett.*, **62**, 597–599.

60 Kronik, L., Burstein, L., Leibovitch, M., Shapira, Y., Gal, D., Moons, E., Beier, J., Hodes, G., Cahen, D., Hariskos, D., Klenk, R., and Schock, H.W. (1995) Band diagram of the polycrystalline CdS/Cu(InGa)Se$_2$ heterojunction. *Appl. Phys. Lett.*, **67**, 1405–1407.

61 Duffy, N.W., Lane, D., Özsan, M.E., Peter, L.M., Rogers, K.D., and Wang, R.L. (2000) Structural and spectroscopic studies of CdS/CdTe heterojunction cells fabricated by electrodeposition. *Thin Solid Films*, **361/362**, 314–320.

62 Duffy, N.W., Peter, L.M., Wang, R.L., Lane, D.W., and Rogers, K.D. (2000) Electrodeposition and characterization of CdTe films for solar cell applications. *Electrochim. Acta*, **45**, 3355–3365.

63 Kampmann, A., Sittinger, V., Rechid, J., and Reineke-Koch, R. (2000) Large area electrodeposition of Cu(In,Ga)Se$_2$. *Thin Solid Films*, **361/362**, 309–313.

64 Rensmo, H., Keis, K., Lindström, H., Södergren, S., Solbrand, A., Hagfeldt, A., Lindquist, S.-E., Wang, L.N., and Muhammed, M. (1997) High light-to-energy conversion efficiencies for solar cells based on nanostructured ZnO electrodes. *J. Phys. Chem. B*, **101**, 2598–2601.

65 Bedja, I., Kamat, P.V., Hua, X., Lappin, A.G., and Hotchandani, S. (1997) Photosensitization of nanocrystalline ZnO films by bis(2,2'-bipyridine)(2,2'-bipyridine-4,4'-dicarboxylic acid) ruthenium(II). *Langmuir*, **13**, 2398–2403.

66 Rao, T.N. and Bahadur, L. (1997) Photoelectrochemical studies on dye-sensitized particulate ZnO thin-film photoelectrodes in nonaqueous media. *J. Electrochem. Soc.*, **144**, 179–185.

67 Fessenden, R.W., and Kamat, P.V. (1995) Rate constants for charge injection from excited sensitizer into SnO_2, ZnO, and TiO_2 semiconductor nanocrystallites. *J. Phys. Chem.*, **99**, 12902–12906.

68 Quintana, M., Edvinsson, T., Hagfeldt, A., and Boschloo, G. (2007) Comparison of dye-sensitized ZnO and TiO_2 solar cells: studies of charge transport and carrier lifetime. *J. Phys. Chem. C*, **111**, 1035–1041.

69 Minami, T. (2005) Transparent conducting oxide semiconductors for transparent electrodes. *Semicond. Sci. Technol.*, **20**, S35–S44.

70 Neumann, C., Lautenschläger, S., Graubner, S., Sann, J., Volbers, N., Meyer, B.K., Bläsing, J., Krost, A., Bertram, F., and Christen, J. (2007) Homoepitaxy of ZnO: from the substrates to doping. *Phys. Stat. Sol. (b)*, **244** (4), 1451–1457.

71 Kawano, T., Yahiro, J., Maki, H., and Imai, H. (2006) Epitaxial growth of wurtzite ZnO crystals in an aqueous solution system. *Chem. Lett.*, **35** (4), 442–444.

72 Loewenstein, T., Sann, J., Neumann, C., Meyer, B.K., and Schlettwein, D. (2008) Influence of molecular adsorbates on the structure of electrodeposited nanocrystalline ZnO. *Phys. Stat. Sol. (a)*, **205**, 2382–2387.

73 Peulon, S. and Lincot, D. (1996) Cathodic electrodeposition from aqueous solution of dense or open-structured zinc oxide films. *Adv. Mater.*, **8** (2), 166–170.

74 Peulon, S. and Lincot, D. (1998) Mechanistic study of cathodic electrodeposition of zinc oxide and zinc hydroxychloride films from oxygenated aqueous zinc chloride solutions. *J. Electrochem. Soc.*, **145**, 864–874.

75 Izaki, M. and Omi, T. (1996) Transparent zinc oxide films prepared by electrochemical reaction. *Appl. Phys. Lett.*, **68**, 2439–2440.

76 Izaki, M. and Omi, T. (1996) Electrolyte optimization for cathodic growth of zinc oxide films. *J. Electrochem. Soc.*, **143** (3), L53–L55.

77 Izaki, M. and Omi, T. (1997) Characterization of transparent zinc oxide films prepared by electrochemical reaction. *J. Electrochem. Soc.*, **144** (6), 1949–1952.

78 Pauporté, T. and Lincot, D. (2001) Hydrogen peroxide oxygen precursor for zinc oxide electrodeposition: I. Deposition in perchlorate medium. *J. Electrochem. Soc.*, **148**, C310–C314.

79 Pauporté, T. and Lincot, D. (2001) Hydrogen peroxide oxygen precursor for zinc oxide electrodeposition: II. Mechanistic aspects. *J. Electroanal. Chem.*, **517** (1–2), 54–62.

80 Pauporté, T., Cortès, R., Froment, M., Beaumont, B., and Lincot, D. (2002) Electrocrystallization of epitaxial zinc oxide onto gallium nitride. *Chem. Mater.*, **14**, 4702–4708.

81 Rao, A.R. and Dutta, V. (2008) Achievement of 4.7% conversion efficiency in ZnO dye-sensitized solar cells fabricated by spray deposition using hydrothermally synthesized nanoparticles. *Nanotechnology*, **19**, 445712.

82 Keis, K., Magnusson, E., Lindstrom, H., Lindquist, S.E., and Hagfeldt, A. (2002) A 5% efficient photo electrochemical solar cell based on nanostructured ZnO electrodes. *Solar Energy Mater. Solar Cells*, **73**, 51–58.

83 Keis, K., Bauer, C., Boschloo, G., Hagfeldt, A., Westermark, K., Rensmo, H., and Siegbahn, H. (2002) Nanostructured ZnO electrodes for dye-sensitized solar cell applications. *J. Photochem. Photobiol. A*, **148**, 57–64.

84 Nusbaumer, H., Moser, J.E., Zakeeruddin, S.M., Nazeeruddin, M.K., and Grätzel, M. (2001) $Co^{II}(dbbiP)_2^{2+}$ complex rivals tri-iodide/iodide redox mediator in dye-sensitized photovoltaic cells. *J. Phys. Chem. B*, **105**, 10461–10464.

85 Fisher, A.C., Peter, L.M., Ponomarev, E.A., Walker, A.B., and Wijayantha, K.G.U. (2000) Intensity dependence of the back reaction and transport of

electrons in dye-sensitized nanocrystalline TiO$_2$ solar cells. *J. Phys. Chem. B*, **104**, 949–958.

86 Pelet, S., Moser, J.E., and Grätzel, M. (2000) Cooperative effect of adsorbed cations and iodide on the interception of back electron transfer in the dye sensitization of nanocrystalline TiO$_2$. *J. Phys. Chem. B*, **104**, 1791–1795.

87 Montanari, I., Nelson, J., and Durrant, J.R. (2002) Iodide electron transfer kinetics in dye-sensitized nanocrystalline TiO$_2$ films. *J. Phys. Chem. B*, **106**, 12203–12210.

88 Cameron, P.J., Peter, L.M., and Hore, S. (2005) How important is the back reaction of electrons via the substrate in dye-sensitized nanocrystalline solar cells? *J. Phys. Chem. B*, **109**, 930–936.

89 Shen, Y., Nonomura, K., Schlettwein, D., Zhao, C., and Wittstock, G. (2006) Photoelectrochemical kinetics of eosin Y-sensitized zinc oxide films investigated by scanning electrochemical microscopy. *Chem. Eur. J.*, **12**, 5832–5839.

90 Shen, Y., Mengesha Tefashe, U., Nonomura, K., Loewenstein, T., Schlettwein, D., and Wittstock, G. (2009) Photoelectrochemical kinetics of Eosin Y-sensitized zinc oxide films investigated by scanning electrochemical microscopy under illumination with different LED. *Electrochim Acta*, **55**, 458–464.

91 Kawano, R. and Watanabe, M. (2005) Anomaly of charge transport of an iodide/tri-iodide redox couple in an ionic liquid and its importance in dye-sensitized solar cells. *Chem. Commun.*, 2107–2109.

92 Zistler, M., Wachter, P., Wasserscheid, P., Gerhard, D., Hinsch, A., Sastrawan, R., and Gores, H.J. (2006) Comparison of electrochemical methods for triiodide diffusion coefficient measurements and observation of non-Stokesian diffusion behaviour in binary mixtures of two ionic liquids. *Electrochim. Acta*, **52**, 161–169.

93 Wachter, P., Zistler, M., Schreiner, C., Berginc, M., Krasovec, U.O., Gerhard, Di., Wasserscheid, P., Hinsch, A., and Gores, H. (2008) Characterisation of DSSC-electrolytes based on 1-ethyl-3-methylimidazolium dicyanamide: Measurement of triiodide diffusion coefficient, viscosity, and photovoltaic performance. *J. Photochem. Photobiol. A*, **197**, 25–33.

94 Keis, K., Lindgren, J., Lindquist, S.E., and Hagfeldt, A. (2000) Studies of the adsorption process of Ru complexes in nanoporous ZnO electrodes. *Langmuir*, **10**, 4688–4694.

95 Yum, J.-H., Jang, S.-R., Walter, P., Geiger, T., Nüesch, F., Kim, S., Ko, J., Grätzel, M., and Nazeeruddin, M.K. (2007) Efficient co-sensitization of nanocrystalline TiO$_2$ films by organic sensitizers. *Chem. Commun.*, **44**, 4680–4682.

96 Cid, J.-J., Yum, J.-H., Jang, S.-R., Nazeeruddin, M.K., Ferrero, E.M., Palomares, E., Ko, J., Grätzel, M., and Torres, T. (2007) Molecular cosensitization for efficient panchromatic dye-sensitized solar cells. *Angew. Chem. Int. Ed.*, **46**, 8358–8362.

97 Silvestri, F., Lopez-Duarte, I., Seitz, W., Beverina, L., Martinez-Diaz, M.V., Marks, T.J., Guldi, D.M., Pagani, G.A., and Torres, T. (2009) A squaraine-phthalocyanine ensemble: towards molecular panchromatic sensitizers in solar cells. *Chem. Commun.*, 4500–4502.

98 Durrant, J.R., Haque, S.A., and Palomares, E. (2004) Towards optimisation of electron transfer processes in dye sensitised solar cells. *Coordin. Chem. Rev.*, **248**, 1247–1257.

99 Haque, S.A., Palomares, E., Cho, B.M., Green, A.N.M., Hirata, N., Klug, D.R., and Durrant, J.R. (2005) Charge separation versus recombination in dye-sensitized nanocrystalline solar cells: the minimization of kinetic redundancy. *J. Am. Chem. Soc.*, **127**, 3456–3462.

100 Fredin, K., Nissfolk, J., Boschloo, G., and Hagfeldt, A. (2007) The influence of cations on charge accumulation in dye-sensitized solar cells. *J. Electroanal. Chem.*, **609**, 55–60.

101 Koops, S.E., O'Regan, B.C., Barnes, P.R.F., and Durrant, J.R. (2009)

Parameters influencing the efficiency of electron injection in dye-sensitized solar cells. *J. Am. Chem. Soc.*, **131**, 4808–4818.

102 Clifford, J.N., Palomares, E., Nazeeruddin, M.K., Grätzel, M., Nelson, J., Li, X., Long, N.J., and Durrant, J.R. (2004) Molecular control of recombination dynamics in dye-sensitized nanocrystalline TiO$_2$ films: free energy vs distance dependence. *J. Am. Chem. Soc*, **126**, 5225–5233.

103 Nazeeruddin, M.K., Kay, A., Rodicio, I., Humphry-Baker, R., Müller, E., Liska, P., Vlachopoulos, N., and Grätzel, M. (1993) Conversion of light to electricity by cis-X2-bis(2,2'-bipyridyl-4,4'-dicarboxylate)ruthenium(II) charge transfer sensitizers (X = Cl$^-$, Br$^-$, I$^-$, CN$^-$ and SCN$^-$) on nanocrystalline TiO$_2$ electrodes. *J. Am. Chem. Soc.*, **115**, 6382–6390.

104 Argazzi, R., Bignozzi, C.A., Heimer, T.A., Castellano, F.N., and Meyer, G.J. (1994) Enhanced spectral sensitivity from ruthenium(II) polypyridyl based photovoltaic devices. *Inorg. Chem.*, **33**, 5741–5749.

105 Nazeeruddin, M.K., Pechy, P., Renouard, T., Zakeeruddin, S.M., Humphry-Baker, R., Comte, P., Liska, P., Cevey, L., Costa, E., Shklover, V., Spiccia, L., Deacon, G.B., Bignozzi, C.A., and Grätzel, M. (2001) Engineering of efficient panchromatic sensitizers for nanocrystalline TiO$_2$-based solar cells. *J. Am. Chem. Soc.*, **123**, 1613–1624.

106 Nazeeruddin, M.K., De Angelis, F., Fantacci, S., Selloni, A., Viscardi, G., Liska, P., Ito, S., Takeru, B., and Grätzel, M.G. (2005) Combined experimental and DFT-TDDFT computational study of photoelectrochemical cell ruthenium sensitizers. *J. Am. Chem. Soc.*, **127**, 16835–16847.

107 Kay, A. and Grätzel, M. (1993) Artificial photosynthesis: 1. Photosensitization of TiO$_2$ solar-cells with chlorophyll derivatives and related natural porphyrins. *J. Phys. Chem.*, **97**, 6272–6277.

108 Cherian, S. and Wamser, C.C. (2000) Adsorption and photoactivity of tetra(4-carboxyphenyl)porphyrin (TCPP) on nanoparticulate TiO$_2$. *J. Phys. Chem. B*, **104**, 3624–3629.

109 Odobel, F., Blart, E., Lagree, M., Villieras, M., Boujtita, H., El Murr, N., Caramori, S., and Bignozzi, C.A. (2003) Porphyrin dyes for TiO$_2$ sensitization. *J. Mater. Chem.*, **13**, 502–510.

110 Campbell, W.M., Burrell, A.K., Officer, D.L., and Jolley, K.W. (2004) Porphyrins as light harvesters in the dye-sensitised TiO$_2$ solar cell. *Coordin. Chem. Rev.*, **248**, 1363–1379.

111 Wang, Q., Carnpbell, W.M., Bonfantani, E.E., Jolley, K.W., Officer, D.L., Walsh, P.J., Gordon, K., Humphry-Baker, R., Nazeeruddin, M.K., and Grätzel, M. (2005) Efficient light harvesting by using green Zn-porphyrin-sensitized nanocrystalline TiO$_2$ films. *J. Phys. Chem. B*, **109**, 15397–15409.

112 Campbell, W.M., Jolley, K.W., Wagner, P., Wagner, K., Walsh, P.J., Gordon, K.C., Schmidt-Mende, L., Nazeeruddin, M.K., Wang, Q., Grätzel, M., and Officer, D.L. (2007) Highly efficient porphyrin sensitizers for dye-sensitized solar cells. *J. Phys. Chem. C*, **111**, 11760–11762.

113 Brumbach, M.T., Boal, A.K., and Wheeler, D.R. (2009) Metalloporphyrin assemblies on pyridine-functionalized titanium dioxide. *Langmuir*, **18**, 10685–10690.

114 Nazeeruddin, M.K., Humphry-Baker, R., Grätzel, M., and Murrer, B.A. (1998) Efficient near IR sensitization of nanocrystalline TiO$_2$ films by ruthenium phthalocyanines. *Chem. Commun.*, 719–720.

115 Nazeeruddin, M.K., Humphry-Baker, R., Grätzel, M., Wöhrle, D., Schnurpfeil, G., Schneider, G., Hirth, A., and Trombach, N. (1999) Efficient near-IR sensitization of nanocrystalline TiO$_2$ films by zinc and aluminum phthalocyanines. *J. Porph. Phthalocyanin.*, **3**, 230–237.

116 He, J.J., Hagfeldt, A., Lindquist, S.E., Grennberg, H., Korodi, F., Sun, L.C., and Akermark, B. (2001)

Phthalocyanine-sensitized nanostructured TiO$_2$ electrodes prepared by a novel anchoring method. *Langmuir*, **17**, 2743–2747.

117 He, J.J., Benko, G., Korodi, F., Polivka, T., Lomoth, R., Akermark, B., Sun, L.C., Hagfeldt, A., and Sundstrom, V. (2002) Modified phthalocyanines for efficient near-IR sensitization of nanostructured TiO$_2$ electrode. *J. Am. Chem. Soc*, **124**, 4922–4932.

118 Reddy, P.Y., Giribabu, L., Lyness, C., Snaith, H.J., Vijaykumar, C., Chandrasekharam, M., Lakshmikantam, M., Yum, J.-H., Kalyanasundaram, K., Grätzel, M., and Nazeeruddin, M.K. (2007) Efficient sensitization of nanocrystalline TiO$_2$ films by a near-IR-absorbing unsymmetrical zinc phthalocyanine. *Angew. Chem. Int. Ed.*, **46**, 373–376.

119 O'Regan, B.C., Lopez-Duarte, I., Martinez-Diaz, M.V., Forneli, A., Albero, J., Morandeira, A., Palomares, E., Torres, T., and Durrant, J.R. (2008) Catalysis of recombination and its limitation on open circuit voltage for dye sensitized photovoltaic cells using phthalocyanine dyes. *J. Am. Chem. Soc.*, **130**, 2906–2907.

120 Ingrosso, C., Curri, M.L., Fini, P., Giancane, G., Agostiano, A., and Valli, L. (2009) Functionalized copper(II)-phthalocyanine in solution and as thin film: photochemical and morphological characterization toward applications. *Langmuir*, **17**, 10305–10313.

121 Cid, J.-J., Garcia-Iglesias, M., Yum, J.-H., Forneli, A., Albero, J., Martinez-Ferrero, E., Vazquez, P., Grätzel, M., Nazeeruddin, M.K., Palomares, E., and Torres, T. (2009) Structure–function relationships in unsymmetrical zinc phthalocyanines for dye-sensitized solar cells. *Chem. Eur. J.*, **15**, 5130–5137.

122 Macor, L., Fungo, F., Tempesti, T., Durantini, E.N., Otero, L., Barea, E.M., Fabregat-Santiago, F., and Bisquert, J. (2009) Near-IR sensitization of wide band gap oxide semiconductor by axially anchored Si-naphthalocyanines. *Energ. Environ. Sci.*, **2**, 529–534.

123 Morandeira, A., Lopez-Duarte, I., O'Regan, B., Martinez-Diaz, M.V., Forneli, A., Palomares, E., Torres, T., and Durrant, J.R. (2009) Ru(II)-phthalocyanine sensitized solar cells: the influence of co-adsorbents upon interfacial electron transfer kinetics. *J. Mater. Chem.*, **19**, 5016–5026.

124 Lukyanets, E.A. (1989) *Electronic Spectra of Phthalocyanines and Related Compounds (in Russian)*, NIITEHIM (NIOPIK), Cherkassy.

125 Stillman, M.J. and Nyokong, T. (1989) Absorption and magnetic circular dichroism spectral properties of phthalocyanines, in *Phthalocyanines: Properties and Applications*, vol. 1 (eds C.C. Leznoff and A.B.P. Lever), Wiley-VCH Verlag GmbH, Weinheim, pp. 133–290.

126 Nazeeruddin, M.K., Zakeeruddin, S.M., Humphry-Baker, R., Jirousek, M., Liska, P., Vlachopoulos, N., Shklover, V., Fischer, C.-H., and Grätzel, M. (1999) Acid–base equilibria of (2,2'-bipyridyl-4,4'-dicarboxylic acid) ruthenium(II) complexes and the effect of protonation on charge-transfer sensitization of nanocrystalline titania. *Inorg. Chem.*, **38**, 6298.

127 Robertson, N. (2008) Catching the rainbow: light harvesting in dye-sensitized solar cells. *Angew. Chem. Int. Ed.*, **47**, 1012–1014.

128 Schlettwein, D., Hesse, K., Gruhn, N.E., Lee, A.P., Nebesny, K.W., and Armstrong, N.R. (2001) Electronic energy levels in individual molecules, thin films and organic heterojunctions of substituted phthalocyanines. *J. Phys. Chem. B*, **105**, 4791–4800.

129 Schlettwein, D. and Armstrong, N.R. (1994) Correlation of frontier orbital positions and conduction type of molecular semiconductors as derived from ups in combination with electrical and photoelectrochemical experiments. *J. Phys. Chem.*, **98**, 11771–11779.

130 Schlettwein, D., Armstrong, N.R., Lee, P.A., and Nebesny, K.W. (1994) Factors which control the n-type or p-type behavior of molecular semiconductor thin films. *Mol. Cryst. Liq. Cryst.*, **253**, 161–171.

131 Chen, S.X., Seki, K., Inokuchi, H., Shi, Z., and Quian, R.Y. (1983) Ultraviolet photoelectron spectra of alpha-polymorphs and beta-polymorphs of copper phthalocyanine. *Bull. Chem. Soc. Jpn.*, **56**, 2565–2568.

132 Sato, N., Seki, K., and Inokuchi, H. (1981) Polarization energies of organic solids determined by ultraviolet photoelectron spectroscopy. *J. Chem. Soc. Faraday Trans. 2*, **77**, 1621–1633.

133 Seki, K. (1989) Ionization energies of free molecules and molecular solids. *Mol. Cryst. Liq. Cryst.*, **171**, 255–270.

134 Kearns, D.R. and Calvin, M. (1961) Solid state ionization potentials of some aromatic organic compounds. *J. Chem. Phys.*, **34**, 2026–2030.

135 Ishii, H., Sugiyama, K., Ito, E., and Seki, K. (1999) Energy level alignment and interfacial electronic structures at organic metal and organic organic interfaces. *Adv. Mater.*, **11**, 605–625.

136 Fahlmann, M. and Salaneck, W.R. (2002) Surfaces and interfaces in polymer-based electronics. *Surf. Sci.*, **500**, 904–922.

137 Witte, G., Lukas, S., Bagus, P.S., and Wöll, C. (2005) Vacuum level alignment at organic/metal junctions: "Cushion" effect and the interface dipole. *Appl. Phys. Lett.*, **87**, 263502.

138 Cahen, D., Kahn, A., and Umbach, E. (2005) Energetics of molecular interfaces. *Mater. Today*, **8**, 32–41.

139 Akaike, K., Kanai, K., Yoshida, H., Tsutsumi, J., Nishi, T., Sato, N., Ouchi, Y., and Seki, K. (2008) Ultraviolet photoelectron spectroscopy and inverse photoemission spectroscopy of [6,6]-phenyl-C_{61}-butyric acid methyl ester in gas and solid phases. *J. Appl. Phys.*, **104**, 023710.

140 Schlettwein, D., Jaeger, N.I., and Wöhrle, D. (1991) Photoelectrochemical investigations of molecular semiconductors: characterization of the conduction type of various substituted porphyrins. *Ber. Bunsenges. Phys. Chem.*, **95**, 1526–1530.

141 Chau, L.K., England, C.D., Chen, S., and Armstrong, N.R. (1993) Visible absorption and photocurrent spectra of epitaxially deposited phthalocyanine thin films: interpretation of exciton coupling effects. *J. Phys. Chem.*, **97**, 2699–2706.

142 Auerhammer, J.M., Knupfer, M., Peisert, H., and Fink, J. (2002) The copper phthalocyanine/Au(100) interface studied using high resolution electron energy-loss spectroscopy. *Surf. Sci.*, **506**, 333–338.

143 Lever, A.B.P., Milaeva, E.R., and Speier, G. (1993) The redox chemistry of metallophthalocyanines in solution, in *Phthalocyanines: Properties and Applications*, vol. 3 (eds C.C. Leznoff and A.B.P. Lever), Wiley-VCH Verlag GmbH, Weinheim, pp. 1–69.

144 Wöhrle, D., and Schmidt, V. (1988) Octabutoxyphthalocyanine, a new electron donor. *J. Chem. Soc. Dalton. Trans.*, 549–551.

145 Giraudeau, A., Louati, A., Gross, M., Andre, J.-J., Simon, J., Su, H., and Kadish, K.M. (1983) Redox properties of octacyano-substituted zinc phthalocyanine ((CN)8PcZn). New charge-transfer complex. *J. Am. Chem. Soc.*, **105**, 2917–2919.

146 Louati, A., El- Meray, M., Andre, J.-J., Simon, J., Kadish, K.M., Gross, M., and Giraudeau, A. (1985) Electrochemical reduction of new, good electron acceptors: the metallooctacyanophthalocyanines. *Inorg. Chem.*, **24**, 1175–1179.

147 Green, J.M., and Faulkner, L.R. (1983) Reversible oxidation and rereduction of entire thin films of transition-metal phthalocyanines. *J. Am. Chem. Soc.*, **105** (10), 2950–2955.

148 Toshima, N., Tominaga, T., and Kawamura, S. (1996) Reversible electrochromism of copper phthalocyanine thin film. *Bull. Chem. Soc. Jpn.*, **69**, 245–253.

149 Toshima, N. and Tominaga, T. (1996) Electrochromism of single transition-metal phthalocyanine thin films: effect of central metal and composite structure. *Bull. Chem. Soc. Jpn.*, **69**, 2111–2122.

150 Schumann, B., Wöhrle, D., and Jaeger, N.I. (1985) Reversible reduction and reoxidation of entire

thin films of octacyanophthalocyanine. *J. Electrochem. Soc.*, **132**, 2144–2149.

151 Nicholson, M.M. (1993) Electrochromism and display devices, in *Phthalocyanines: Properties and Applications*, vol. 3 (eds C.C. Leznoff and A.B.P. Lever), Wiley-VCH Verlag GmbH, Weinheim, pp. 71–117.

152 Daum, P., Lehnhard, J.R., Rolison, D., and Murray, R.W. (1980) Diffusional charge transport through ultrathin films of radiofrequency plasma polymerized vinylferrocene at low temperature. *J. Am. Chem. Soc.*, **102**, 4649–4653.

153 Murray, R.W. (1984) Chemically modified electrodes. *Electroanal. Chem.*, **13**, 191–368.

154 Schroeder, A.H., Kaufman, F.B., Patel, V., and Engler, E.E. (1980) Comparative behavior of electrodes coated with thin films of structurally related electroactive polymers. *J. Electroanal. Chem.*, **113**, 193–208.

155 Martin, C.R., Rubinstein, I., and Bard, A.J. (1982) Polymer films on electrodes: 9. Electron and mass transfer in Nafion films containing tris(2,2′-bipyridine)ruthenium(2+). *J. Am. Chem. Soc.*, **104**, 4817–4824.

156 Oyama, N. and Anson, F. (1980) Factors affecting the electrochemical responses of metal complexes at pyrolytic graphite electrodes coated with films of poly(4-vinylpyridine). *J. Electrochem. Soc.*, **127**, 640–647.

157 Schlettwein, D., Wöhrle, D., and Jaeger, N.I. (1989) Reversible reduction and reoxidation of thin films of tetrapyrazinotetraazaporphyrines. *J. Electrochem. Soc.*, **136**, 2882–2886.

158 Jaeger, N.I., Lehmkuhl, R., Schlettwein, D., and Wöhrle, D. (1994) Observation of a transient structural change during the reversible reduction of a porphyrin thin film electrode. *J. Electrochem. Soc.*, **141**, 1735–1739.

159 Wahlster-Yoshida, R., Yoshida, T., Schlettwein, D., Tsukatani, K., Yanagi, H., Kaneko, M., and Minoura, H. (2000) Electrochromic redox reactions of vapour-deposited thin films of tetrapyridotetraazaporphyrinatozinc(II). *J. Porphyrins Phthalocyanines*, **4**, 112–122.

160 Hesse, K. and Schlettwein, D. (1999) Spectroelectrochemical investigations on the reduction of hexadecafluorophthalocyaninatozinc ($F_{16}PcZn$). *J. Electroanal. Chem.*, **476**, 148–158.

161 Schlettwein, D., Hesse, K., and Oekermann, T. (1999) (Photo-)elektrochemische Untersuchungen an phthalocyaninmodifizierten Elektroden, in *Elektrochemische Reaktionstechnik Und Synthese, Gdch- Monographie Band 14* (eds J. Russow, G. Sandstede, and R. Staab), GDCh, Frankfurt am Main, pp. 216–222.

162 Hansch, C., Leo, A., and Taft, R.W. (1991) A survey of Hammett substituent constants and resonance and field parameters. *Chem. Rev.*, **91**, 165–195.

163 Lever, A.B.P. (1993) Derivation of metallophthalocyanine redox potentials via Hammett parameter analysis. *Inorg. Chim. Acta*, **203**, 171–174.

164 Zecevic, S., Simic-Glavaski, B., Yeager, E., Lever, A.B.P., and Minor, P.C. (1985) Spectroscopic and electrochemical studies of transition metal tetrasulfonated phthalocyanines: V. Voltammetric studies of adsorbed tetrasulfonated phthalocyanines (MTsPc) in aqueous solutions. *J. Electroanal. Chem.*, **196**, 339–358.

165 L'Her, M. and Pondaven, A. (2003) Electrochemistry of phthalocyanines, in *The Porphyrin Handbook*, vol. 16 (eds K.M. Kadish, K.M. Smith, and R. Guilard), Academic Press, pp. 117–169.

166 Nyokong, T. (2006) Electrodes modified with monomeric M-N_4 catalysts for the detection of environmentally important molecules, in *N4-Macrocyclic Metal Complexes: Electrocatalysis, Electrophotochemistry & Biomimetic Electroanalysis* (eds F. Bedioui, J. Zagal, and J. Dodelet), Springer, New York, pp. 315–361.

167 Yeager, E. (1984) Electrocatalysts for O_2 reduction. *Electrochim. Acta*, **29**, 1527–1537.

168 Savy, M., Coowar, F., Riga, J., Verbist, J.J., Bronoel, G., and Besse, S. (1990) Investigation of O_2 reduction in alkaline media on macrocyclic chelates

impregnated on different supports: influence of the heat treatment on stability and activity. *J. Appl. Electrochem.*, **20**, 260–268.

169 Collin, J.P., and Sauvage, J.P. (1989) Electrochemical reduction of carbon dioxide mediated by molecular catalysts. *Coordin. Chem. Rev.*, **93**, 245–268.

170 Buck, T., Bohlen, H., Wöhrle, D., and Schulz-Ekloff, G. (1993) Influence of substituents and ligands of various cobalt(II) porphyrin derivatives coordinatively bonded to silica on the oxidation of mercaptan. *J. Mol. Catal.*, **80**, 253–267.

171 Yoshida, T., Kamato, K., Tsukamoto, M., Iida, T., Schlettwein, D., Wöhrle, D., and Kaneko, M. (1995) Selective electrocatalysis for CO_2 reduction in the aqueous phase using cobalt phthalocyanine/poly-4-vinylpyridine modified electrodes. *J. Electroanal. Chem.*, **385**, 209–225.

172 Schlettwein, D. and Yoshida, T. (1998) Electrochemical reduction of substituted cobalt phthalocyanines adsorbed on graphite. *J. Electroanal. Chem.*, **441**, 139–146.

173 Darwent, J.R., Douglas, P., Harriman, A., Porter, G., and Richoux, M.-C. (1982) Metal phthalocyanines and porphyrins as photosensitizers for reduction of water to hydrogen. *Coordin. Chem. Rev.*, **44**, 83–126.

174 Ferraudi, G. (1989) Photochemical properties of metallophthalocyanines in homogeneous solution, in *Phthalocyanines, Properties and Applications*, vol. 1 (eds C.C. Leznoff and A.B.P. Lever), Wiley-VCH Verlag GmbH, Weinheim, pp. 291–340.

175 Sobbi, A.K., Wöhrle, D., and Schlettwein, D. (1993) Photochemical stability of various porphyrins in solution and as thin film electrodes. *J. Chem. Soc. Perkin Trans. 2*, 481–488.

176 Mack, J. and Stillman, M.J. (2003) Electronic structure of metal phthalocyanine and porphyrin complexes from analysis of the UV-visible absorption and magnetic circular dichroism spectra and molecular orbital calculations, in *The Porphyrin Handbook*, vol. 16 (eds K.M. Kadish, K.M. Smith, and R. Guilard), Academic Press, pp. 43–116.

177 Nyokong, T. (2007) Effects of substituents on the photochemical and photophysical properties of main group metal phthalocyanines. *Coordin. Chem. Rev.*, **251**, 1707–1722.

178 Ishii, K. and Kobayashi, N. (2003) The photophysical properties of phthalocyanines and related compounds, in *The Porphyrin Handbook*, vol. 16 (eds K.M. Kadish, K.M. Smith, and R. Guilard), Academic Press, pp. 1–42.

179 Kasuga, K. (1996) Chemical fixation and photoreduction of carbon dioxide catalyzed by metal phthalocyanine derivatives, in *Phthalocyanines, Properties and Applications*, vol. 4 (eds C.C. Leznoff and A.B.P. Lever), Wiley-VCH Verlag GmbH, Weinheim, pp. 201–217.

180 Snow, A. (2003) Phthalocyanine aggregation, in *The Porphyrin Handbook*, vol. 17 (eds K.M. Kadish, K.M. Smith, and R. Guilard), Academic Press, pp. 129–176.

181 Schlettwein, D. (2006) Photoelectrochemical reactions at phthalocyanine electrodes, in *N4-Macrocyclic Metal Complexes: Electrocatalysis, Electrophotochemistry & Biomimetic Electroanalysis* (eds F. Bedioui, J. Zagal, and J.-P. Dodelet), Springer, New York, pp. 467–515.

182 Schlettwein, D., Jaeger, N.I., and Oekermann, T. (2003) Photoelectrochemical reactions at phthalocyanine electrodes, in *The Porphyrin Handbook*, vol. 16 (eds K.M. Kadish, K.M. Smith, and R. Guilard), Academic Press, pp. 247–283.

183 Memming, R. (1996) *Semiconductor Electrochemistry*, Wiley-VCH Verlag GmbH, Weinheim, pp. 81–150.

184 Nozik, A.J. and Memming, R. (1996) Physical chemistry of semiconductor–liquid interfaces. *J. Phys. Chem.*, **100**, 13061–13078.

185 Shumov, Y.S. and Heyrovsky, M. (1975) The relation between catalytic and photoelectro-chemical properties of

phthalocyanine films. *J. Electroanal. Chem.*, **65**, 469–471.

186 Sevastyanov, V.I., Alferov, G.A., Asanov, A.N., and Komissarov, G.G. (1975) Photovoltaic effect in the films of pigments contacting with electrolyte. *Biofizika*, **20**, 1004–1009.

187 Meier, H., Albrecht, W., Tschirwitz, U., Zimmerhackl, E., and Geheeb, N. (1977) Zum photovoltaischen Effekt am System Organischer Halbleiter/ Elektrolyt. *Ber. Bunsenges. Phys. Chem.*, **81**, 592–597.

188 Meshitsuka, S. and Tamaru, K. (1977) Photoelectrocatalysis by metal phthalocyanine evaporated films in the oxidation of oxalate ion. *J. Chem. Soc. Faraday Trans. 1*, **73**, 236–242.

189 Meshitsuka, S. and Tamaru, K. (1977) Spectral distributions of photoelectrochemical reactions over metal phthalocyanine electrodes. *J. Chem. Soc. Faraday Trans. 1*, **73**, 760–767.

190 Minami, N., Watanabe, T., Fujishima, A., and Honda, K.-I. (1979) Photoelectrochemical study on copper phthalocyanine films. *Ber. Bunsenges. Phys. Chem.*, **83**, 476–481.

191 Tachikawa, H. and Faulkner, L.R. (1978) Electrochemical and solid state studies of phthalocyanine thin film electrodes. *J. Am. Chem. Soc.*, **100**, 4379–4385.

192 Leempoel, P., Fan, F.-R.F., and Bard, A.J. (1983) Semiconductor electrodes: 50. Effect of mode of illumination and doping on photochemical behavior of phthalocyanine films. *J. Phys. Chem.*, **87**, 2948–2955.

193 Fan, F.-R. and Faulkner, L.R. (1979) Phthalocyanine thin films as semiconductor electrodes. *J. Am. Chem. Soc.*, **101**, 4779–4787.

194 Simon, J. and Andre, J.-J. (1985) *Molecular Semiconductors: Photoelectrical Properties and Solar Cells*, Springer, Berlin, pp. 103–149.

195 Schlettwein, D. (2001) Electronic properties of molecular organic semiconductor thin films, in *Supramolecular Photosensitive and Electroactive Materials* (ed. H. Nalwa), Academic Press, San Diego, pp. 211–338.

196 Santerre, F., Cote, R., Veilleux, G., Saint-Jacques, R.G., and Dodelet, J.P. (1996) Highly photoactive molecular semiconductors: determination of the essential parameters that lead to an improved photoactivity for modified chloroaluminum phthalocyanine thin films. *J. Phys. Chem.*, **100**, 7632–7645.

197 Klofta, T.J., Sims, T.D., Pankow, J.W., Danziger, J., Nebesny, K.W., and Armstrong, N.R. (1987) Spectroscopic and photoelectrochemical studies of trivalent metal phthalocyanine thin films: the role of gaseous dopants (oxygen and hydrogen) in determining photoelectrochemical response. *J. Phys. Chem.*, **91**, 5651–5659.

198 Klofta, T., Rieke, P., Linkous, C., Buttner, W.J., Nanthakumar, A., Mewborn, T.D., and Armstrong, N.R. (1985) Tri- and tetravalent phthalocyanine thin film electrodes: comparison with other metal and demetallated phthalocyanine systems. *J. Electrochem. Soc.*, **132**, 2134–2143.

199 Buttner, W.J., Rieke, P.C., and Armstrong, N.R. (1985) The gold/ GaPc-Cl/ferri, ferrocyanide/GaPc-Cl/ platinum photoelectrochemical cell. *J. Am. Chem. Soc.*, **107**, 3738–3739.

200 Rieke, P.C. and Armstrong, N.R. (1984) Light-assisted, aqueous redox reactions at chlorogallium phthalocyanine thin-film photoconductors: dependence of the photopotential on the formal potential of the redox couple and evidence for photoassisted hydrogen evolution. *J. Am. Chem. Soc.*, **106**, 47–50.

201 Mezza, T.M., Linkous, C.L., Shepard, V.R., and Armstrong, N.R. (1981) Improved photoelectrochemical efficiencies at methalocyanine-modified SnO_2 electrodes. *J. Electroanal. Chem.*, **124**, 311–320.

202 Klofta, T., Buttner, W.J., and Armstrong, N.R. (1986) Effect of crystallite size and hydrogen and oxygen uptake in the photoelectrochemistries of thin films of chlorogallium phthalocyanine. *J. Electrochem. Soc.*, **133**, 1531–1532.

203 Klofta, T.J., Danziger, J., Lee, P.A., Pankow, J., Nebesny, K.W.,

and Armstrong, N.R. (1987) Photoelectrochemical and spectroscopic characterization of thin films of titanyl phthalocyanine: comparisons with vanadyl phthalocyanine. *J. Phys. Chem.*, **91**, 5646–5651.
204 Yanagi, H., Douko, S., Ueda, Y., Ashida, M., and Wöhrle, D. (1992) Improvement of photoelectrochemical properties of chloroaluminum phthalocyanine thin films by controlled crystallization and molecular orientation. *J. Phys. Chem.*, **96**, 1366–1372.
205 Karmann, E., Schlettwein, D., and Jaeger, N.I. (1996) Photoelectrochemical oxidation of 2-mercaptoethanol at the surface of octacyanophthalocyanine thin film electrodes. *J. Electroanal. Chem.*, **405**, 149–158.
206 Yanagi, H., Tsukatani, K., Yamaguchi, H., Ashida, M., Schlettwein, D., and Wöhrle, D. (1993) Semiconducting behavior of substituted tetraazaporphyrin thin films in photoelectrochemical cells. *J. Electrochem. Soc.*, **140**, 1942–1948.
207 Karmann, E., Meyer, J.-P., Schlettwein, D., Jaeger, N.I., Anderson, M., Schmidt, A., and Armstrong, N.R. (1996) Photoelectrochemical effects and (photo)conductivity of "n-type" phthalocyanines. *Mol. Cryst. Liq. Cryst.*, **283**, 283–291.
208 Yanagi, H., Kanbayashi, Y., Schlettwein, D., Wöhrle, D., and Armstrong, N.R. (1994) Photochemical investigations on naphthalocyanine derivatives in thin films. *J. Phys. Chem.*, **98**, 4760–4766.
209 Schlettwein, D., Karmann, E., Oekermann, T., and Yanagi, H. (2000) Wavelength-dependent switching of the photocurrent direction at the surface of molecular semiconductor electrodes based on orbital-confined excitation and transfer of charge carriers from higher excited states. *Electrochim. Acta*, **45**, 4697–4704.
210 Oekermann, T., Schlettwein, D., Jaeger, N.I., and Wöhrle, D. (1999) Influence of electron withdrawing substituents on photoelectrochemical surface phenomena at phthalocyanine thin film electrodes. *J. Porphyrins Phthalocyanin.*, **3**, 444–452.
211 Schlettwein, D., Wöhrle, D., Karmann, E., and Melville, U. (1994) Conduction type of substituted tetraazaporphyrins and perylene tetracarboxylic acid diimides as detected by thermoelectric power measurements. *Chem. Mater.*, **6**, 3–6.
212 Meyer, J.-P., Schlettwein, D., Wöhrle, D., and Jaeger, N.I. (1995) Charge Transport in thin films of molecular semiconductors as investigated by measurements of thermoelectric power and electrical conductivity. *Thin Solid Films*, **258**, 317–324.
213 Meyer, J.-P. and Schlettwein, D. (1996) Influence of central metal and ligand system on conduction type and charge carrier transport in phthalocyanine thin films. *Adv. Mater. Opt. Electron.*, **6**, 239–244.
214 Schlettwein, D., Oekermann, T., Jaeger, N.I., Armstrong, N.R., and Wöhrle, D. (2003) Interfacial trap states in junctions of molecular semiconductors. *Chem. Phys.*, **285**, 103–112.
215 Hiller, S., Schlettwein, D., Armstrong, N.R., and Wöhrle, D. (1998) Influence of surface reactions and ionization gradients on junction properties of F_{16}PcZn. *J. Mater. Chem.*, **8**, 945–954.
216 Oekermann, T., Schlettwein, D., and Wöhrle, D. (1997) Characterization of N,N'-dimethyl-3,4,9,10-perylenetetracarboxylic acid diimide and phthalocyaninatozinc(II) in electrochemical photovoltaic cells. *J. Appl. Electrochem.*, **27**, 1172–1178.
217 Schlettwein, D. and Jaeger, N.I. (1993) Identification of the mechanism of the photoelectrochemical reduction of oxygen on the surface of a molecular semiconductor. *J. Phys. Chem.*, **97**, 3333–3337.
218 Shepard, V.R. and Armstrong, N.R. (1979) Electrochemical and photoelectrochemical studies of copper and cobalt phthalocyanine-tin oxide electrodes. *J. Phys. Chem.*, **83**, 1268–1276.
219 Jaeger, C.D., Fan, F.-R.F., and Bard, A.J. (1980) Semiconductor electrodes:

26. Spectral sensitization of semiconductors with phthalocyanine. *J. Am. Chem. Soc.*, **102**, 2592–2598.

220 Giraudeau, A., Fan, F.-R.F., and Bard, A.J. (1980) Semiconductor electrodes: 30. Spectral sensitization of the semiconductors titanium oxide (n-TiO_2) and tungsten oxide (n-WO_3) with metal phthalocyanines. *J. Am. Chem. Soc.*, **102**, 5137–5142.

221 Yanagi, H., Chen, S.Y., Lee, P.A., Nebesny, K.W., Armstrong, N.R., and Fujishima, A. (1996) Dye-sensitizing effect of TiOPc thin film on n-TiO_2 (001) surface. *J. Phys. Chem.*, **100**, 5447–5451.

222 Taira, S., Miki, T., and Yanagi, H. (1999) Dye-sensitization of n-TiO_2 single-crystal electrodes with vapor-deposited oxometal phthalocyanines. *Appl. Surf. Sci.*, **143**, 23–29.

223 Armstrong, N.R., Nebesny, K.W., Collins, G.E., Lee, P.A., Chau, L.K., Arbour, C., and Parkinson, B.A. (1991) O/I-MBE: formation of highly ordered phthalocyanine/semiconductor junctions by molecular-beam epitaxy: photoelectrochemical characterization. *Proc. SPIE*, **1559**, 18–26.

224 Chau, L.K., Arbour, C., Collins, G.E., Nebesny, K.W., Lee, P.A., England, C.D., Armstrong, N.R., and Parkinson, B.A. (1993) Phthalocyanine aggregates on metal dichalcogenide surfaces: dye sensitization on tin disulfide semiconductor electrodes by ordered and disordered chloroindium phthalocyanine thin films. *J. Phys. Chem.*, **97**, 2690–2698.

225 Chau, L.K., Osburn, E.J., Armstrong, N.R., O'Brien, D.F., and Parkinson, B.A. (1994) Dye sensitization with octasubstituted liquid crystalline phthalocyanines. *Langmuir*, **10**, 351–353.

226 Deng, H., Mao, H., Liang, B., Shen, Y., Lu, Z., and Xu, H. (1996) Aggregation and the photoelectric behavior of tetrasulfonated phthalocyanine adsorbed on a TiO_2 microporous electrode. *J. Photochem. Photobiol. A*, **99**, 71–74.

227 Hodak, J., Quinteros, C., Litter, M.I., and San Roman, E. (1996) Sensitization of TiO_2 with phthalocyanines: 1. Photo-oxidations using hydroxoaluminium tricarboxymonoamidephthalocyanine adsorbed on TiO_2. *J. Chem. Soc. Faraday Trans.*, **92**, 5081–5088.

228 Yanagi, H., Ohoka, Y., Hishiki, T., Ajito, K., and Fujishima, A. (1997) Characterization of dye-doped TiO_2 films prepared by spray-pyrolysis. *Appl. Surf. Sci.*, **113**, 426–431.

229 Fang, J., Wu, J., Zhang, X., Mao, H., Shen, Y., and Lu, Z. (1997) Fabrication, characterization and photovoltaic study of a GaTSPc-CdS/TiO_2 particulate film. *J. Mater. Chem.*, **7**, 737–740.

230 Robertson, N. (2006) Optimizing dyes for dye-sensitized solar cells. *Angew. Chem. Int. Ed.*, **45**, 2338–2345.

231 Hara, K., Sato, T., Katoh, R., Furube, A., Ohga, Y., Shinpo, A., Suga, S., Sayama, K., Sugihara, H., and Arakawa, H. (2003) Molecular design of coumarin dyes for efficient dye-sensitized solar cells. *J. Phys. Chem. B*, **107**, 597–606.

232 Wang, Z.-S., Hara, K., Dan-oh, Y., Kasada, C., Shinpo, A., Suga, S., Arakawa, H., and Sugihrara, H. (2005) Photophysical and (photo) electrochemical properties of a coumarin dye. *J. Phys. Chem. B*, **109**, 3907–3914.

233 Hara, K., Wang, Z.-S., Sato, T., Furube, A., Katoh, R., Sugihara, H., Dan-oh, Y., Kasada, C., Shinpo, A., and Suga, S. (2005) Oligothiophene-containing coumarin dyes for efficient dye-sensitized solar cells. *J. Phys. Chem. B*, **109**, 15476–15482.

234 Hara, K., Kurashige, M., Ito, S., Shinpo, A., Suga, S., Sayama, K., and Arakawa, H. (2003) Novel polyene dyes for highly efficient dye-sensitized solar cells. *Chem. Commun.*, 252–253.

235 Hara, K., Sato, T., Katoh, R., Furube, A., Yoshihara, T., Murai, M., Kurashige, M., Ito, S., Shinpo, A., Suga, S., and Arakawa, H. (2005) Novel conjugated organic dyes for efficient dye-sensitized solar cells. *Adv. Funct. Mater.*, **15**, 246–252.

236 Kitamura, T., Ikeda, M., Shigaki, K., Inoue, T., Anderson, N.A., Ai, X.,

Lian, T., and Yanagida, S. (2004) Phenyl-conjugated oligoene sensitizers for TiO$_2$ solar cells. *Chem. Mater.*, **16**, 1806–1812.

237 Alex, S., Santhosh, U., and Das, S. (2005) Dye sensitization of nanocrystalline TiO$_2$: enhanced efficiency of unsymmetrical versus symmetrical squaraine dyes. *J. Photochem. Photobiol A*, **172**, 63–71.

238 Horiuchi, T., Miura, H., Sumioka, K., and Uchida, S. (2004) High efficiency of dye-sensitized solar cells based on metal-free indoline dyes. *J. Am. Chem. Soc.*, **126**, 12218–12219.

239 Thomas, K.R., Lin, J.T., Hsu, Y.-C., and Ho, K.-C. (2005) Organic dyes containing thienylfluorene conjugation for solar cells. *Chem. Commun.*, 4098–4100.

240 Chen, Y.-S., Li, C., Zeng, Z.-H., Wang, W.-B., Wang, X.-S., and Zhang, B.-W. (2005) Efficient electron injection due to a special adsorbing group's combination of carboxyl and hydroxyl: dye-sensitized solar cells based on new hemicyanine dyes. *J. Mater. Chem.*, **15**, 1654–1661.

241 Yoshida, T. and Schlettwein, D. (2004) Electrochemical self-assembly of oxide/dye composites, in *Encyclopedia of Nanoscience and Nanotechnology*, vol. 2 (ed. A. Nalwa), American Scientific Publishers, Stevenson Ranch, pp. 819–836.

242 Yoshida, T. and Minoura, H. (2000) Electrochemical self-assembly of dye-modified zinc oxide thin films. *Adv. Mater.*, **12**, 1219–1222.

243 Yoshida, T., Miyamoto, K., Hibi, N., Sugiura, T., Minoura, H., Schlettwein, D., Oekermann, T., Schneider, G., and Wöhrle, D. (1998) Self assembled growth of nanoparticulate porous ZnO thin film modified by 2,9,16,23-tetrasulfophthalocyanatozinc(II) by one-step electrodeposition. *Chem. Lett.*, **7**, 599–600.

244 Yoshida, T., Tochimoto, M., Schlettwein, D., Schneider, G., Wöhrle, D., Sugiura, T., and Minoura, H. (1999) Self-assembly of zinc oxide thin films modified with tetrasulfonated metallophthalocyanines by one-step electrodeposition. *Chem. Mater.*, **11**, 2657–2667.

245 Schlettwein, D., Oekermann, T., Yoshida, T., Tochimoto, M., and Minoura, H. (2000) Photoelectrochemical sensitisation of ZnO-tetrasulfophthalocyaninatozinc composites prepared by electrochemical self-assembly. *J. Electroanal. Chem.*, **481**, 42–51.

246 Oekermann, T., Yoshida, T., Schlettwein, D., Sugiura, T., and Minoura, H. (2001) Photoelectrochemical properties of ZnO/tetrasulfophthalocyanine hybrid thin films prepared by electrochemical self-assembly. *Phys. Chem. Chem. Phys.*, **3**, 3387–3392.

247 Yoshida, T., Terada, K., Schlettwein, D., Oekermann, T., Sugiura, T., and Minoura, H. (2000) Electrochemical self-assembly of nanoporous ZnO/eosin Y thin films and their sensitized photoelectrochemical performance. *Adv. Mater.*, **12**, 1214–1217.

248 Okabe, K., Yoshida, T., Sugiura, T., and Minoura, H. (2001) Electrodeposition of photoactive ZnO/xanthene dye hybrid thin films. *Trans. Mater. Res. Soc. Jpn.*, **26**, 523–526.

249 Yoshida, T., Oekermann, T., Okabe, K., Schlettwein, D., Funabiki, K., and Minoura, H. (2002) Cathodic electrodeposition of ZnO/eosinY hybrid thin films from dye added zinc nitrate bath and their photoelectrochemical characterizations. *Electrochemistry*, **70**, 470–487.

250 Pauporte, T., Yoshida, T., Goux, A., and Lincot, D. (2002) One-step electrodeposition of ZnO/eosinY hybrid thin films from a hydrogen peroxide oxygen precursor. *J. Electroanal. Chem.*, **534**, 55–64.

251 Yoshida, T., Yoshimura, J., Matsui, M., Sugiura, T., and Minoura, H. (1999) Self-assembly of ZnO/tetrabromophenol blue mixed thin film by one-step electrodeposition and its sensitized photoelectrochemical performance. *Trans. Mater. Res. Soc. Jpn.*, **24**, 497–500.

252 Karuppuchamy, S., Nonomura, K., Yoshida, T., Sugiura, T., and Minoura,

H. (2002) Cathodic electrodeposition of oxide semiconductor thin films and their application to dye-sensitized solar cells. *Solid State Ionics*, **151**, 19–27.

253 Karuppuchamy, S., Yoshida, T., Sugiura, T., and Minoura, H. (2001) Self-assembly of ZnO/riboflavin 5′-phosphate thin films by one-step electrodeposition and its characterization. *Thin Solid Films*, **397**, 63–69.

254 Nonomura, K., Yoshida, T., Schlettwein, D., and Minoura, H. (2003) One-step electrochemical synthesis of $ZnO/Ru(dcbpy)_2(NCS)_2$ hybrid thin films and their photoelectrochemical properties. *Electrochim. Acta*, **48**, 3071–3078.

255 Schubert, U., Lorenz, A., Kundo, N., Stuchinskaya, T., Gogina, L., Salanov, A., Zaikonovskii, V., Maizlish, V., and Shaposhnikov, G.P. (1997) Cobalt phthalocyanine derivatives supported on TiO_2 by sol-gel processing: 1. Preparation and microstructure. *Chem. Ber./Recl.*, **130**, 1585–1589.

256 Schneider, G., Wöhrle, D., Spiller, W., Stark, J., and Schulz-Ekloff, G. (1998) Photooxidation of 2-mercaptoethanol by various water soluble phthalocyanines in aqueous alkaline solution under irradiation with visible light. *Photochem. Photobiol.*, **60**, 333–342.

257 Schlettwein, D., Oekermann, T., Yoshida, T., Sugiura, T., Minoura, H., and Wöhrle, D. (2002) Electrochemically self-assembled ZnO/dye electrodes: preparation and time-resolved photoelectrochemical measurements. *Proc. SPIE*, **4465**, 113–122.

258 Michaelis, E., Nonomura, K., Schlettwein, D., Yoshida, T., Minoura, H., and Wöhrle, D. (2004) Hybrid thin films of ZnO with porphyrins and phthalocyanines prepared by one-step electrodeposition. *J. Porph. Phthalocyanin.*, **8**, 1366–1375.

259 Nonomura, K., Loewenstein, T., Michaelis, E., Wöhrle, D., Oekermann, T., Yoshida, T., Minoura, H., and Schlettwein, D. (2006) Photoelectrochemical characterization of electrodeposited ZnO thin films sensitized by porphyrins and/or phthalocyanines. *Phys. Chem. Chem. Phys.*, **8**, 3867.

260 de Jongh, P.E., Meulenkamp, E.A., Vanmaekelbergh, D., and Kelly, J.J. (2000) Charge carrier dynamics in illuminated, particulate ZnO electrodes. *J. Phys. Chem. B*, **104**, 7686–7693.

261 Yoshida, T., Pauporte, T., Lincot, D., Oekermann, T., and Minoura, H. (2003) Cathodic electrodeposition of ZnO/eosin Y hybrid thin films from oxygen-saturated aqueous solution of $ZnCl_2$ and eosin Y. *J. Electrochem. Soc.*, **150**, C608–C615.

262 Goux, A., Pauporte, T., Yoshida, T., and Lincot, D. (2006) Mechanistic study of the electrodeposition of nanoporous self-assembled ZnO/eosin Y hybrid thin films: effect of eosin concentration. *Langmuir*, **22**, 10545–10553.

263 Boeckler, C., Feldhoff, A., and Oekermann, T. (2007) Electrodeposited zinc oxide/phthalocyanine films: an inorganic/organic hybrid system with highly variable composition. *Adv. Funct. Mater.*, **17**, 3864–3869.

264 Pauporte, T., Yoshida, T., Cortes, R., Froment, M., and Lincot, D. (2003) Electrochemical growth of epitaxial eosin/ZnO hybrid films. *J. Phys. Chem. B*, **107**, 10077–10082.

265 Komatsu, D., Zhang, J., Yoshida, T., and Minoura, H. (2007) Electrochemical growth of ZnO/eosin Y hybrid thin film. *Trans. Mater. Res. Soc. Jpn.*, **32**, 417–420.

266 Yoshida, T., Iwaya, M., Ando, H., Oekermann, T., Nonomura, K., Schlettwein, D., Wöhrle, D., and Minoura, H. (2004) Improved photoelectrochemical performance of electrodeposited ZnO/EosinY hybrid thin films by dye re-adsorption. *Chem. Commun.*, 400–401.

267 Loewenstein, T., Nonomura, K., Yoshida, T., Michaelis, E., Wöhrle, D., Rathousky, J., Wark, M., and Schlettwein, D. (2006) Efficient sensitization of mesoporous electrodeposited zinc oxide by cis-bis(isothiocyanato)bis(2,2′-bipyridyl-4,4′-dicarboxylato)-ruthenium(II). *J. Electrochem. Soc.*, **153**, A699–A704.

268 Nonomura, K., Loewenstein, T., Michaelis, E., Wöhrle, D., Yoshida, T., Minoura, H., and Schlettwein, D. (2006) Photoelectrochemical characterisation and optimisation of electrodeposited ZnO thin films sensitised by porphyrins and phthalocyanines. *Phys. Chem. Chem. Phys.*, **33**, 3867–3875.

269 Nonomura, K., Komatsu, D., Yoshida, T., Minoura, H., and Schlettwein, D. (2007) Dependence of the photoelectrochemical performance of sensitised ZnO on the crystalline orientation in electrodeposited ZnO thin films. *Phys. Chem. Chem. Phys.*, **9**, 1843–1849.

270 Matsui, M., Ito, A., Kotani, M., Kubota, Y., Funabiki, K., Jin, J., Yoshida, T., Minoura, H., and Miura, H. (2009) The use of indoline dyes in a zinc oxide dye-sensitized solar cell. *Dyes Pigments*, **80**, 233–238.

271 Oekermann, T., Yoshida, T., Minoura, H., Wijayantha, K.G.U., and Peter, L.M. (2004) Electron transport and back reaction in electrochemically self-assembled nanoporous ZnO/dye hybrid films. *J. Phys. Chem. B*, **108**, 8364–8370.

272 Oekermann, T., Yoshida, T., Boeckler, C., Caro, J., and Minoura, H. (2005) Capacitance and field-driven electron transport in electrochemically self-assembled nanoporous ZnO/dye hybrid films. *J. Phys. Chem. B*, **109**, 12560–12566.

273 Idowu, M., Loewenstein, T., Hastall, A., Nyokong, T., and Schlettwein, D. (2010) Photoelectrochemical characterization of electrodeposited ZnO thin films sensitized by octacarboxy metallophthalocyanine derivatives. *J. Porph. Phthalocyanin.*, **14**, 142–149.

274 Singh, V., Saswat, B., and Kumar, S. (2005) Low temperature deposition of Indium tin oxide (ITO) films on plastic substrates. *MRS Symp. Proc.*, **869**, D2.9.1–D2.9.6.

275 Ikegami, M., Miyoshi, K., Miyasaka, T., Teshima, K., Wie, T.C., Wan, C.C., and Wang, Y.Y. (2007) Platinum/titanium bilayer deposited on polymer film as efficient counter electrodes for plastic dye-sensitized solar cells. *Appl. Phys. Lett.*, **90**, 153122.

276 Peccell products information, Peccell Technologies, Inc. Yokohama, Japan (http://www.hs-kr.com/pds/peccell_products_en.pdf, accessed 2 October 2009).

277 Hara, H., Hanada, T., Shiro, T., and Yatabe, T. (2004) Properties of indium zinc oxide thin films on heat withstanding plastic substrate. *J. Vac. Sci. Technol. A*, **22**, 1726–1729.

278 Miyake, A., Yamada, T., Makino, H., Yamamoto, N., and Yamamoto, T. (2008) Effect of substrate temperature on structural, electrical and optical properties of Ga-doped ZnO films on cycro olefin polymer substrate by ion plating deposition. *Thin Solid Films*, **517**, 1037–1041.

279 Gimpel, S., Möhring, U., Müller, H., Neudeck, A., and Scheibner, W. (2004) Textile-based electronic substrate technology. *J. Ind. Textiles*, **33**, 179–189.

280 Möhring, U., Gimpel, S., Neudeck, A., Scheibner, W., and Zschenderlein, D. (2006) Conductive, sensorial and luminescent features in textile structures, in *Proceedings, 3rd International Forum on Applied Wearable Computing, Bremen, Germany, March 15–16* (ed. H. Kenn), Italian National Agency for New Technologies, Energy and Sustainable Economic Development (http://spring.bologna.enea.it/ifawc/2006/Proceedings.htm, accessed 2 October 2009).

281 Post, R., Orth, M., Russo, P., and Gershenfeld, N. (2000) E-broidery: design and fabrication of textile-based computing. *IBM Syst. J.*, **39**, 840–860.

282 Catrysse, M., Puers, R., Hertleer, C., Van Langenhove, L., van Egmond, H., and Matthys, D. (2004) Towards the integration of textile sensors in a wireless monitoring suit. *Sens. Actuators A*, **114**, 302–311.

283 Jiang, H., Sakurai, S., and Kobayashi, K. (2009) Fabrication and enhanced performance of a dye-sensitized solar cell with a ClO_4^-–poly(3,4-ethylenedioxythiophene)/TiO_2/FTO counter electrode. *Electrochem. Solid-State Lett.*, **12**, F13–F16.

284 Fan, X., Wang, F., Chu, Z., Chen, L., Zhang, C., and Zou, D. (2007) Conductive mesh based flexible dye-sensitized solar cells. *Appl. Phys. Lett.*, **90**, 073501.

285 Fan, X., Chu, Z., Wang, F., Zhang, C., Chen, L., Tang, Y., and Zou, D. (2008) Wire-shaped flexible dye-sensitized solar cells. *Adv. Mater.*, **20**, 592–595.

286 Fan, X., Chu, Z., Chen, L., Zhang, C., Wang, F., Tang, Y., Sun, J., and Zou, D. (2008) Fibrous flexible solid-type dye-sensitized solar cells without transparent conducting oxide. *Appl. Phys. Lett.*, **92**, 113510.

287 Ramier, J., Plummer, C.J.G., Leterrier, Y., Månson, J.-A.E., Eckert, B., and Gaudiana, R. (2008) Mechanical integrity of dye-sensitized photovoltaic fibers. *Renewable Energy*, **33**, 314–319.

288 Loewenstein, T., Hastall, A., Mingebach, M., Zimmermann, Y., Neudeck, A., and Schlettwein, D. (2008) Textile electrodes as substrates for the electrodeposition of porous ZnO. *Phys. Chem. Chem. Phys.*, **10**, 1844–1847.

289 Rudolph, M., Loewenstein, T., Arndt, E., Zimmermann, Y., Neudeck, A., and Schlettwein, D. (2009) Pulsed electrodeposition of porous ZnO on Ag-coated polyamide filaments. *Phys. Chem. Chem. Phys.*, **11**, 3313–3319.

290 Loewenstein, T., Rudolph, M., Mingebach, M., Strauch, K., Zimmermann, Y., Neudeck, A., Sensfuss, S., and Schlettwein, D. (2010) Textile-compatible substrate electrodes with electrodeposited ZnO: a new pathway to textile-based photovoltaics. *Chem. Phys. Chem.*, **11**, 783–788.

5
Thin-Film Semiconductors Deposited in Nanometric Scales by Electrochemical and Wet Chemical Methods for Photovoltaic Solar Cell Applications

Oumarou Savadogo

5.1
Introduction

The world net electricity generation has been estimated to increase 77% from 18 trillion kilowatt hours (kWh) in 2006 to 31.8 trillion kWh in 2030 with a value of 23.2 trillion kWh in 2015[1)]. On the other hand, the world market capacity of solar photovoltaic power systems escalated from 1.3 gigawatts (GW) in 2001 to 15.2 GW in 2008 for systems which have been installed. In particular, market installations reached a record high of 5.95 GW in 2008 corresponding to a growth of 110% from 2007. The contribution of solar energy to the world net electricity generation is estimated to be 6% by 2050.

The production of photovoltaic (PV) modules is still based mainly on crystalline silicon (Si) (94%) while 4% of modules are based on thin-film amorphous Si solar cells and 2% are polycrystalline compound solar cells based on CdTe and $CuIn_2Se$ [1]. Despite the tremendous progress in all aspects of production of Si-based solar cells and the rapid decrease of production cost for PV modules from $5 per peak watt at the beginning of the 1990s to $2.5 per peak watt in 2009, or $0.7 per kWh, this remains effectively too high. Subsidies from some government policies and/or the carbon dioxide market to increase the utilization of clean energy for sustainable development can contribute to reduce the PV energy cost to $0.25–0.40 per kWh during the first year of the system installation. This cost is similar to that of classic energy where energy cost is higher than $0.25–0.30 per kWh. For economic viability of this energy without subsidies, the development of ultralow-cost PV systems is one of the important issues to ensure a smooth transition to sustainable energy development. According to a recent US Department of Energy study in the USA [2], a major research effort is needed to close the huge gap between the current use of solar energy and its enormous underdeveloped potential. One of the identified thrusts of the

---

1) Sources 2006: Energy Information Administration, (Energy International Annual 2006 (June-December 2008):www.eia.doe.gov/iea, 2009. Projections: World Energy Projections Plus (2009).

required research is to bridge this gap by "more efficient solar cells created using nanotechnologies".

The main idea of thin films is to use mostly low-cost materials (glass, metals, and plastics) and very little high-cost semiconductors. This is because the few-micrometers-thick semiconductor thin films deposited on a substrate surface take typically about $2\text{--}6\,\mathrm{g\,m^{-2}}$, so even very expensive semiconductors (say, \$1000 per kg) can cost very little at this level leading to an energy cost of a few cents per kWh [1].

The significant challenge for solar light conversion to electricity or fuel is to develop PV, photoelectrochemical, or photochemical systems that exhibit combinations of efficiency and capital cost per unit area that result in a total power cost of less than \$0.50 per peak watt hydrogen produced by PV electrolysis. What is needed is to bring hydrogen cost down to \$0.05 per kWh, which is similar to the present cost of hydrogen from steam reforming of natural gas. A cost level of \$0.50 per peak watt would make solar light conversion very attractive for large-scale application of solar light conversion devices. A total solar cell cost of $\$130\,\mathrm{m^{-2}}$ with a light conversion efficiency of 50% are necessary to reach this peak watt cost. There is no current solar cell technology which meets these two requirements. From basic scientific principles, these requirements are feasible but the development of related commercial devices is facing major scientific as well as engineering development challenges. Based on the above discussion on the related cost reduction of thin-film solar cell systems, the development of devices based on nanoscale films should help respond to these challenges. Thin-film materials and nanomaterials are based on material layers the thicknesses of which range from monolayers of nanometers to several micrometers. Semiconductor materials based on electronic and solar devices and optical coatings are the main applications benefiting from thin-film fabrication.

Polycrystalline thin films can be prepared by chemical vapor deposition (CVD), as well as vacuum evaporation [3, 4], sputtering [5], molecular beam epitaxy [6], layer-wise chemisorptions [7], chemical vapor deposition [8, 9] and liquid-phase atomic layer expitaxy [10], physical vapor deposition (PVD) [11], spray pyrolysis [12], molecular beam epitaxy [13–15], low-pressure metal organic vapor-phase epitaxy (LPMOVE) [16, 17], successive ionic layer adsorption and reaction (SILAR) [18], pulsed layer deposition (PLD) [19], traveling heater (TH) [20], radiofrequency diode sputtering (RDS) [21], chemical wet deposition, and electrodeposition [22–25]. Each thin-film deposition technique has its own advantages and disadvantages. But chemical bath deposition (CBD), chemical solution deposition (CSD) or chemical wet deposition (WD), and electrodeposition are easier process methods for depositing thin-film semiconductors.

This chapter describes (i) materials and composite materials fabrication including some fundamental considerations of the fabrication conditions of some thin films; (ii) the development of thin films deposited with additives; (iii) the development of thin-film solar systems; and (iv) perspectives on thin-film solar cell technologies.

5.2
Materials and Composite Materials Fabrication

5.2.1
Fundamental Considerations

Three approaches can be used to prepare uniform films of semiconductor nanoparticles based on wet chemistry for solar applications. The first, CDB, takes place in a solution containing both atomic components (e.g., Cd^{2+} and Se^{2-}) from cation and anion sources with complexing agents in an electrolyte with controlled pH and temperature, or by chemical reaction of a second component (e.g., chalcogen or halogen) from its ion source following an electrodeposited layer of metal particles. In favorable cases, particularly when there is a good epitaxial match between the substrate and the semiconductor lattice, well-crystallized films can be formed. The second method, electrodeposition, involves particle self-assembly, in which preformed colloidal particles are attached layer-by-layer to a growing surface from cation and anion sources mixed in an electrolyte including complexing agents if necessary. In this case, the quality of the film depends on the properties of the nanoparticles used, as well as the attachment chemistry and, of course, the experimental conditions including current and/or potential control. The third method, sol–gel fabrication, is a direct layer-by-layer synthesis from the chemical components (e.g., anion and cation) of the semiconductor. This method offers the potential for achieving the finest control over the film growth process. The sol–gel method typically involves a two-step chemisorption/chemical activation cycle. One component is adsorbed or reacted chemically with molecules on the surface, but the reaction is self-limiting at the extent of a single monolayer. The chemisorbed monolayer is then activated in the second step, by reaction with an appropriate reagent or by redox reactions in the liquid phase.

5.2.1.1 Chemical Bath Deposition

Chemical bath deposition is a technique in which thin semiconductor films are deposited on substrates immersed in dilute solutions containing metal ions and a source of hydroxide, telluride, sulfide, selenide, etc., ions. One of the first chemically deposited semiconductors, reported in 1869, was a PbS thin film [26]. During the ensuing 140 years, CBD has been used to deposit films of metal sulfides, selenides, and oxides, and various other compounds. While it is a well-known technique in a few specific areas (notably photoconductive lead salt detectors, photoelectrodes, and, more recently, thin-film solar cells), it is by and large an under-appreciated technique.

The more recent interest in all things nano has provided a boost for CBD, since it is a low-temperature, solution-phase (almost always aqueous) technique, which often results in very small crystal sizes. This is evidenced by the existence of size quantization commonly found in CBD semiconductor films. The intention of this review is to provide an overview of how the technique has been used to fabricate nanocrystalline semiconductor films, as well as some of the properties of these films. Since CBD films are usually porous, surface effects can be very important with the result that various surface-dependent properties as well as surface modifications must be considered in film preparation. For CBD films used as photoelectrodes in photoelectrochemical cells and solid-state solar cells, it has been found that film structure, morphology, patterning, porosity, crystal shape, and method of preparation may have important effects on system performances and lifetime. The equipment, materials, and chemicals needed are relatively modest and less costly compared to the other methods (low-cost vessels, aqueous solutions containing few chemicals at very low concentrations, and substrates on which deposition is carried out). However, in comparison to the other thin-film technologies, the CBD method requires improved reproducibility in the properties of the materials that are obtained.

Accordingly, the number of materials produced by this method has increased steadily to include CdS ($Eg = 2.4–2.6\,eV$), $FeCdS_3$ ($Eg = 2.30\,eV$), $AgAlS_2$ ($Eg = 2.54\,eV$), CdTe ($Eg = 1.4–1.5\,eV$), CdSe ($Eg = 1.7–1.8\,eV$), ZnS ($Eg = 3.2–3.6\,eV$), ZnO ($Eg = 3–3.3\,eV$), ZnSe ($Eg = 2.70\,eV$), ZnTe ($Eg = 2.25\,eV$), PbS ($Eg = 0.37\,eV$), PbSe ($Eg = 0.27\,eV$), PbTe ($Eg = 0.29–0.4\,eV$), SnS ($Eg = 1.2–1.6\,eV$), Bi_2S_3 ($Eg = 1.70–2.35\,eV$), Bi_2Se_3 ($Eg = 0.8–1.2\,eV$), Sb_2S_3 ($Eg = 1.6–2.48\,eV$), Sb_2Se_3 ($Eg = 1.46\,eV$), Cu_2S ($Eg = 1.7\,eV$), CuS ($Eg = 2.2–2.4\,eV$), and CuSe ($Eg = 2.0–2.2\,eV$) among others. The ability to produce a wide range of materials is due primarily to the feasibility of producing multilayer films by this technique followed by annealing. The methods of CBD and electrodeposition most used in practice involve an aqueous solution. These chemical and electrochemical processes proceed through complicated and usually incompletely understood reaction mechanisms that involve the co-deposition of each element of the semiconductor material. Different types of wet chemical synthesis techniques are based on precipitation [27, 28], sol–gel [29], and colloidal systems [30, 31].

The solubility product (K_s) is the key parameter for understanding the basic aspects of CBD of thin-film semiconductors [12, 23]. The solubility product of a material is defined as the product of the concentrations of the ions of this material dissolved in a given electrolyte. In general the dissolution of an A_xB_y material is expressed according to

$$A_xB_y \rightleftarrows xA^{n+} + yB^{m-} \tag{5.1}$$

K_s is defined from the dissolution of this compound as

$$K_s = [A^{n+}]^x[B^{m-}]^y \tag{5.2}$$

This indicates that K_s increases with the solubility of the salt.

We may keep in mind that K_s is a thermodynamic parameter derived from the free energies of formation (ΔG^0) of the species involved in the dissolution of the solid compound $A_xB_y(s)$ into its ions in solution ($xA^{n+}(aq)$ and $yB^{m-}(aq)$) at thermodynamic equilibrium according to

$$A_xB_y(s) \rightleftarrows xA^{n+}(aq) + yB^{m-}(aq) \tag{5.3}$$

for which the free energy of dissolution at equilibrium is given by

$$\Delta G^0 = x\Delta G^0(xA^{n+}(aq)) + y\Delta G^0(yB^{m-}(aq)) \tag{5.4}$$

Since

$$\Delta G^0 = -RT \ln K_s \qquad (5.5)$$

then

$$K_s = e^{-\Delta G^0/RT} \qquad (5.6)$$

As a thermodynamic parameter, K_s will indicate the range of ion concentrations in which the compound precipitates. This serves to define what concentrations of ions are required in the deposition electrolyte to ensure (or avoid) compound precipitation. For example, for CdTe the K_s value is 10^{-42}, and therefore

$$[Cd^{2+}][Te^{2-}] = 10^{-42} \text{ M} \qquad (5.7)$$

This product means that if we want the dissolution of the solid to occur, we may consider situations where the concentration of each ion is equal. But for the formation of the solid compound, we may consider the product of the ion concentrations. For example, if the electrolyte contains 0.05 M of tellurium ions and 10^{-41} M of Cd^{2+}, CdTe will not precipitate, in principle, because the ion product is smaller than K_s; for an electrolyte containing 0.5 M of tellurium ions and 10^{-41} M of Cd^{2+}, CdTe will precipitate, in principle, because the ion product is greater than K_s.

Because K_s is a thermodynamic value, it may happen that, for kinetic reasons, the ion product which is supposed to result in precipitation may not actually result in solid formation. That is, the K_s value should be regarded as an indicator for conditions under which solid formation is, in principle, feasible.

In almost all cases, CBD is a technique for controlling the homogeneous precipitation of water-insoluble compounds and their solid solutions. Thus, for depositing thin films of a compound M_nX_m, a solution of M^{n+} ions, with a complexing agent (or ligand) L added to it, is prepared. Ligands that are generally used are NH_3, CN^-, ethylenediaminetetraacetic acid (EDTA), triethanolamine (TEA), and trisodium citrate, among others. The formation of complex ions $[M(L)_i]^+$ is the key factor in reaction control since they avoid the immediate precipitation of the metal ions in the solution when the precipitating anions are added to it. Moreover, because most of these processes are carried out in an alkaline medium, the metal-ion complexation process also avoids precipitation of OH^-, thus making film deposition possible. The precipitating agent of the complex is a compound which, upon hydrolysis, slowly generates the anions in the solution. For example, sulfur ions (S^{2-}) can be generated from thiourea or thioacetamide (TAM), and selenium ions (Se^{2-}) from Na_2SeO_3 or H_2SeO_3. The cations are, of course, generated by decomposition of the complex ions:

$$[M(L)_i]^{m+} \rightleftarrows M^{n+} + iL \qquad (5.8)$$

In the case of CdS, complex anion decomposition is described by

$$Cd(L)_n^{2+} \rightleftarrows Cd^{2+} + nL \qquad (5.9)$$

when the solution is heated, for example. In all cases, compound nucleation starts when the ionic product $[M^{n+}]^m[X^{m-}]^n$ exceeds the solubility product. It may continue to deposit slowly as a film in the immersed deposited ions in an aqueous ammonia medium on the substrates. For two non-interfering, independent complexing agents used to complex the two cations, the ions dissociates in an aqueous solution to give metal ions according the reactions [32]

$$M(A)_n^{2+} \rightleftarrows M^{2+} + nA \qquad (5.10)$$

and

$$M(B)_n^{2+} \rightleftarrows M^{2+} + nB \qquad (5.11)$$

When one complexing agent is used, only one of the above equations correctly applies; however, in the report cited, two non-interfering complexing agents were used.

Since thiourea has a higher dissociation constant, the fraction of S^{2-} ions in the solution is expected to be more than the fraction of thiourea in the solution. As is the case in an atom-by-atom deposition process, the solubility conditions of multicomponents in an ion-by-ion condensation process are relaxed [32].

Depending on the deposition experimental conditions, CBD can occur in the bulk solution by homogeneous nucleation, or on a substrate by heterogeneous nucleation. Undesired homogeneous nucleation is characterized by precipitate formation in the deposition electrolyte, which is to be minimized because it is detrimental to formation of the surface film. Heterogeneous nucleation occurs when ions are adsorbed onto the substrate, creating a site for reactions and promoting further growth of the film. The energy required for the homogeneous reaction is higher than that required to form an interface between the substrate and the ions. In optimal experimental conditions, heterogeneous nucleation is energetically preferred over homogeneous nucleation. In general heterogeneous nucleation is the first nucleation process to occur during CBD.

The different basic mechanism pathways shown above clearly suggest that the kinetics of CBD can vary from one chemical system to another depending on the material to be deposited and the experimental conditions. But in all cases the variation of the film thickness with time (e.g., Figure 5.1 shows the film growth of Sb_2S_3 [33]) is characterized by three main stages: induction, linear growth, and termination. The behavior of these stages depends on the reaction pathways. During the induction process, no film is deposited on the substrate. After a certain time of reaction, linear growth of the film with time is observed on the substrate, where most of the film is deposited, followed by a termination stage where no film growth is observed. In many cases, the color of the films is used to determine the starting of the linear growth stage because during the induction period, no film is deposited and the electrolyte is colorless. In the case, for example, of CdS formation, the yellow color of the film appears when its deposition on the substrate starts. During this linear stage some precipitates of CdS also form in the electrolyte, and its translucence is reduced and it becomes progressively cloudier. This corresponds to the occurrence of both homogeneous and heterogeneous nucleation processes during film growth on the substrate. For this system, the termination stage is characterized by a dark brown color of the electrolyte with large clusters of precipitates. The film-covered substrate is generally left in the electrolyte until the termination is over. Optimized experimental conditions are those

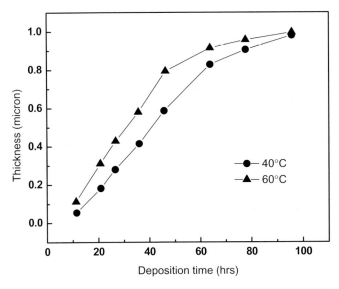

Figure 5.1 The variation of film thickness with time during the CBD of Sb$_2$S$_3$ [33].

which promote heterogeneous nucleation, leading to high-quality films with very low defect density.

As typical examples, we present in the following the mechanisms related to the CBD of cadmium sulfide and of antimony sulfide.

5.2.1.1.1 Case of CdS

Chemical bath deposition of CdS thin films was first reported by Mokrushin and Tkachev [34] and later by many other authors [35–41]. The basic mechanism of film formation was supposed to be either (i) an ion-by-ion condensation of Cd^{2+} and S^{2-} on the substrate from an aqueous basic medium, containing thiourea and cadmium ions in the form of a complex species, as indicated in Equation 5.11, or (ii) the result of the adsorption of colloidal particles of CdS onto the substrate surface. CdS film formation may be achieved using one of the following three complexing methods [42].

1) The tetraamine complex method [35–40], in which film formation follows the reaction

$$[Cd(NH_3)_4]^{2+} + SC(NH_2)_2 + 2OH^- \rightarrow CdS + 4NH_3 + OC(NH_2)_2 + H_2O \quad (5.12)$$

2) The cyano complex method [39]:

$$[Cd(CN)_4]^{2+} + SC(NH_2)_2 + 2OH^- \rightarrow CdS + 4CN^- + OC(NH_2)_2 + H_2O \quad (5.13)$$

3) The TEA (C$_6$H$_{12}$N(OH)$_3$) complex method [41], in which film formation is based on the reaction

$$[Cd(TEA)]^{2+} + SC(NH_2)_2 + 2OH^- \rightarrow CdS + TEA + OC(NH_2)_2 + H_2O. \quad (5.14)$$

It has been shown that the TEA process leads to high-quality films [43–45]. The mechanism involving the CBD of CdS thin films from the ammonia–thiourea system have been studied *in situ* by means of the quartz crystal microbalance technique (QCM) [25]. The formation of CdS was assumed to result from the decomposition of adsorbed thiourea molecules via the formation of an intermediate surface complex with cadmium hydroxide. This mechanism is different from the dissociation mechanism involving the formation of free sulfide ions in solution, and which had previously been reported [46–49]. Thus, the influence of growth parameters such as bath temperature, deposition rate, bath composition, etc., on various film properties has been studied [37, 39, 41, 50, 51], and the main parameters which determine the quality of the films were deduced. The chemical deposition of CdS thin films generally consisted of the decomposition of thiourea in an alkaline solution containing a cadmium salt. The deposition process was based on the slow release of Cd^{2+} and S^{2-} ions in solution which then condensed on an ion-by-ion basis on the substrate. The reaction process for the formation of CdS may be described by the following steps [25, 35, 36, 43, 52–54].

1) Ammonium ion formation:

$$NH_3 + H_2O \rightleftarrows HN_4^+ + OH^-; \quad K = 1.8 \times 10^{-5} \tag{5.15}$$

2) Cadmium salt reaction with the anions to form the complex compound:

$$Cd^{2+} + 2OH^- \rightleftarrows Cd(OH)_2; \quad K = 1.88 \times 10^{14} \tag{5.16}$$

$$Cd(Ac)_2 + 4NH_3 \rightleftarrows [Cd(NH_3)_4](Ac)_2 \tag{5.17}$$

3) Diffusion of the complex ion, OH^-, and thiourea on the catalytic surface of CdS. In the case where thiourea is the S^{2-} source in an alkaline medium, the sulfide ions are released as follows [25, 37, 52, 55, 56]:

$$SC(NH_2)_2 + 3OH^- \rightleftarrows 2NH_3 + CO_3^{2-} + HS^- \tag{5.18}$$

$$HS^- + OH^- \rightarrow S^{2-} + H_2O \tag{5.19}$$

In the case where H_2S is the S^{2-} source in an acid medium, the dissociation proceeds as follows [51]:

$$H_2S + H_2O \rightleftarrows HS^- + H_3O^+; \quad K_1 = 1 \times 10^{-7} \tag{5.20}$$

$$HS^- + H_2O \rightleftarrows S^{2-} + H_3O^+; \quad K_2 = 10^{-14} \tag{5.21}$$

4) Formation of CdS:

$$Cd(NH_3)_4^{2+} + S^{2-} \rightleftarrows CdS + 4NH_3; \quad K_3 = 7.1 \times 10^{28}. \tag{5.22}$$

Thus, the deposition of CdS occurs when the ionic product of $[Cd^{2+}]$ and $[S^{2-}]$ exceeds the solubility product (K_s) of CdS. The very low value of the solubility product of CdS (1.4×10^{29} at $25\,°C$) [52] implies that CdS precipitation can take place even at the lowest S^{2-} ion concentration that it is possible to obtain in solution. They are many ways of obtaining suitable complex species of Cd^{2+} ions in solution, but most of the literature dealing with the chemical bath deposition process is based on the tetraamine complex method where ammonia is the complexing agent and the hydroxide source, and NH_4Ac/NH_3 serves as a buffer. Thus, the global reaction for the process is given by

$$Cd(NH_3)_4^{2+} + SC(NH_2)_2 + 4OH^- \rightarrow CdS + 6NH_3 + CO_3^{2-} + H_2O \quad (5.23)$$

At a given temperature, the rate of formation of CdS is determined by the concentration of Cd^{2+} provided by $Cd(NH_3)_4^{2+}$ and the concentration of S^{2-} from the hydrolysis of $(NH_2)_2CS$. The rate of hydrolysis of $(NH_2)_2CS$ depends upon the pH and the temperature of the electrolyte. For example, at 80 °C, the rate constant of hydrolysis is at least twice as high at a pH of 13 (3.8×10^{-3}) than at a pH of 13.7 (8.2×10^{-3}). These constants increase significantly at 100 °C to 1×10^{-2} and 2.5×10^{-2} at a pH of 13 and 13.7, respectively [57]. From the various equations indicated above, the presence of an ammonium salt in the electrolyte will increase the concentration of $Cd(NH_3)_4^{2+}$ and reduce the concentration of Cd^{2+} and S^{2-} and, as a result, decrease the rate of CdS formation. By contrast, an increase in the concentration of ammonia will increase the pH of the solution, promoting the formation of S^{2-}. This will also increase the concentration of $Cd(NH_3)_4^{2+}$ and reduce the concentration of Cd^{2+}, and thus the rate of CdS formation.

As a result of these considerations, the rate of deposition of CdS may be controlled by varying the concentration of ammonia as well as the ammonium salt and the temperature of the electrolyte.

Solution-grown CdS films can occur on the substrate (heterogeneous deposition) or in solution (homogeneous precipitation). Homogeneous precipitation makes poorly performing films because they are formed from the adsorption of CdS particles, and not by ion-ion deposition directly onto the surface. The homogeneous process may be suppressed by choosing the conditions such that the formation of CdS occurs at low rates, such as low concentrations of $Cd(Ac)_2$ and $(NH_2)_2CS$, high concentrations of NH_3 and NH_4Ac, low temperature, vigorous stirring, etc. The heterogeneous process may be optimized by adequate preparation of the substrate surface for nucleation by cleaning and etching the conductive glass surface. For example, an indium tin oxide (ITO) glass substrate may be cleaned in ultrasonic baths of acetone, methanol, and isopropanol, followed by isopropanol vapor degreasing, a 60 s ultrasonic rinse in 50% HCl, and then a 120 s rinse in flowing deionized water. This nucleation is more effective on a crystalline substrate than on an amorphous substrate. As a result, film quality is strongly affected by the initial nucleation process. Thus, a SnO_2-coated glass substrate with a smooth surface may lead to discontinuous films, whereas a glass surface etched with hydrofluoric acid will significantly improve the continuity of films. The resistance of CdS films may be controlled by the addition of appropriate dopant salts. For example, during electrodeposition, indium chloride ($InCl_3$), boric acid (H_3BO_3), cuprous iodide (CuI), and cuprous chloride (CuCl), among others, may be used as the donor dopant. But the solubility of the dopant salt in the deposition electrolyte may lower the efficiency of the doping process. Based on the above considerations, CdS chemical deposition is influenced by various parameters. However, the CdS film properties may be improved by surface treatment activation in an appropriate electrolyte. For example, a CdCl$_2$ solution saturated in methanol may be used to activate the CdS surface before annealing in air at 200–400 °C for 5–30 min. By contrast, the stability of the deposited films still needs to be improved. This lack of stability is the key problem to be solved in the development of CdS-based thin-film solar cells. As a result, attempts to entirely eliminate the CdS layer in CIS/CdS solar cells are under way.

5.2.1.1.2 Case of Sb$_2$S$_3$

The steps of the mechanism related to the formation of Sb$_2$S$_3$ films are based on: (i) slow release of Sb^{3+} and S^{2-} ions from their respective sources, (ii) their respective precipitation into an aqueous ammonia medium, and (iii) their subsequent condensation on the substrate. The following successive chemical equations for the formation of the films have been suggested [33].

1) Complexation of the antimony cation:

$$SbO^+ + nTEA \rightarrow SbO^+(TEA)_n \qquad (5.24)$$

2) The tautomerization reaction of TAM:

$$\underset{CH_3-C-NH_2}{\overset{S}{\|}} \rightleftharpoons \underset{CH_3-C=NH}{\overset{SH}{|}} \qquad (5.25)$$

3) The formation of the SH$^-$ species:

$$\underset{CH_3-C=NH}{\overset{SH}{|}} + OH^- \rightleftharpoons \underset{CH_3-C=NH}{\overset{OH}{|}} + SH^- \qquad (5.26)$$

4) The formation of S^{2-}:

$$SH^- + OH^- \rightleftharpoons S^{2-} + H_2O \qquad (5.27)$$

5) The formation of the Sb$_2$S$_3$ films according to:

$$2SbO^+(TEA)_n + 3S^{2-} + 2H_2O \rightleftharpoons Sb_2S_3 + nTEA + 4OH^- \qquad (5.28)$$

The tautomerization reactions (5.25) should proceed from left to right for the next step of the reaction (5.26).

In the case of antimony sulfide [33], we have shown that the thickness of the n-Sb$_2$S$_3$ thin film increased with the concentration of the metal cations, for example potassium antimonyl tartrate (PAT) (Figure 5.2). It also increased with the anion sources, for example TAM (CH$_3$CSNH$_2$) (Figure 5.3) and the concentration of the complexing agent, for example ammonium (Figure 5.4). It was shown that the rate of the film deposition increased regularly or linearly up to a concentration of (i) 0.25 M PAT, (ii) 0.2 M TAM, and (iii) 1.4 M NH$_4$OH. All these curves are also characterized by three main stages: induction, linear growth, and termination. It was also shown that the rate of the film growth is slow at 40 °C (1.89 Å min^{-1}), and

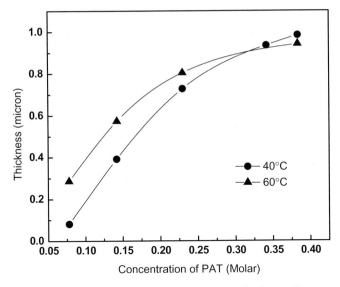

Figure 5.2 Effect of PAT concentration on the growth of Sb_2S_3 films at two different temperatures [33].

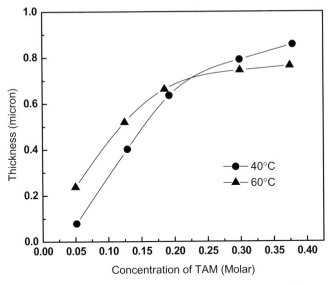

Figure 5.3 Effect of TAM concentration on the growth of Sb_2S_3 films at two different temperatures [33].

the growth rate is higher at 60 °C (2.8 A min^{-1}) in the linear stage region of the kinetics. This was explained on the basis of higher rates of release of Sb^{2+} and S^{2-} ions into the solution. These results are in agreement with the various points addressed above.

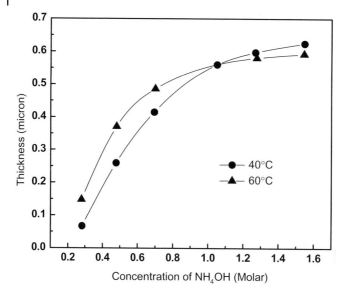

Figure 5.4 Effect of NH$_4$OH concentration on the growth of Sb$_2$S$_3$ films at two different temperatures [33].

5.2.1.1.3 Case of FeCdS$_3$ and AgAlS$_2$

The reaction bath for the deposition of FeCdS$_3$ (band gap = 2.30 eV) [58a] contained 5 ml of 0.2 M ferrous nitrate, 5 ml of 0.2 M cadmium chloride, 2 ml of 14 M ammonia, 2 ml of 0.01 M EDTA, 10 ml of 0.1 M thiourea, and 21 ml of distilled water which were added in that order and a deposition time of 20 hours was allowed for. The pH after mixing was controlled in the range 9–10 using drops of ammonia. If, during deposition, the semiconductor cations and anions precipitate rapidly by homogeneous reaction, then a thin film cannot form on the substrate immersed in the solution. However, if the homogeneous precipitation reaction is slow, which the additives like NH$_3$ and EDTA promote, then thin solid films of neutral atoms could form on the substrate. The complexing agents used slow down the precipitation action and enables the formation of FeCdS$_3$ according to the following steps:

$$Fe(NO_3)_3 \cdot 9H_2O + EDTA \rightleftarrows [Fe(EDTA)]^{2+} + (NO_3)^- \quad (5.29)$$

$$[Fe(EDTA)]^{2+} \rightleftarrows Fe^{2+} + EDTA \quad (5.30)$$

$$CdCl_2 + 4NH_3 \rightleftarrows [Cd(NH_3)_4]^{2+} + 2Cl^- \quad (5.31)$$

$$[Cd(NH_3)_4]^{2+} \rightleftarrows Cd^{2+} + 4NH_3 \quad (5.32)$$

$$(NH_2)_2CS + OH^- \rightleftarrows CH_2N_2 + H_2O + HS^- \quad (5.33)$$

$$HS^- + OH^- \rightleftarrows H_2O + S^{2-} \quad (5.34)$$

$$Fe^{2+} + Cd^{2+} + 3S^{2-} \rightleftarrows FeCdS_3 \quad (5.35)$$

Sulfide ions are released by the hydrolysis of thiourea, but Fe^{2+} and Cd^{2+} ions form ferrous–EDTA complex and tetraamine–cadmium complex ions by combining with EDTA and NH_3 respectively in the pH range of 9–10. The $[Fe(EDTA)]^{2+}$ and $[Cd(NH_3)_4]^{2+}$ complexes adsorb on the glass, then a heterogeneous nucleation and growth takes place by ionic exchange of reaction S^{2-} ions. This process is referred to as an ion-by-ion process and in this way brownish (reddish) yellow $FeCdS_3$ was deposited on a glass slide in the form of a transparent, uniform, and adherent thin film.

Similarly, non-interfering, independent complexing agents with a pH adjustment can be used to complex the two cations as shown in the following case [58b]. Thin films of $AgAlS_2$ (Band gap of 2.54 eV) [58b] have been prepared by the CBD technique from aqueous solutions of $AgNO_3 \cdot 5H_2O$, $Al_2(SO_4)_3 \cdot 14H_2O$, and $(NH_2)_2CS$ and in which sodium hydroxide solution was used as pH adjuster and EDTA as complexing agent. The thin films were coated on commercial microscope glass substrates by the reaction of a solution of 0.2 M $AgNO_3 \cdot 5H_2O$, 1.0 M $(NH_2)_2SC$, 0.1 M $Al_2(SO_4)_3 \cdot 14H_2O$, 1.0 M EDTA, and 2.0 M NaOH and enables the formation of $AgAlS_2$ according to the following steps:

$$AgNO_3 \cdot 5H_2O + (EDTA) \rightleftarrows [Ag(EDTA)]^+ + (NO_3)^- \tag{5.36}$$

$$Ag(EDTA)^+ \rightleftarrows Ag^+ + EDTA \tag{5.37}$$

$$Al_2(SO_4)_3 \cdot 14H_2O + EDTA \rightleftarrows 2[Al(EDTA)]^{3+} + 3(SO_4)^{2-} \tag{5.38}$$

$$(NH_2)_2CS + OH^- \rightleftarrows CH_2N_2 + H_2O + HS^- \tag{5.39}$$

$$2HS^- + 2OH^- \rightleftarrows 2H_2O + S^{2-} \tag{5.40}$$

$$Ag^+ + Al^{3+} + 2S^{2-} \rightleftarrows AgAlS_2 \tag{5.41}$$

The reaction bath was allowed to stand for about 44 hours and thereafter the substrate was removed, washed with distilled water, and dried in open air at room temperature.

5.2.1.2 Electrodeposition

In principle, both anodic and cathodic deposition may be used for the creation of II–VI compounds. It has been shown that CdS can be made by the anodic deposition of sulfur from a solution containing S^{2-} ions on a cadmium anode [59]. We have applied a similar method to CdS creation [60]; however, the stoichiometry of the deposit is not easy to regulate. By contrast, if each element of a semiconducting material (e.g., Cd and Te for CdTe, or Cd and S for CdS) can be deposited cathodically, it may be possible to deposit this semiconductor cathodically. Cathodic deposition can occur either by deposition of the individual components in the required ratio in a multi-step process, or by deposition resulting from discharge and decomposition of a complex containing the two components. The former process has been proposed for the electrodeposition of CdS [61], CdSe and Ag_2S [62], CdTe [63], $CuInSe_2$ [64], and Sb_2S_3 [65]. The latter process has been proposed for the cathodic deposition of Ni_3S_2 from a nickel thiosulfate solution [66]:

$$Ni_3(S_2O_3)_3 + 6e^- \rightleftarrows Ni_3S_2 + S_2O_3^{2-} + 2SO_3^{2-} \tag{5.42}$$

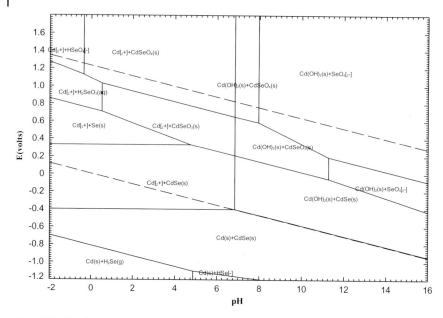

Figure 5.5 Pourbaix diagram showing corrosion, passivation, and immunity domains of CdSe [23].

The two different cathodic deposition processes can be differentiated by studying the current efficiency of deposition and the secondary products formed in solution [67]. The Pourbaix diagrams for CdS, CdTe, CuInSe$_2$, CdSe, and Sb$_2$S$_3$ based on thermodynamic considerations (Figures 5.5–5.9 respectively) show clearly that the voltage–pH region where stability of the corresponding materials can be obtained is very limited. In particular, thermodynamics considerations for the case of CdTe and CuInSe$_2$ electrodeposition are discussed below.

5.2.1.2.1 The Case of CdTe

The thermodynamic electrochemical reactions for the electrodeposition of CdTe and the corresponding potentials on the hydrogen scale are [63, 68–89] as follows.
For

$$TeO_2 + 4H^+ + 4e^- \rightleftarrows Te + 2H_2O \tag{5.43}$$

or

$$HTeO_2^+ + 3H^+ + 4e^- \rightleftarrows Te + 2H_2O \tag{5.44}$$

$$V_{TeO_2/Te} = 0.560 + 0.015 \log[HTeO^+] - 0.045 pH \tag{5.45}$$

The equilibria associated with CdTe stability include

$$Te + Cd^{2+} + 2e^- \rightleftarrows CdTe \tag{5.46}$$

$$V_{Te/CdTe} = 0.074 + 0.030 \log[Cd^{2+}] \tag{5.47}$$

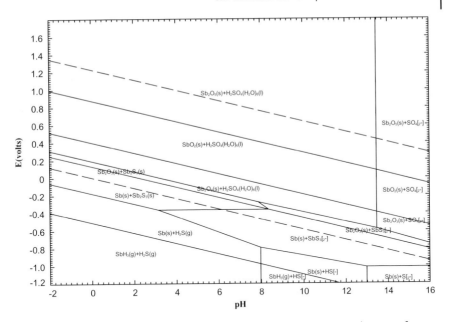

Figure 5.6 Pourbaix diagram showing corrosion, passivation, and immunity domains of Sb_2S_3 [23].

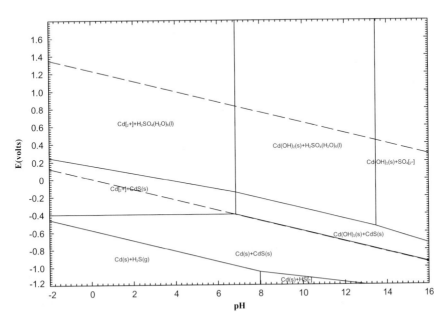

Figure 5.7 Pourbaix diagram showing corrosion, passivation, and immunity domains of CdS [23].

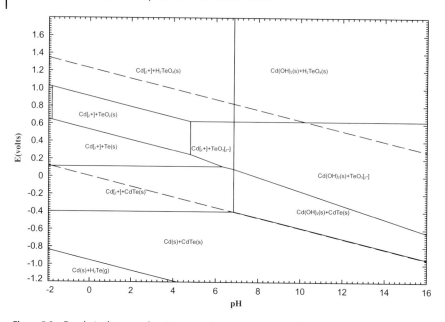

Figure 5.8 Pourbaix diagram showing corrosion, passivation, and immunity domains of CdTe [23].

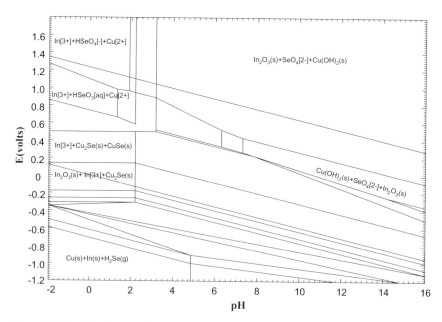

Figure 5.9 Pourbaix diagram showing corrosion, passivation, and immunity domains of CuInSe$_2$ [23].

$$Cd^{2+} + 2e^- \rightleftarrows Cd \qquad (5.48)$$

$$V_{Cd^{2+}/Cd} = -0.4025 + 0.059 \log[Cd^{2+}] \qquad (5.49)$$

$$Te + 2H^+ + 2e^- \rightleftarrows H_2Te \qquad (5.50)$$

$$V_{Te/H_2Te} = -0.740 - 0.0295 \log[H_2Te] - 0.059 pH \qquad (5.51)$$

$$CdTe + 2H^+ + 2e^- \rightleftarrows Cd + H_2Te \qquad (5.52)$$

$$V_{CdTe/H_2Te} = -1.217 - 0.0295 \log[H_2Te] - 0.059 pH \qquad (5.53)$$

For

$$Cd + Te \rightleftarrows CdTe \qquad (5.54)$$

the free energy of reaction may be as much as $-99.7\,kJ\,mol^{-1}$ [68]. In practice, CdTe is electrodeposited essentially from aqueous solutions [63, 68, 69]. Due to its complexity, various reaction mechanisms have been proposed in the literature. The majority of authors [72, 73, 82, 85, 87] assume that, in the first mechanism, the first step is the four-electron reduction of $HTeO_2^+$ to Te, according to Equations 5.43 and 5.44, followed by Equation 5.46.

The second mechanism [77] assumes that reactions (5.43) and (5.44) are followed by reaction (5.50), which produces H_2Te corresponding to a direct six-electron reduction:

$$HTeO_2^+ + 5H^+ + 6e^- \rightleftarrows H_2Te + 2H_2O \qquad (5.55)$$

which then reacts with Cd^{2+} in a chemical step to precipitate CdTe:

$$H_2Te + Cd^{2+} \rightleftarrows CdTe + 2H^+ \qquad (5.56)$$

The third mechanism involves a solid-state reaction between electrodeposited Cd and Te (5.37). A reaction mechanism which involves the presence of free sites with adsorbed species such as Cd^{2+}, $HTeO_2^{2+}$, Te, and CdTe on the surface has been proposed [86].

This is similar to the first mechanism described above, where surface sites, s, are involved. The first step is the reduction of $HTeO_2^+$ at a site:

$$HTeO_2^+, s + 3H^+ + 4e^- \rightleftarrows Te, s + 2H_2O \qquad (5.57)$$

The Te,s is then reduced to Te^{2-}, forming CdTe with Cd^{2+}, according to

$$Te, s + Cd^{2+} + 2e^- \rightleftarrows CdTe, s \qquad (5.58)$$

If reaction (5.58) is not fast enough, the following competing reaction will occur, which leads to the formation of an excess of Te in the films:

$$Te, s \rightleftarrows Te + s \qquad (5.59)$$

This model was developed to explain the decrease of limiting current when excess $CdSO_4$ is added to the solution. A model based on the assumption that Cd atoms are only incorporated at Te sites and Te atoms at Cd sites has been developed [71] to explain the case where the substrate is completely covered with CdTe. This last model supports the third mechanism of a solid-state reaction between electrodeposited Cd and Te. The first mechanism was obtained for electrodes that strongly adsorbed $HTeO_2^+$ [90–92]. Subsequently, Cd^{2+} reacts with the electrochemically crystallized Te via reaction (5.46).

The second mechanism appears to be important only for electrode surfaces that are more electrocatalytic than graphitic carbon. Thus, this route has been identified on Pt [67] and Au [90]. This direct six-electron route plays a dominant role in CdSe electrodeposition chemistry [93, 94]. The competition between the different reactions will control the overall composition of the film from stoichiometric tellurium- or cadmium-rich films [70], depending on both the nature of the substrate and the applied potential. For example, a sensitization (cathodic reduction) of thin oxide showed an improvement in the stoichiometry, optical properties (transmission), quantum efficiency, etc., of the films, but CdS was found to be an appropriate substrate for CdTe electrodeposition. The type (n-type or p-type) of the CdTe films depends on the electrodeposition potential. The electrolyte composition may have an important effect on film quality (uniformity, adherence, freedom from pinholes, etc.). Sulfate-based electrolytes furnish better quality films than chloride-based electrolytes. An appropriate Cl^-/SO_4^{2-} ratio may give good-quality films. After heat treatment, CdTe surface preparation with bromine/methanol etching, followed by etching in $(K_2Cr_2O_7)/H_2SO_4$, removes surface CdO layers to leave a highly Te-rich surface. Subsequent treatment by a reducing agent leaves an essentially oxide-free, Te-rich surface with a Cd:Te (reduced) ratio which can be 0.3. This surface must yield a good ohmic contact when Au is deposited onto it. However, there is still a fundamental problem of stability with CdTe grown on CdS, which is due to (i) the formation of $CdTe_{1-x}S_x$ at the interface, (ii) the degradation of the CdTe backcontact due to atmospheric interactions and/or corrosion, (iii) the diffusion of impurities in the films, and (iv) the lack of stability of the films themselves. These points have to be addressed to allow further improvement and long term stability of CdS/CdTe thin-film solar cells for practical applications.

5.2.1.2.2 The Case of CuInSe$_2$

The conditions for electrodeposition of $CuInSe_2$ are first and foremost related to the basic individual electrochemical reactions involved in the deposition of Cu, In, and Se, and to their corresponding Nernst equations, listed below [64, 95].

$$Cu^{2+} + 2e^- \rightleftarrows Cu \tag{5.60}$$

$$V_{Cu^{2+}/Cu} = V^0_{Cu^{2+}/Cu} + (RT/2F)\ln(a_{Cu^{2+}}/a_{Cu}) = 0.34 + 0.0295 \log(a_{Cu^{2+}}/a_{Cu}) \tag{5.61}$$

$$In^{3+} + 3e^- \rightleftarrows In \tag{5.62}$$

$$V_{In^{3+}/In} = V^0_{In^{3+}/In} + (RT/3F)\ln(a_{In^{3+}}/a_{In}) = -0.34 + 0.0197 \log(a_{In^{3+}}/a_{In}) \tag{5.63}$$

$$HSeO_2^+ + 4H^+ + 4e^- + OH^- \rightleftarrows H_2SeO_3 + 4H^+ + 4e^- \rightleftarrows Se + 3H_2O \tag{5.64}$$

$$V_{H_2SeO_3/Se} = V^0_{H_2SeO_3/Se} + (RT/4F)\ln(a_{HSeO_2^+}/a_{Se}) + (3RT/4F)\ln(C_{H^+})$$
$$= 0.74 + 0.0148 \log(a_{HSeO_2^+}/a_{Se}) - 0.0433 \text{pH} \tag{5.65}$$

where the electrode potential V is with respect to the normal hydrogen electrode, a_{Cu2+}, a_{In3+}, and a_{HSeO2+} are the activities of the respective ions in the solution, and a_{Cu}, a_{In}, and a_{Se} are the activities of the respective atoms in the electrodeposits, which are equal to 1 for pure elemental deposition.

According to the Nernst equations, the electrode potentials for Se and Cu are more positive than that of In. Thus, the deposition of Se and Cu will, of course, precede the deposition of In. Consequently, for the simultaneous deposition of Cu, In, and Se, we may adjust the pH and the concentration of the electrolyte such that the electrode potential of all the individual deposits may come closer to each other. For example, a higher concentration of In makes the electrode potential of In closer to that of Cu and Se. The presence of Cu in the electrolyte may enhance the deposition of Cu with Se.

By contrast, the dependence of the electrode potentials on the pH of the electrolyte may be used to determine the immunity and passivity domains of Cu, In, and Se. To obtain pure corrosion-free electrodeposits the experimental conditions should be permitted to be in the respective immunity domains of Cu, In, and Se. Thus, the governing factors in the co-deposition of the alloy compounds at any given temperature and current density are (i) the standard electrode potential of the electrode/electrolyte, (ii) cathodic polarization caused by a difference in deposition potentials, (iii) relative ion concentrations, (iv) the potential of the electrode oxide, and (v) hydrogen overpotential on the electrodeposited cathode. The effect of each parameter can be predicted, but if two or more conditions are varied simultaneously, it is more difficult to estimate the magnitude of the changes. Consequently, for the direct deposition of $CuInSe_2$, the pH of the electrolyte should lie between 0.5 and 9.5 and the deposition potential should preferably lie between −0.05 and −0.75 V vs. NHE. But these considerations are purely thermodynamic, because they ignore the polarization phenomena and modifications caused by the co-deposition of another element. For example, a decrease in the concentration of an ion does not change the equilibrium potential greatly, but can introduce significant concentration polarization. Thus, the use of dilute electrolytes generally leads to the formation of powdery or dentritic layers which are unusable in most practical applications. To avoid this problem, deposition potentials may be shifted by adding an appropriate complexing agent to the electrolyte. The stoichiometry of the film depends upon the solute concentration, the pH of the electrolyte, electrolysis current density, and deposition time. A careful analysis of the electrodeposition potential curve appears to be the best way to determine the ideal experimental conditions for the preparation of various definite compounds like $CuInSe_2$ by electrolytic co-deposition.

5.2.1.3 Sol–Gel Method

The sol–gel method is a wet chemical technique which is used for the fabrication of materials (ceramics, semiconductors, composites, etc.) starting from a chemical electrolyte which acts as the precursor for a gel with an integrated network of either particles or polymers. Precursors are metal alkoxides and metal chlorides which can undergo various forms of hydrolysis and polycondensation reactions [96].

Classically, this method allows the synthesis of materials (typically metal oxides like ZnO, TiO_2, Nb_2O_3, WO_3, etc.) at almost room temperature using preparation techniques different from the process of fusion of oxides. In the case of metal oxide fabrication, reactions involve connection of the metal centers with oxo (M–O–M) or hydroxo (M–OH–M) bridges, resulting in metal-oxo or metal-hydroxo polymerization in solution. Accordingly, the hydrolysis and the condensation reactions lead to the formation of a new phase (sol) according to (Figure 5.10) [96]:

$$M-O-R + H_2O \rightarrow M-OH + R-OH \text{ (hydrolysis)} \tag{5.66}$$

$$M-OH + HO-M \rightarrow M-O-M + H_2O \text{ (water condensation)} \tag{5.67}$$

$$M-O-R + HO-M \rightarrow M-O-M + R-OH \text{ (alcohol condensation)} \tag{5.68}$$

In sol–gel synthesis, a soluble precursor molecule is hydrolyzed to form a dispersion of solid colloidal particles, typically having diameters of a few hundred nanometers, suspended in a liquid (the sol phase). Further reaction causes bonds to form between the sol particles, which then condense to form a new phase (the gel phase) in which solid macromolecules are immersed in a liquid phase (solvent). The gel is then typically heated to yield the desired material. From the hydrolysis and condensation of the sol process it may be possible to get: (i) uniform particles, (ii) gels which can give (iii) aerogels from solvent extraction, (iv) xerogels from gelation or evaporation of the sol, (v) dense films from heat treatment of the xerogel, (vi) various types of fibers from the initial sol–gel mixture, and (vii) xerogels from the solvent evaporation of the gel, the evaporation of the xerogel leading to the formation of a dense ceramic. It is also possible to incorporate in the samples some soft doping agents. For systems involving colloids, the volume fraction of particles (or particle density) is so low that a significant amount of fluid must be removed to get the gel. This can be done by sedimentation, centrifugation, and phase separation [97].

Sol–gel chemistry has been developed to the point where it is a powerful approach for preparing initially inorganic materials such as glasses and ceramics [98] and also has been successfully applied to the synthesis of CdS, ZnS [98, 99], oxides based on Ti, Zr, Al, V, Mn, Co, Zn, W, B, etc. [100], and mixed Ti, Ta oxides [101] as thin films. This method for the synthesis of inorganic materials has a number of advantages over more conventional synthetic procedures. For example, high-purity materials can be synthesized at a lower temperature. In addition, homogeneous multicomponent systems can be obtained by mixing precursor solutions; this allows for easy chemical doping of the materials prepared. Finally, the rheological properties of the sol and the gel can be utilized in processing the material, for example by dip coating of thin films, spinning of fibers, etc. The concepts of sol–gel synthesis and template preparation of nanomaterials can be used to yield new routes for preparing nanostructures of semiconductors and other inorganic materials. This can be accomplished by conducting sol–gel synthesis within the pores of various micro- and nanoporous

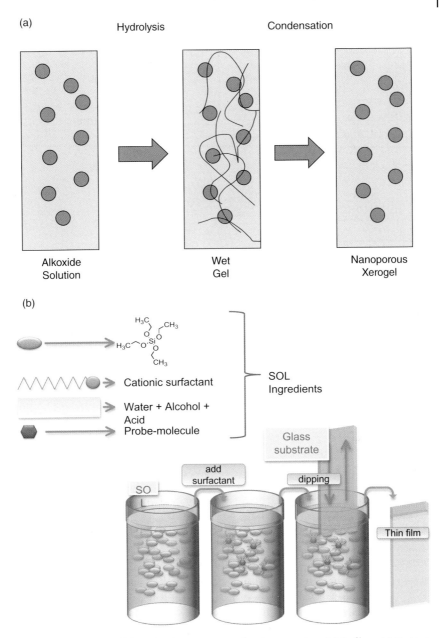

Figure 5.10 Overview of the sol–gel process. (a) Principle of hydrolysis and condensation of sol–gel process: case of alkoxide sol hydrolysis to wet gel condensation to give nanoporous xerogel. (b) Different steps of sol–gel process to get thin films. (c) Various steps from precursor dissolution to materials processing to obtain thin-film coating, powder, and dense ceramic materials from the sol–gel method.

(c)

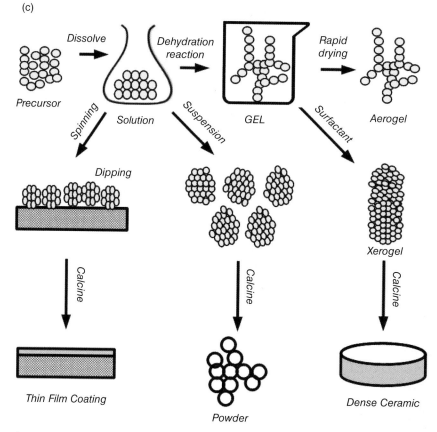

Figure 5.10 Continued

systems where monodisperse tubules and fibrils of the desired material can be obtained.

The relative composition and ratio of the solvent, the complexing agent, and the inorganic precursor have an important effect on the kinetics of the material preparation, and the various physicochemical properties [102]. For example, in the case of ZnO deposition using $Zn(CH_3COO)_2 \cdot 2H_2O$ as precursor in an ethanolic solution containing monoethanolamine (MEA) as complexing agent, where the ratio $[H_2O]/[Zn^{2+}] = 2$, the hydrolysis and condensation of the Zn^{2+} cation are relatively slow due to the low quantities of water. It was shown that MEA acts as a complexing agent and also retards the condensation of Zn^{2+}. The pH of the electrolyte was increased due to the presence of the MEA, promoting the formation of ZnO. The role of the acetate is to complex Zn^{2+} which is in competition with the MEA complexation [102]. Complex chemical relationships have been proposed [102] (Figure 5.11).

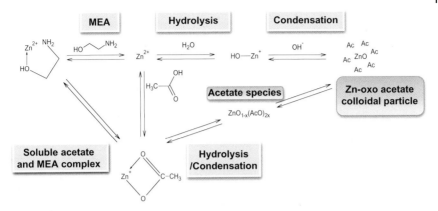

Figure 5.11 The complex chemical relationships of the main species and the chemical equilibria taking place in the initial solutions of ZnO formation by the sol–gel method [102].

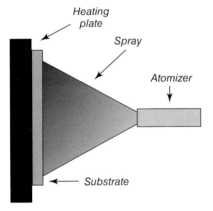

Figure 5.12 Schematic diagram of spray pyrolysis equipment [103].

5.2.1.4 Other Wet Methods

5.2.1.4.1 Spray Pyrolysis Methods

For solar cell materials and technology, spray pyrolysis is used for thin-film preparation. Film deposition based on this method consists of spraying a metal salt solution onto a heated substrate (Figure 5.12). The solution is usually made by dissolving salts of the constituent atoms of the desired compound in aqueous medium. The salt droplets impact on the substrate surface, spread into a disk-shaped feature, and undergo thermal decomposition. The shape and size of the disk depends on the momentum and volume of the droplets, as well as the substrate temperature. The film is then composed of overlapping disks of metal salt being converted into oxides on the heated substrate [103]. Typical spray pyrolysis equipment consists of an atomizer, precursor solution, substrate heater, and temperature controller. The following atomizers are usually used in spray pyrolysis: air blast (the liquid is exposed to a stream of air) [104], ultrasonic (ultrasonic frequencies produce the short wavelengths necessary for fine atomization) [105], and electrostatic (the liquid is exposed to a high electric field) [106]. Spray pyrolysis is a processing technique being considered in research to prepare thin and thick films, ceramic coatings, and powders. Unlike many other film deposition techniques, spray pyrolysis represents a very simple and relatively cost-effective processing method (especially with regard to equipment costs). It offers an extremely easy technique for preparing films of any composition. It does not require high-quality substrates or chemicals. The method has been employed for the deposition of dense porous films, and for powder production [103] including multilayered films. It has been also used for the production of materials in the glass industry [107] and in solar cell production [108].

Various parameters can influence the quality of the films. Among them, the co-solvent, the temperature, and the precursor solution are the main parameters that can have important effects on the film characteristics.

The deposition temperature is involved in all the processes mentioned above, except in the aerosol generation. Consequently, the substrate surface temperature is the main parameter that determines the film morphology and properties. By increasing the temperature, the film morphology can change from a cracked to a porous microstructure. Accordingly, the properties (optical, electrical, morphological, etc.) of deposited films can be varied and thus controlled by changing the deposition temperature.

It has been shown that the following processes will occur with increasing substrate temperature [109] (Figure 5.13). At lower temperature, droplets of the co-solvent and the precursor may splash onto the substrate and decompose (case A). When the temperature increases (case B), the solvent is evaporated during the flight of the droplet and gives a dry precipitate which is deposited on the substrate followed by some decomposition. At still higher temperature (case C), there is

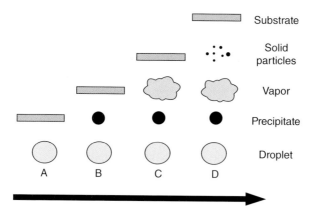

Figure 5.13 Description of the deposition processes initiated with increasing substrate temperature [103].

evaporation of the solvent before the droplet reaches the substrate, but in this case the solid precipitate is not decomposed; rather, it firstly melts, and then vaporizes. The resulting vapor diffuses to the substrate surface according to a CVD process. At the highest temperature of the process (case D), the droplet vaporizes before it can reach the surface. It is considered that solid particles are formed during a chemical reaction in this vapor phase. It is known that the films formed under the conditions of cases A and D are poorly or completely non-adherent. Adherent films are obtained when the process temperature used corresponds to case C. However, this case can rarely occur because, in general, the deposition temperature is too low to allow the vaporization of the droplet, which just decomposes without melting and vaporizing. In practice, processes corresponding to case B can allow the formation of high-quality films.

A variant of the spray pyrolysis method based on electrostatic spray-assisted vapor deposition (ESAVD) can also be used [110] in which the mixed co-solvent and precursor electrolyte is atomized by an electric field. The morphology of the thin films obtained from this method is very dependent on the process temperature (Figure 5.14) [110]. For example, in the case of CdS deposition [110] amorphous films are obtained below 300 °C (process I in Figure 5.14). Cadmium

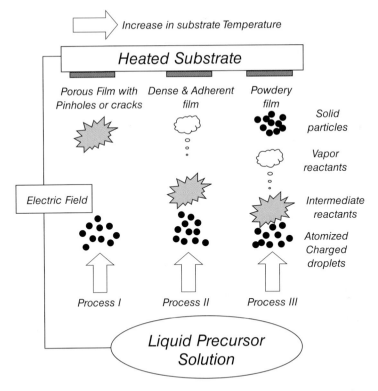

Figure 5.14 Schematic diagram of the proposed electrostatic spray-assisted vapor deposition mechanisms for different process conditions [110].

chloride ($CdCl_2$) and thiourea ((NH_2)$_2$CS) were used in the study cited. In particular $CdCl_2$ and (NH_2)$_2$CS with a molar ratio of 1:1 were dissolved in H_2O/C_2H_5OH solvent to produce 0.005 M homogeneous solution. In this case the films contain pinholes and cracks. At temperatures above 450 °C, the droplets of the mixed electrolyte are vaporized and decompose before the deposition of powder films on the substrate (process III in Figure 5.14). At intermediate temperatures, for example between 300 and 450 °C, the two types of the above film morphologies can be obtained. For an optimized temperature between 400 and 450 °C, a dense film with good adhesion (process II in Figure 5.14) can be obtained, corresponding to a heterogeneous CVD reaction.

Another method derived from spray pyrolysis is the ultrasonic spray pyrolysis method which consists of spraying a solution on a heated substrate. The apparatus of the deposition system typically used in this work is shown in Figure 5.15 [111]. Ultrasonic spray pyrolysis involves spraying a solution on a heated substrate. An electrical substrate heater and copper plate can be used to heat the substrate. Substrate temperature can be controlled by an appropriate thermocouple. This provides thermal energy for solvent evaporation and atomic rearrangement. The chemical electrolyte which contains in particular the precursor particles and the co-solvent is atomized into the stream of fine droplets (30–60 µm in diameter) via an ultrasonic spray head (containing a 58 kHz transducer). The aerosol droplets, generated by vibration of the transducer, are transported by a carrier gas to the heated surface of the substratre. The quality of the prepared films depends on, among others: (i) the electrolyte flow rate, (ii) the carrier gas pressure/flow rate, (iii) the distance between the ultrasonic spray head and the substrate, and (iv) the deposition time. Typical flow rates are around $1\,\text{ml}\,\text{min}^{-1}$ and the carrier gas

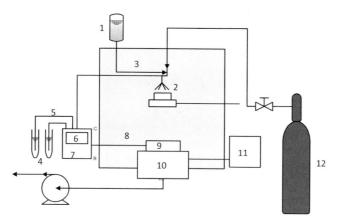

Figure 5.15 Ultrasonic spray pyrolysis system [111]: (1) container, (2) ultrasonic spray head, (3) flow meter, (4) water–ice bath for reference temperature, (5) thermocouple wire, (6) ultrasonic generator, (7) voltmeter for thermocouple, (8) iron–constantan thermocouple, (9) glass substrates, (10) heater, (11) current source, (12) nitrogen gas tank.

pressure can be 0.1 to 0.9 bar. Argon or nitrogen can be used as carrier gas and the deposition time is typically less than an hour. The optimum selection of experimental conditions also depends on the type and thickness of film.

5.2.1.4.2 Chemical Vapor Deposition (CVD)

The CVD method involves a chemical reaction in a vapor phase which results in deposition of a solid on a heated surface. Various PVD processes such as ion plating, sputtering, molecular beam, evaporation, and epitaxy might be also included in CVD processes [112]. The various CVD methods are powerful processes used for the fabrications of a wide variety of thin-film materials including solar cell materials and semiconductor materials for electronic applications, as well as the manufacture of coatings, powders, fibers, and monolithic components. There are two types of CVD reactors, the differential reactor and the starved reactor, according to the value of the flow rate (F_r) defined as

$$F_r = [\text{reactant out}]/[\text{reactant in}] \qquad (5.69)$$

If $F_r = 1$, we have a differential reactor which is characterized by a constant composition through the reactor. This type of reactor seems to be more useful for research activities.

If $F_r \ll 1$, we have a starved reactor and large gradients or fast reactions occur. This type of reactor is typically used for industrial applications. A schematic of the typical classic CVD process (Figure 5.16) [113] shows that the complete reactor is

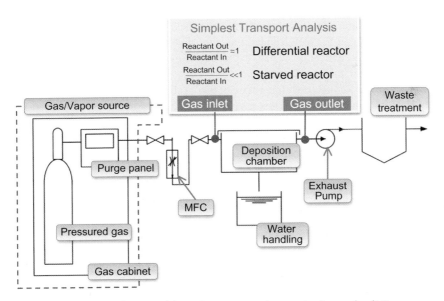

Figure 5.16 Schematic diagram of the CVD process method: (i) gas measurement and metering; (ii) transport of molecules by gas flow and diffusion; (iii) transport of heat by convection, conduction, and radiation; (iv) chemical reactions in the gas phase and at the surfaces; (v) plasma formation and behavior [113].

composed of: (i) a gas vapor system, (ii) a gas metering system (gas inlet and outlet determination, etc), (iii) a deposition chamber including a sample holding system, (iv) a gas transport analysis system, (v) an exhaust pump, and (vi) a waste treatment system. Accordingly for the proper management of a CVD process, it is important to develop and understand the following aspects related to the system: (i) the chemical reactions involved during the film deposition process, (ii) the mechanism of the transport of the species involved in these reactions, (iii) the mode of heat transport, (iv) the gas metering and flow measurements, and (v) the conditions of plasma formation and behavior and the characterization of the obtained films.

Based on molecular kinetics of gases colliding on a surface, the number of molecules impinging on a plane per square centimeter each second is [114]

$$J = n(8RT/\pi M)^{1/2}/4 \tag{5.70}$$

This results in the maximum possible deposition flux (J_{Max}) for a given partial pressure of precursor:

$$J_{Max} = (3.5 \times 10^{22} \times P)/(MT)^{1/2} \tag{5.71}$$

where n is the molar volume N/V, R the universal gas constant (8.3 J mol^{-1} K^{-1}), T the absolute temperature (K), M the molar mass (g mol^{-1}), P is in torr, and J is in molecules cm^{-2} s^{-1}.

Atmospheric pressure chemical vapor deposition (APVD) and jet vapor deposition of (JVD) of CdTe for high-efficiency solar cells have been developed by in an international project of the National Renewable Energy Laboratory [115]. All APCVD samples prepared for comparisons in this study had common processing steps through to CdTe deposition. Substrates used to make these cells were commercially available 3 × 5 inch soda-lime glass slides coated with SnO$_2$:F. A 2000 Å layer of CdS was sputter-deposited on these substrates, then pre-treated with CdCl$_2$ at 450 °C for 50 minutes in nitrogen. CdTe was deposited by APCVD at about 0.2 µm min^{-1} at a substrate temperature of 580 °C. The 3–5 µm thick CdTe was then treated with CdCl$_2$ for 35 minutes at 410 °C, followed by a 10 second etch in a 0.1% Br$_2$/methanol solution. Back contacts were formed by diffusing a thin layer (30 Å) of Cu into the CdTe, etching the surface to remove excess Cu, and then evaporating Au. APCVD samples used in these studies were either as-deposited, with CdCl$_2$ treatment, or with Au/Cu contacts.

5.2.1.4.3 Liquid-Phase Deposition (LPD)

The first experiment on the method of liquid-phase deposition (LPD) to produce thin films was reported by Nagayama et al. [116]. The method consists of depositing metal oxide thin films onto immersed substrates via a ligand exchange equilibrium reaction of metal–fluorocomplex ions and a fluorine-consuming reaction by the addition of fluorine scavengers, such as boric acid or aluminum metal. In particular, it has been reported that SiO$_2$ thin films can be produced by this new chemical method of LPD [116]. The advantages of this method are its low-temperature processing, its simplicity of operation, and inexpensive equipment requirements. It was found that the films prepared by the LPD method possess dense structure and good chemical stability. Also, the LPD method can be applied for large-area thin-film preparation which may involve using complex shapes, or on various kinds of substrates. As an example, TiO_2 films have been synthesized by LPD using ammonium hexafluorotitanate and boric acid as treatment solution, with $[TiF_6]^{2-}$ to H_3BO_3 molar ratios of 1:2 to 1:4 as the most suitable for the formation of these thin films [117–123]. The clean substrates are immersed into the treatment solution for 48 hours. The reaction is summarized as follows [124]:

$$[TiF_6]^{2-} + nH_2O \Leftrightarrow [TiF_{6-n}(OH)_n]^{2-} + nHF \qquad (5.72)$$

$$H_3BO_3 + 4HF \Leftrightarrow HBF_4 + 3H_2O \qquad (5.73)$$

An example of a general flow chart of a TiO_2 thin-film preparation which is often used for material processing is shown in Figure 5.17 [123]. The disadvantages of the LPD method are: (i) films can only be prepared on substrates with OH^- radicals and (ii) the growth rate of the films is slow. This method is very attractive in theoretical and industrial areas of research [125–135], in part owing to fundamental understanding of the deposition mechanism [136, 137], and in part to the

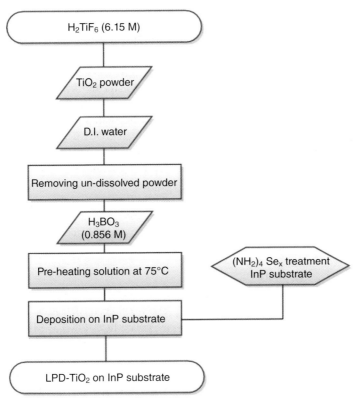

Figure 5.17 Flow chart for the preparation of TiO_2 films by the LPD method (adapted from [123]).

versatility of the method for the formation of other oxides, including those of Ti, Sn, Zr, V, Cd, Zn, Ni, and Fe [138–145].

5.2.1.4.4 Solution Deposition by Successive Ionic-Layer Adsorption and Reaction (SIAR)

This process, introduced in the mid-1980s [146], is intended to grow polycrystalline or epitaxial thin films of water-insoluble ionic or ionocovalent compounds, C_mA_n, by heterogeneous chemical reaction at the solid–solution interface between adsorbed C^{n+} cations and A^{m-} anions. The process involves an alternate immersion of the substrate in a solution containing a soluble salt of the cation of the compound to be grown and then in a solution containing a soluble salt of the anion (Figure 5.18) [146]. The substrate supporting the growing film is rinsed in high-purity deionized water after each immersion. Polycrystalline and epitaxial thin films of ZnS and CdS have been deposited on different substrates following this process at room temperature [146]. This method has also been used for the deposition of ZnO [147] and Cu_2O [148]. A recent review of this approach for gas sensing has been published [149]. This method has a high potential for thin-film semiconductor deposition for solar cell applications. Its utilization for such applications should be emphasized [150].

5.2.1.4.5 Electroless Deposition

Electroless deposition takes place by an autocatalytic redox process, in which the cation of the metal to be deposited is reduced by a soluble reductant at the surface

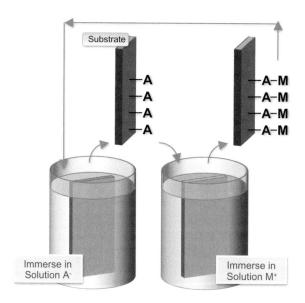

Figure 5.18 Schematic diagram of SILAR. M^+ refers to a cation (usually a metallic precursor); A^- an anion; X-L a precursor compound, where X is the nonmetallic species (generally a group VI element) and L a ligand.

of the metal feature being formed, or at the surface of catalysts used to initiate the deposition [151, 152]. To occur, the redox process requires a catalyst. In the case of noncatalytic substrates, the surface first has to be activated with a metal catalyst such as palladium [153]. Selective deposition can be achieved either by the selective deactivation of a catalytic substrate or by the selective activation of a nonreactive surface by a catalyst [154]. Electroless deposition can operate at ambient temperatures and pressures, and represents a simple and economic method of thin-film deposition. This technique has been used to deposit thin films of β-FeOOH and Fe_3O_4 on amine-functionalized silicon that had been immersed in Pd salt solution for sensitization, dimethylamine borane complex (DMAB; CH_2NHBH_3) in the deposition serving as the reducing agent [155]. .Further studies of the structure of the support–semiconductor interface produced could therefore provide information regarding both the fundamental growth mechanism and the viability of such a technique for manufacturing processes of thin-film semiconductors for solar cell applications.

5.2.2
Preparation of Active Materials

5.2.2.1 Preparation by Chemical Deposition

Several approaches based on CBD have been used for the development of thin-film semiconductors for solar cell applications [22–24] of typically 0.02–1 μm thickness. From these references and those therein, the binary compounds that have been fabricated by CBD include sulfides (CdS, Cu_xS, CoS, PbS, Sb_2S_3, PdS_2, SnS, SnS_2, ZnS, NiS, TlS In_2S_3, HgS, MnS, MoS, Bi_2S_3, Ag_2S, As_2S_3, etc.) and selenides (CdSe, Cu_xSe, CoSe, PbSe, Sb_2Se_3, SnSe, $SnSe_2$, ZnSe, NiSe, HgSe, MoSe, Bi_2Se_3, Ag_2Se, As_2S_3). In some cases metal hydroxide thin films are deposited and subsequently converted into oxide films [156]. To obtain good-quality and cheaper semiconductor thin films, several optimized experimental conditions have been developed and the influence of various factors such as temperature, concentration, stirring, and others has been studied. More fundamentals studies related to thin-film chemical deposition have also been published [157, 158]

Studies of the kinetic parameters of thin-film deposition rates and performance are very important to develop appropriate films for solar cell applications. For example, substrate spacing and thin-film yield in CBD of semiconductor thin films have been studied [159] (Figure 5.19). Based on a mathematical model, it was proposed and experimentally verified in the case of CdS thin films that the film thickness reaches an asymptotic maximum with increasing substrate separation. The model takes into account the following several factors:

1) The effect of initial concentration (C_i) of the metal ions M^{n+} in the bath, which influences the kinetics of film formation through a first-order dissociation of metal–ligand (M–L) complex [ML_y] (rate constant, k), leading to the formation of metal chalcogenide M_qX_r. It was assumed that chalcogenide (X) ions are readily available in the bath.

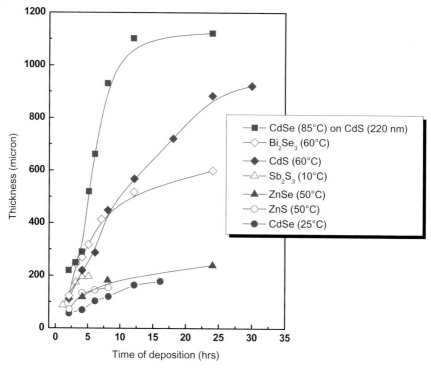

Figure 5.19 Film thickness versus duration of deposition for different semiconductor thin films obtained by the CBD technique [159].

2) Precipitate formation (γ), which is nonlinear with concentration and temperature (T).

3) A geometry factor, which takes into account the ratio of substrate area to volume of the solution used (s).

4) An induction period (t_i) for the setting up of the relevant chemical equilibrium in the bath and the commencement of formation of insoluble compound M_qX_r.

5) An induction period (t_c) for the formation of the catalytic surface on the substrate for film formation. The quantity (C_f) of the soluble metal ions M^{2+} utilized in the formation of the thin film of insoluble semiconductor metal chalcogenide compound M_qX_r at the end of a duration t of the deposition process is:

$$C_f(t) = C_i[1-\exp(-\alpha s)](1-\gamma)\times[1-\exp(-t/t_c)]\{1-\exp[-k(t-t_i)]\} \qquad (5.74)$$

The interested reader is referred to [159] for a detailed explanation of the model, which is summarized in table 1 of that work.

This approach involves considerations of surface catalysis which occurs prior to thin-film deposition. In this model the induction period for formation of a catalytic surface for thin-film growth t_c is the probe of the surface catalysis. From Equation 5.74, it can be seen that there is no film deposition when no catalytic surface is created on the substrate surface, that is, $t_c \to \infty$, and facile thin-film formation when a catalytic surface is provided, $t_c \to 0$. On the other hand, when $\gamma \to 0$, no precipitate formation occurs in the electrolyte and a catalytic surface is provided on the substrate. The model is based on only the effect of the concentration of metal ions and not on each of the components (anion concentrations, supporting electrolyte, pH adjustment, etc.), all of which play an important role in the growth rate and film thickness as indicated elsewhere [23, 25, 65].

5.2.2.1.1 CdS and CdSe Preparation

Chemical deposition of CdS may be a highly reproducible and controllable technique which yields CdS layers with excellent optoelectronic and solar cell properties. The deposition of CdS films has been carried out on various substrates using different solution compositions [35–38, 40, 42, 160–170]. The electrical conductivity and thickness of deposited In-doped CdS from an ammoniacal solution of $InCl_3$, $CdSO_4$, and thiourea were found to vary with the concentration on $InCl_3$ in solution [171]. However, due to the fact that $In(OH)_3$ is insoluble in water and that, unlike $Cd(OH)_2$, $In(OH)_3$ is also insoluble in ammonia, the incorporation of In into CdS is unlikely. The deposition of CdS films from an aqueous solution containing cadmium acetate ($Cd(Ac)_2$), thirourea (($NH_2)_2CS$), ammonia (NH_4OH), and ammonium acetate (NH_4Ac) has shown that the addition of boron into the solution reduces the resistivity of CdS films. A minimum in dark resistivity was observed at a BO_3^{2-}/Cd^{2+} ratio of 0.001, as the boron concentration increased and approached that of undoped films [53]. Optimum conditions for better overall morphological and optical properties have been investigated [172]. Better conditions for good morphological and optical properties were: $[Cd^{2+}]$, (0.1–3) $\times 10^{-3}$ M; thiourea, (3–5) $\times 10^{-3}$ M; buffered solution, 20×10^{-3} M; pH 9.0–9.5; temperature, 85 °C; and time, 13–30 minutes. Significantly lower precipitated particulate density was obtained using the buffered solutions under optimum conditions and the films were stoichiometric, grew a strong (111) or (002) texture, and had a direct bandgap of 2.40 eV. CdS thin-film photoconductivity has been determined for films prepared from aqueous chemical baths containing a TEA complex of Cd^{2+} ions and thiourea, mixed in different nonequimolar ratios [43, 45]. This led to the conclusion that the TEA complex method results in the production of high-quality films. The chemical deposition of CdS from solutions at 50–70 °C containing citratocadmium complexes and thiourea has been reported [173]. The films showed an optical transmittance of about 80%, a dark conductivity of 10^{-8} cm^{-1}, and photosensitivity of about 106–107 at 1 kW m^{-1}. The films were converted to n-type by annealing in air at 400–500 °C for an hour. It has also been shown that the impurity phase in the CdS films involves a mixture of cadmium oxide (CdO) [174]. Analysis of the films and the powders obtained using thiourea and TAM as sulfurizing agents showed that the impurity phase is predominantly present when thiourea is used in the chemical bath. The CBD of CdS layers, using the ammonia process, has been studied by combined *in situ* quartz crystal microbalance and electrochemical impedance techniques [46]. The film was shown to have, in general, a duplex structure with an inner compact layer and an outer porous layer, growing at longer reaction times. An excess of thiourea was found to be very favorable for obtaining total coverage of the substrate with a minimum thickness of the CdS film (about 30 nm on an unactivated Au substrate). For various optoelectronic or solar cell applications, it is vital to determine the conditions of stability of CdS thin films related to the mechanism of their deposition.

The cathodic electrodeposition of CdS from aqueous solutions of Cd^{2+} and $S_2O_3^{2-}$ at 90 °C has been investigated [67, 175–183]. The electrodeposition was carried out using a standard three-electrode configuration in galvanostatic mode. CdS deposition was studied on Pt wire and ITO electrodes. The reference electrode was saturated calomel and the counterelectrode was a spectrographically pure carbon rod of geometric area of 16 cm². The CdS electrolyte was prepared by heating a 0.2 M $CdCl_2$ solution to 90 °C, followed by adding, immediately prior to commencing CdS deposition, sufficient $Na_2S_2O_3 \cdot 5H_2O$ to produce 0.009 M $S_2O_3^{2+}$ in solution. The deposition potential was −0.695 V, giving a current density of the order of 0.06–0.10 mA cm^{-2}, which subsequently decreased. The potential was then adjusted to maintain the current density in the range 0.035–0.045 mA cm^{-2}. The deposition charge over a one-hour period was approximately 5.4 C (about 0.135 C m^{-2}). A lower pH increased the rate of the electrochemical process as shown by the increase in CdS film thickness as the pH decreased. A pH of about 3.5 yielded the greatest thickness [175]. CdS was electrodeposited from a stirred solution at 90 °C in which the electrolyte was 0.2 M Cd^{2+} and 0.01 M $S_2O_3^{2+}$ with the pH adjusted to 2 [64, 184, 185]. The deposition potential was +40 mV from the measured Cd deposition potential. It was evident that electrodeposited CdS thin films were of better quality and more adherent than those obtained with CBD [64].

Electrodeposition is less used than CBD for the fabrication of CdS thin films. This may be due to the lower rate of film formation by electrodeposition. Recently, thin films of CdS were prepared by CBD using potassium nitrilotriacetate (K_3NTA or NTA) as the complexing agent for Cd and thiourea as the S source [186–188]. Most crucial is pH adjustment, which governs the formation of a cadmium hydroxy species and the start of thiourea desulfuration [188], while the influence of illumination is only marginal. It was shown that the use of NTA for the deposition of CdS, where it serves as the complexing agent for Cd, eliminates the problems of volatility and toxicity of the commonly used ammonia. Moreover, the NTA-based CBD leads to smaller sizes of the prepared nanocrystals. Their size can be increased by subsequent thermal annealing. This can be used for top growth on a variety of substrates including coated glass, semiconductors, metals, polymers, and cloth.

Other typical deposition conditions for CdS have been reported [24, 173]. They are based on the use of 30 ml of 0.1 M $Cd(CH_3COO)_2 \cdot 2H_2O$, 8–12 ml of 1 M sodium citrate, 15 ml of 1.5 M NH_4OH, and 5–10 ml of 1 M thiourea, with deionized water to make the volume up to 100 ml. The deposition is allowed to proceed at 50–70 °C for up to 12 hours. Very smooth and uniform coatings of CdS are formed except at long durations of deposition [24, 173].

Thin films of CdSe were prepared by CBD using the NTA complexing agent for Cd and sodium selenosulfate (Na_2SeSO_3) as the Se source [186, 189–191]. Careful attention is needed for the preparation of Na_2SeSO_3, which is done by dissolving elemental Se powder in Na_2SO_3 [189], because this step is usually the major source of growth irreproducibility. It is also very important to maintain a constant illumination while films are growing, because the preparation of CdSe is very sensitive to the intensity and the spectral content of the incident light. One of the classic methods for CdSe preparation, reported by us [23, 192] and others [193, 194], is based on 30 ml of 0.1 M $Cd(NO_3)_2 \cdot 4H_2O$, 12 ml of 1 M sodium citrate, 1.2 ml of 30% (about 15 M) NH_4OH, and 0.4 g of N,N-dimethylselenourea dissolved in 30 ml of freshly prepared 0.01 M Na_2SO_3, with the volume made up to 100 ml with water. Deposition may be made at room temperature (24 hours) or at temperatures up to 60 °C (8 hours); for prolonged deposition periods there is a risk of the film peeling from the glass substrates.

5.2.2.1.2 Cu_xS and Cu_xSe Preparation

Thin films of Cu_xS possess near ideal solar control characteristics with a bandgap of 1.2–2.2 eV [23, 24, 195]. Accordingly, Cu_xS thin films have received particular attention in the development of the experimental conditions for their CBD. A large variety of Cu_xS (with x from 1 to 2) films have been deposited by chemical methods including Cu_2S films [196]. Cu_xS thin films with a wide range of sheet resistances (r) and optical transmittances, indicating different composition x, have been obtained from chemical baths constituted of copper(II) chloride, TEA, and thiourea at appropriate pH (10–12). Depending on the deposition parameters, a range of resistances from 30 Ω to 1 MΩ, a range of transmittances (at 500 nm) from 1 to 65, and a range of color of reflected daylight (golden yellow, purple, blue, green, etc.) can be obtained [197]. The films have been found to be stable with respect to electrical and optical properties on storage under ambient. The deposition of Cu_xS films is made possible by the presence of the Cu^+/Cu^{2+} and S^{2-} ions available in the chemical bath. The S^{2-} ions are easily provided through the hydrolysis of thiourea in an alkaline or ammoniacal (normally of pH 10–12) bath, as reported for CdS CBD [198]. To get Cu_xS films with x closer to 2 (typically 1.75 and 2), Cu^+ salts such as CuCl or CuCN which are highly insoluble in water and acidic media (pH about 3.5) are used following their topotaxial formation from CdS and $CdCl$ [199]:

$$CdS + 2CuCl \rightarrow Cu_xS + CdCl_2 \tag{5.75}$$

This method is different from the classic CBD. To enhance the direct CBD rate, the equilibrium reaction [200]

$$2Cu^+ \rightarrow Cu + Cu^{2+} \tag{5.76}$$

could be displaced in either direction through complexation, depending on the complexing agent. For example, the equilibrium constant $K = [Cu^{2+}]/[Cu^+]$ is approximately 10^5 for the chelating ethylenediamine (EN), which will drive the reaction forward; for the nonchelating ammonia, K is approximately 10^{-2}, which can promote the reverse reaction, thus giving rise to $[Cu(NH_3)_2]^+$. Formation of $[Cu_4(SC(NH_3)_2)]_6^{4+}$ and $[Cu_4(SC(NH_3)_2)]_6^{4+}$ complex ions from aqueous solutions of copper(II) salt and thiourea has been reported at acidic pH (=1). Those complexes may affect the film deposition kinetics [201].

Copper sulfide (Cu_2S) thin films were deposited using a modified chemical deposition method. The preparative conditions such as concentration, pH of cationic and anionic precursors, adsorption, reaction and rinsing time durations, complexing agent, etc., were optimized to get stoichiometric Cu_2S thin films [202]. This method is based on successive immersion and rinsing after each immersion of the substrate into separate cation and anion precursor solutions with ion-exchange water to avoid homogeneous precipitation. A schematic representation for the deposition of Cu_2S thin films by a modified chemical method is shown in Figure 5.20 [202], which indicates that it is different from the classic method because the cationic precursor is separated from the anionic precursor by an ion-exchange water bath. The chemicals used as the cationic and the anionic precursors in such a method are, respectively: (i) copper(II) sulfate pentahydrate ($CuSO_4 \cdot 5H_2O$) solution complexed with a mixture of 2 N TEA and 2 N hydrazine hydrate (HH), the

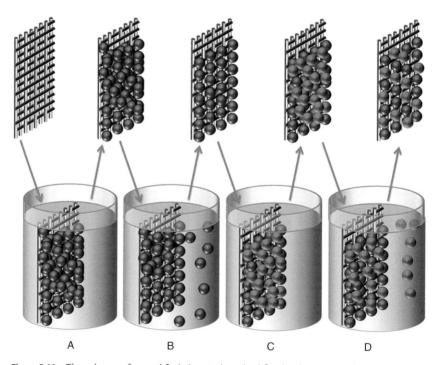

Figure 5.20 The scheme of a modified chemical method for the deposition of Cu_2S films onto glass substrates: A, cationic precursor (copper(II) sulfate pentahydrate); B, ion-exchange water; C, anionic precursor (sodium sulfide); D, ion-exchange water [202].

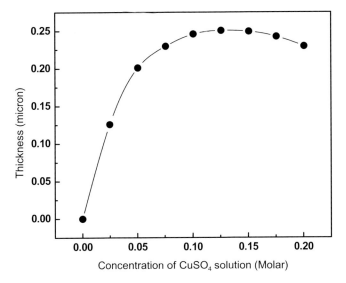

Figure 5.21 Plot of film thickness against concentration of CuSO$_4$ solution for 20 deposition cycles for Cu$_2$S thin film [202].

pH of this solution being adjusted to about 5; and (ii) sodium sulfide (Na$_2$S·H$_2$O) solution with pH ≈ 12. A concentration of 0.05 M for the sodium sulfide solution as anion source and highly purified deionized water as ion-exchanging water were used. The variation of the film thickness against concentration of CuSO$_4$·5H$_2$O solution for 20 deposition cycles (Figure 5.21) showed that the film thickness increased with copper ion concentration. The maximum thickness value at a concentration of 0.12 M was interpreted as the formation of an outer porous layer as the film peeled off from the substrate. Similarly the Cu$_2$S film thickness was found to increase with the number of deposition cycles (Figure 5.22) for optimized concentrations of CuSO$_4$·5H$_2$O (0.12 M) and sodium sulfide (0.05 M), with a maximum film thickness of 0.44 μm.

It is well established that copper selenide usually exists as the copper(I) form (CuSe or Cu$_3$Se$_2$) [202–205]. Copper(II) selenide in Cu$_3$Se$_2$ form is often reported as an impurity phase along with CuSe [204, 205]. Face-centered cubic (FCC) structures of Cu$_x$Se thin films have been chemically prepared using copper acetate and sodium selenosulfate solution onto Pt(111) substrate at 75.2 °C [206]. Copper selenide (Cu$_2$Se or Cu$_{2-x}$Se) films have also been deposited using tetraamine copper and sodium selenosulfate solution onto glass substrates in an alkaline medium at 55 °C [207]. The deposition time was 35 minutes. The optical bandgap was found to be 1.39–1.44 eV. The films were p-type. Cu$_{2-x}$Se films have also been deposited using CuCl$_2$ and Na$_2$SeSO$_3$ solutions complexed with TEA and NH$_4$OH onto glass substrate in an alkaline medium at 90–95 °C for 30 minutes [208]. The optical bandgap was found to be 1.2 eV. The electrical resistivity was of the order of 10^{-1} Ω cm. The interphase conversion of chemically prepared Cu$_{2-x}$Se and

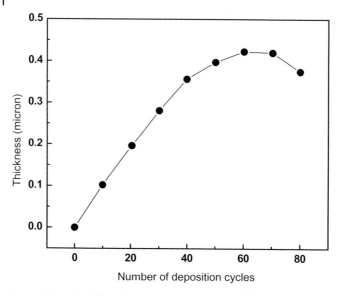

Figure 5.22 Plot of film thickness against number of deposition cycles for Cu_2S thin film [202].

tetragonal Cu_3Se_2 thin films has been studied [209]. This interphase conversion was found to be cyclic. The bandgap of the film was varied from 2.27 to 2.84 eV. Cu_3Se_2 and CuSe thin films have also been deposited using copper sulfate and sodium selenosulfate solutions at 60 °C on polyester substrate [210]. These various results indicate clearly that the film deposition conditions have important effects on film properties (composition, bandgap, structure, conductivity, etc.) and the resulting performance of the films in solar cell applications. Accordingly, special attention must be paid optimizing experimental conditions for thin-film solar cells applications.

In addition, there are other less expensive and simple modified chemical methods that are suitable for the deposition of metal chalcogenide thin films [203]. These are mainly based on the adsorption and reaction of ions from solutions and rinsing between every immersion step with deionized water to avoid homogeneous precipitation in the solution. As an example, stoichiometric copper(I) selenide (Cu_2Se) thin films have been deposited onto glass substrate using this modified chemical method. The deposition conditions such as concentration and pH of cationic and anionic precursor solutions, immersion and rinsing times, number of immersions, etc., were optimized for Cu_2Se films [203]. The cationic precursor was a solution of 0.1 M copper(II) sulfate ($CuSO_4 \cdot 5H_2O$) with the pH adjusted to 2 by adding tartaric acid. The anionic precursor was 0.05 M sodium selenosulfate (Na_2SeSO_3) solution with a pH of 12. The solutions were placed into separate beakers of 50 ml capacity each. An ample quantity of distilled water was used for the rinsing step. The deposition was carried out at room temperature (300 K) from unstirred baths. The concentration, pH, and temperature of precursor solutions and the time for adsorption, reaction, and rinsing are important parameters.

One deposition cycle consists of: (i) immersion of the glass substrate in the cationic precursor solution for 30 s to allow copper ions to adsorb on the surface of the glass substrate; (ii) rinsing the glass in double-distilled water for 50 s to remove loosely bound or excess copper ions from the surface; (iii) immersion of the glass in the anionic precursor solution for 20 s which allows selenium ions to react with pre-adsorbed copper ions to form Cu_2Se; and (iv) rinsing the sample in double-distilled water for 50 s which helps the separation of the unreacted or unadsorbed or excess selenium ions or powdery Cu_2Se from the material surface. Several tens of cycles can be performed depending of the desired thickness of the sample. As an example, for 65 cycles, copper(I) selenide film of terminal thickness about 330 μm was obtained [203]. In the study described above, the formation of Cu_2Se was attributed to the presence of organic complexing agent (tartaric acid) with sulfate ions. This system acts as a reducing agent and releases Cu^+ ions instead of Cu^{2+} ions from the $CuSO_4$ solution. The formation of stoichiometric Cu_2Se involves the following steps.

1) For the copper(II) sulfate precursor, in the presence of organic complexing agents, the sulfate anion, having reducing properties, can reduce Cu^{2+} into Cu^+ ions [210]:

$$[Cu(\text{tartaric acid})^{2+}] \rightarrow Cu^+ + \text{tartaric acid} \tag{5.77}$$

2) Hydrolysis of sodium selenosulfate takes place in solution, which releases selenide ions:

$$Na_2SeSO_3 + OH^- \rightarrow Na_2SO_4 + HSe^- \tag{5.78}$$

$$HSe^- + OH^- \rightarrow H_2O + Se^{2-} \tag{5.79}$$

3) Cu^+ ions react with Se^{2-} ions to form copper(I) selenide:

$$2Cu^+ + Se^{2-} \rightarrow Cu_2Se \tag{5.80}$$

Therefore Cu_2Se deposition is possible from tartaric acid-complexed Cu^{2+} ions from the bath.

These results show clearly that the CBD of Cu_xS and Cu_xSe can be achieved for thin-film solar cell applications. The basic solution for Cu_xS deposition could be [211]: 5 ml of 1 M $CuCl_2 \cdot 3H_2O$, 4 ml of TEA, 8 ml of 30% NH_4OH, 10 ml of 1 M NaOH, 6 ml of 1 M thiourea, with water making up the remainder. Deposition was carried out at room temperature for best results; the films peel from glass substrates during prolonged deposition, although peeling may be prevented by using ZnS substrate films [212].

For Cu_xSe thin films, one of the basic electrolytes for CBD could be [24]: 10 ml of 0.5 M $CuSO_4 \cdot 5H_2O$ solution, 1.5 ml of 30% NH_4OH, 12 ml of 0.4 M Na_2SeSO_3 solution (prepared by refluxing 4 g of selenium powder in 100 ml of 1 M Na_2SO_3 solution for about 3 hours), with water making up the remainder. At room temperature, films of approximately 0.2 μm thick were deposited in 7 hours. The value

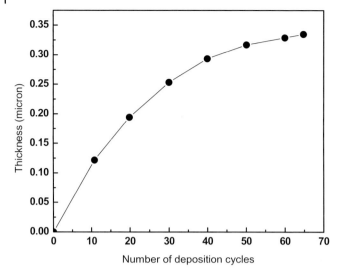

Figure 5.23 Plot of film thickness against number of deposition cycles for Cu_2Se thin film deposited onto glass substrate [203].

of x may depend on the exact bath composition and may even gradually change with the duration of deposition (Figure 5.23). If one uses modified chemical deposition [203], thicker films can be obtained.

5.2.2.1.3 CuInS$_z$ Preparation

The CBD of CuInS$_z$ thin films [213–216] assumes the formation of Cu ions in aqueous solutions of copper(II) salts in the overall presence of TEA, thiourea, and aqueous ammonia. The formation of Cu$_x$S (low chalcocite: 85%; djurleite: 10%; Cu$_2$O: 5%) by sulfuration of thin copper films in thiourea solution at 60–80 °C also indicates the stabilization of Cu in the presence of thiourea [217]. In the work described here, a copper(II) salt, CuCl·23H$_2$O, was used. The chemical baths were based on a 0.5 mol l^{-1} solution of CuCl$_2$/2½H$_2$O, approximately 9.4 mol l^{-1} (as supplied) TEA, 30% ammonia (aq.), 1 mol l^{-1} solution of thiourea, and distilled water. Different volume compositions of the above chemicals can be employed. Copper(II) salts have been used to generate Cu(I) ions, as in chemically deposited CuInS films [213–216], or to generate Cu^{2+} ions, as in chemically deposited CuSe thin films [218, 219]. The complexation efficiency, mechanism, and effect on the film deposition rate depend, of course, on the type of complexing agent and its interaction with water including the simultaneous presence of different complex ions in the deposition bath [219, 220]. The bath composition is not the only limiting parameter of the film deposition performance. The substrate properties are important parameters which may have a significant impact on the film deposition properties. The film substrate is a nucleation center which acts as a catalytic surface promoting the growth and adherence of particular atoms/ions [221, 222]. This fundamental aspect is one of the most important issues to be addressed in the development of adherent and performing CBD thin films for solar cell applications.

5.2.2.1.4 CdTe Preparation

The preparation of CdTe by CBD is not common. This is due the oxidative instability of aqueous solutions containing the telluride ion. Very few reports are available in the literature that describe the preparation of hexagonal CdTe by CBD [223–226]. It has been shown that, in aqueous solution, CBD of CdTe thin films is based on chemical reaction of $CdSO_4$ complexed with TEA, NaOH, and ammonia with sodium tellurosulfite. As an example, the deposition of CdTe can be achieved using 10 ml of (1 M) $CdSO_4$ complexed with TEA, NaOH, and sufficient amount of ammonia at 75 °C [223]. The sodium tellurosulfite was synthesized from tellurium powder in a sodium sulfite solution by a reflux action. The decomposition of sodium tellurosulfite was made possible in an aqueous alkaline medium containing $CdSO_4$ complexed with TEA that allowed control of Cd^{2+} ions. The principle of the decomposition process is to provide a slow generation of Te^{2-} ions in such a way that the solubility product is just exceeded. The CdTe thin-film growth was found to occur preferentially at the surface of the substrate, where the free energy of nucleation is small [224–226]. As indicated in Section 5.2.1.1, for any CBD, for the reaction to take place, the ionic product of the species must exceed the solubility product. The kinetics of film formation can be understood from the following reaction sequence [226]:

$$Na_2TeSO_3 + OH^- \rightarrow Na_2SO_4 + HTe^- \quad (5.81)$$

$$HTe^- + OH^- \rightarrow H_2O + Te^{2-} \quad (5.82)$$

$$Cd(TEA)_n^{2+} + Te^{2-} \rightarrow CdTe + n(TEA) \quad (5.83)$$

The preparation of CdTe:CdS clusters, which used a chemical deposition method, has been reported [227]. To increase the utility of CBD, tellurocarbonyls have been synthesized recently [228, 229]. The successful preparation of thin films of CdTe by an exchange reaction is described by the following equation [230]:

$$Cd(OH)_2 + TeQ^* \rightarrow CdTe + 2OH^- \quad (5.84)$$

where TeQ* describes the source of tellurium. The prepared TeQ solution contains an excess of sulfinic acid (Rongalite) which acts as a reducing agent. Firstly, cadmium hydroxide was prepared from the following solution: 5 ml of 0.5 M $Cd(NO_3)_2$ in 30% H_2O_2, 15 ml of aqueous 1 M sodium citrate solution, 15 ml of 0.5 M KOH in 30% H_2O_2, 8 ml of 30% H_2O_2, and 37 ml of distilled water. This allows the formation of a $Cd(OH)_2$ film of 1 mm thickness after 2 hours of successive deposition at room temperature and drying in air for 24 hours. Secondly, the white $Cd(OH)_2$ film was immersed in TeQ solution in a beaker at 70 °C, which changed to a brown color and gradually intensified in color to the dark brown color of CdTe. Accordingly, the steps required for the preparation of CdTe deposits on glass are summarized as follows: (i) a film of $Cd(OH)_2$ on glass is first prepared; (ii) a solution containing a source of dissolved tellurium, designated as TeQ, is prepared; and (iii) the $Cd(OH)_2$ film on glass is immersed in the TeQ solution where it is converted to CdTe. This opens the way for the development of new approaches for the preparation of CdTe using CBD. This is an important issue related to the development of low-cost, high-surface-area CdS/CdTe thin-film solar cells. Significant efforts must be devoted to this approach for other tellurium based solar cell materials.

5.2.2.1.5 Other Chalcogenide Materials Preparation

As indicated above for cadmium sulfides and selenides, CBD is a well-known deposition process for some other chalcogenides such as sulfides and selenides of Zn, In, Co, Hg, Pb, Sb_2S_3 and $Sb_2Bi_{2-x}S_3$ [231]. The deposition of metal cholcogenide films is based of following properties of thiosulfate/selenosulfate [232].

1) The reducing ability of the agents, for example thiosulfate as in the following half reaction:

$$2S_2O_3^{2-} \rightarrow S_4O_6^{2-} + 2e^- \qquad (5.85)$$

2) The complexing ability of these agents, whereby metal–thiosulfate and metal–selenosulfate complexes are formed.

3) The ability of these agents to gradually release sulfide/selenide ions upon hydrolytic decomposition, both in acidic media:

$$S_2O_3^{2-} + H^+ \rightarrow S + HSO_3^- \qquad (5.86)$$

$$S + 2e^- \text{ (from reaction 5.85)} \rightarrow S^{2-} \qquad (5.87)$$

and in alkaline media:

$$S_2O_3^{2-} + OH^- \rightarrow HS^- + SO_4^- \qquad (5.88)$$

$$HS^- + OH^- \rightarrow S^{2-} + H_2O \qquad (5.89)$$

Then the chalcogenide ions will combine with the metal ions released from the thiosulfate/selenosulfate complexes, and upon hydrolysis precipitating the corresponding chalcogenides. Due to the differences in the stability of the metal–chalcogenide complexes initially formed, the optimal concentrations, pH, and temperatures may vary from one chalcogenide to another. Optimal experimental conditions must be established for each system.

Other chalcogenide semiconductor materials have been studied extensively due to their potential use in photoconductive devices and solar cells [233–235]. Their wide range of bandgaps (from 0.8 to 3.5 eV) allows the possibility to develop optimized thin-film solar cells based on appropriate buffers (n-type ZnS, ZnO, In_2S_3, Sb_2S_3, etc.) [33, 223, 236–240] which may replace CdS in $Cu(InGa)(SSe)_2$ (CIGS)-based solar cells to avoid toxic cadmium and realize more environmentally friendly PV technology. In addition to the CIGS active materials, chemically deposited PbS–CuS and ZnS–CuS coatings can result after interfacial diffusion in materials such as $Pb_xCu_yS_z$ and $Zn_xCu_yS_z$ with p-type conductivities and are stable up to a temperature of 300 °C [4]. Annealing of Bi_2S_3 (bismuthinite)–CuS (covellite) coating at temperatures of 250–300 °C leads to the formation of a new compound, Cu_3BiS_3 (wittichenite), with p-type conductivity [24, 241]. The following other active materials can be developed using this approach [242]: $CuSbSe_2$ (E_g = 0.83 eV with a hole mobility of $5\,cm^2\,V^{-1}\,s^{-1}$), $AgSbS_2$ (E_g = 1.73 eV with a hole mobility of $1500\,cm^2\,V^{-1}\,s^{-1}$), $AgSbSe_2$ (E_g = 0.58), $AgBiS_2$ (E_g = 0.9 eV), Cu_3SbS_4 (E_g = 0.46 eV), Cu_3SbSe_4 (E_g = 0.32 with a hole mobility of $50\,cm^2\,V^{-1}\,s^{-1}$), Ag_2SnSe_3 (E_g = 0.81 eV with a hole mobility of $910\,cm^2\,V^{-1}\,s^{-1}$), Cu_2SnSe_3 (E_g = 0.96 eV with a hole mobility of $870\,cm^2\,V^{-1}\,s^{-1}$), Cu_2SnS_3 (E_g = 0.91 eV with a hole mobility of $605\,cm^2\,V^{-1}\,s^{-1}$), and $PbSnS_3$ (E_g = 1.05 eV). They can be produced by sold-state reaction during deposition, for example:

$$2CuS + SnS \rightarrow Cu_2SnS_3 \; (E_g = 0.91 \text{ eV}) \tag{5.90}$$

$$PbS + SnS_2 \rightarrow PbSnS_3 \; (E_g = 1.05 \text{ eV}) \tag{5.91}$$

$$Bi_2S_3 + Ag_2S \rightarrow 2AgBiS_2 \; (E_g = 0.9 \text{ eV}) \tag{5.92}$$

$$Sb_2S_3 + Ag_2S \rightarrow 2AgSbS_2 \; (E_g = 1.73 \text{ eV}) \tag{5.93}$$

Some other combinations are possible. In all cases the base CuS [211], SnS [243], PbS [244], Bi_2S_3 [245], Ag_2S [246], and Sb_2S_3 [23, 33] semiconductor films are produced by CBD. It has also been reported that, in the case of SnS and SnS_2 [247], controlling the bath composition and temperature is essential to produce good film composition. Further, new solar cell materials can be fabricated by combining CBD with thermal evaporation or sputtering of metals. This is, for example, the case for $Sb_2S_3 + 3CuS + 3Cu \rightarrow 2Cu_3SbS_3$. The development of new solar cell/solar energy materials which may be prepared through the interfacial diffusion of ions in multilayer chemically deposited thin films during annealing could be improved upon, consequently increasing the technology opportunities of these new materials.

Indium sulfide (In_2S_3) is an important material for PV applications [239, 240]. Due to its band value (2–2.8 eV), it can replace CdS in CIGS-based solar cells to avoid toxic Cd and realize more environmentally friendly PV technology. Results for the development of Cd-free buffer layers for CIGS thin-film solar cells from different laboratories have shown that In_2S_3, with optimal physical properties, can meet the requirements for use as window material or buffer layer for these PV structures [248].

CBD of In_2S_3 thin films can be achieved on commercial conduction ITO glass substrates immersed in baths containing $InCl_3$ (0.0083 M), a carboxylic acid (0.1 M), and TAM (0.1 M). The pH of the final solution was adjusted accordingly by the dropwise addition of NaOH or HCl (5 M). The temperature of the solution was maintained at 80 °C for a total deposition time of 30 minutes [248]. As for the classic CBD process, the formation of In_2S_3 films is based on the slow release of In^{3+} and S^{2-} ions in an acidic medium and their subsequent condensation on the substrate when the ionic product exceeds the solubility product K_{ps} ($=10^{-73}$) [249]. Sulfide ions are provided by the hydrogen sulfide produced during the TAM hydrolysis in dilute acid solutions [250]. Finally, the hydrogen sulfide is dissociated to give rise to the sulfide ions needed for In_2S_3 precipitation. The mechanism of film formation using CBD is not clear; however, the following tentative reactions steps which are very similar to other classic-based semiconducting materials using CBD reactions have been suggested [251]:

$$CH_3-CS-NH_2 + H^+ + 2H_2O \Leftrightarrow CH_3COOH + H_2S + NH_4^+ \quad (5.94)$$

$$H_2S \Leftrightarrow HS^- + H^+ \quad (5.95)$$

$$HS^- \Leftrightarrow S^{2-} + H^+ \quad (5.96)$$

$$2In^{3+} + 3S^{2-} \Leftrightarrow In_2S_3 \quad (5.97)$$

CBD-prepared indium hydroxy sulfide $In(OH)_xS_z$ has also been used as buffer layer in CIGS solar cells [251–253]. It was shown that the films can be prepared from an acidic bath (pH 2.2–2.5) containing 0.025 M indium(III) chloride, 0–0.1 M acetic acid, and 0.05–0.3 M TAM at 70 °C. The reaction solution was stirred in a 20 ml quartz beaker and the deposition was made on the desired substrates, followed by washing thoroughly with distilled water and drying in air. Some films were also air-annealed for 10 minutes at 300 and 400 °C. The presence of oxygen in the films has been found to be the origin of their optical properties which make them good candidates to substitute for CdS in thin-film solar cells as buffer layer. The films have a n-type electrical conductivity and their optical bandgap is about 2.8 eV (higher than that obtained with In_2S_3 which has a bandgap of around 2 to 2.45 eV).

CBD of ZnO and ZnS has been developed because ZnS and ZnO exhibit respective bandgaps of 3.2–3.5 and 2.7 eV [33, 223, 236–240] which means they are interesting buffer layers for CIGS solar cells. It has been reported that the CBD of ZnS and ZnO is complicated because attempts to grow ZnO directly on glass and some other substrates led in most cases to no growth or to irreproducible growth. Moreover, the deposition parameters were very narrow [254–256] and post-annealing was required [22, 257]. In the past decade, CBD of Zn-based buffer layers has attracted attention. Because the properties of ZnO strongly depend on its morphology and microstructure, it is essential to control precisely the size, shape, microstructure, and crystallinity of ZnO for its application in PV solar cells or other photoelectrical devices. The precise control of ZnO crystal evolution in the CBD process is an important issue for cell performance optimization and the mechanism of the formation of CBD ZnO crystals must be elucidated. Improved mechanistic understanding is crucial for the development of high-efficiency solar cells based on CIGS [258–262] and other related materials. The solution chemistry of CBD of ZnO can be summarized as follows [263]. In an aqueous solution, metal cations M^{z+} are solvated to aquo-ions, typically $[M(OH_2)_n]^{z+}$. The $M-OH_2$ bond is polarized which facilitates deprotonation of the coordinated water. In dilute solutions, a range of monomeric species exist such as $[M(OH_2)_{n-p}(OH)_p]^{(z-p)+}$ and other hydroxyl species $[M(OH)_n]$ (it is often customary to omit the water); ultimately, oxoanions are formed. In order to form the polynuclear species that subsequently develop into metal oxide particles, reactions involving condensation must occur.

Two important processes have been recognized. Olation is the formation of an "ol" bridge by reaction of a hydroxo- and aquo-species:

$$M-OH + M-OH_2 \rightarrow M-OH-M + H_2O \quad (5.98)$$

Oxolation leads to an "oxo" bridge by the dehydration of hydroxo-species:

$$M_2-(OH)_2 \rightarrow M-O-M + H_2O \quad (5.99)$$

Because zinc hydroxide is amphoteric, its complexation by OH_2 can lead to soluble species such as "$[Zn(OH)_3]^{2-}$" and "$[Zn(OH)_4]^-$" and hence "zinc hydroxide" is more soluble in basic solution. A simple consideration of the solubility product (10^{-16}) would suggest this solubility of ZnO in alkaline solution [264]. In the CBD of ZnO, ligands are employed to keep the free zinc ion concentration low. Raising the bath temperature promotes some dissociation of the zinc complex, leading to controlled supersaturation of the free metal ion. Zinc is a labile metal ion in aqueous solution and equilibria within stirred solutions are generally attained quickly. Thermodynamic modeling of solutions representing the initial states of deposition baths is a useful aid to understanding these systems. Understanding the initial states of the deposition will sustain the development of large-area ZnO thin films for solar cell applications.

The first CBD ZnS-based buffer fabrication was done with an efficiency level of 9–10% [262]. In the CBD of ZnS buffer, thiourea is generally used as the sulfur source and ammonia as complexing agent [259]. The quality of the ZnS films, and their solar performance, is controlled by the concentration of ammonia. In particular, depending on this concentration, co-deposition of ZnS, ZnO, and $Zn(OH)_2$ may occur resulting in many different possible compositions and phases (ZnS, ZnO, and $Zn(OH)_2$) for the CBD process [261]. Optimal film compositions must be identified before large-scale preparation of these films for solar cell applications because CBD of ZnS is more complex and difficult than that of CdS [262]. Although significant efforts have been undertaken to make CIGS-based cells free of Cd, until now, such cells usually have lower efficiency and less reproducible behavior than their Cd-containing counterparts. The continued investigation of the deposition conditions leading to superior film properties for the CBD of ZnS in basic aqueous ammonia solutions would be worthwhile.

A simple CBD method was employed to deposit ZnS thin films onto glass substrates using thiourea as sulfide ion source and zinc acetate as zinc ion source in an alkaline bath. For the preparation of ZnS thin films, 0.2 M zinc acetate solution was mixed with an equal volume of 0.2 M thiourea, and ammonia solution was added slowly to form the complex; the pH then was adjusted between 9 and 10 at 303 K [232].

ZnS thin films of different thicknesses were prepared by CBD using thiourea and zinc acetate as S^{2-} and Zn^{2+} sources, respectively. Thermo-EMF measurement indicated that films prepared by this method are of n-type. ZnS thin films were prepared by decomposition of thiourea in an alkaline solution containing a zinc salt. The reaction mechanism for deposition of ZnS films has been reported. In aqueous solution, zinc acetate dissociates to give Zn^{2+} ions. Hydrolysis of ammonia in water gives OH$^-$ ions, which forms a complex of $Zn(NH_3)_4^{2+}$. Thiourea in alkaline medium acts as an S^{2-} ion source. In short, the reaction for the process is:

$$Zn(NH_3)_4^{2+} + SC(NH_2)_2 + 2OH^- \rightarrow ZnS \downarrow + 4NH_3 + CH_2N_2 + 2H_2O \quad (5.100)$$

CBD was used to deposit ZnO and ZnS thin films with the use of an aqueous medium involving Zn salt, ammonium sulfate, aqueous ammonia, and thiourea [265]. Observations of the physical and chemical properties of the grown layers as a function of ammonia concentration were reported. Rapid growth of nanostructured ZnO films on fluorine-doped tin oxide (FTO) glass substrates was obtained. ZnO films crystallized in a wurtzite hexagonal structure and, with a very small quantity of $Zn(OH)_2$ and ZnS phases, were obtained for an ammonia concentration ranging from 0.75 to 2.0 M. Flower-like and columnar nanostrucured ZnO films were deposited in two ammonia concentration ranges, respectively, one between 0.75 and 1.0 M and the other between 1.4 and 2.0 M. ZnS films were formed with a high ammonia concentration of 3.0 M. The formation mechanisms of ZnO, $Zn(OH)_2$, and ZnS phases in the CBD process were discussed. The technique can be used to directly and rapidly grow nanostructured ZnO film photoanodes, and the resulting structures have been demonstrated for potential applications of CBD nanostructured ZnO films for PV cells. Some of the well-established experimental conditions which may allow good ZnS thin films are [212]: 5 ml of 1 M zinc sulfate, 4.4 ml of NH_3/NH_4Cl (pH 10), 5.4 ml of 50% TEA, and 2 ml of 1 M TAM, with deionized water to make up to 100 ml by volume. Depositions were made at 50 °C for about 6 hours or at room temperature for about 20 hours to obtain ZnS films of 0.2 μm thickness. Another Zn semiconductor, ZnSe (E_g = 2.7 eV), can be fabricated using the following conditions [266]: 35 ml of 0.1 M zinc acetate, 16 ml of 0.8 M sodium citrate, 5 ml of 7.4 M ammonium hydroxide, and 20 ml of 0.07 M N,N-dimethylselenourea, with the volume made up to 100 ml with deionized water. Depositions may be made at room temperature or in an oven at temperatures up to 60 °C.

Antimony sulfide (Sb_2S_3) is another interesting material for solar cell films because the value of its bandgap (E_g = 2.48 eV) makes it suitable as a buffer layer for CIGS thin-film solar cells. One of the important difficulties in the use of antimony salts in aqueous media arises from the strong tendency of antimony salts to hydrolyze in water solutions and precipitate as insoluble hydroxy salts [33]. Clear solutions can be obtained only in strongly acidic media or by using complexing agents. As an example, solid antimony(III) chloride was dissolved in a small volume cold water or glacial acetic acid and then complexed by sodium thiosulfate [33, 232, 267]. The following bath parameters were found to be optimal in the case of glacial acetic acid: 3.540 g of $SbCl_3$ was dissolved in 7–8 ml of glacial acetic acid and a 1.0 M aqueous solution of sodium thiosulfate was slowly introduced with constant stirring until a clear solution was obtained. As much as 100 ml of 1.0 M sodium thiosulfate may be needed. Then 400–450 ml of cold (10–15 °C) distilled water was added and the substrates were introduced into the bath. For practical considerations, it was found that vertically supported substrates had films deposited on both sides, while horizontally mounted substrates had films deposited on the side facing the bottom of the bath container. For good adherence, it was found that the substrates needed to be left in the bath for at least 5 and up to 24 hours, depending on the desired thickness.

The advantages of CBD for forming Sb_2S_3 and Bi_2S_3 compounds are also related to the possibility of annealing of multilayer thin-film semiconductors like Bi_2S_3–CuS [241] and Sb_2S_3–CuS [268] to form films of ternary composition, Cu_3BiS_3 and $CuSbS_4$, respectively. We consider that the possibility of creating new materials through interfacial diffusion of atoms in chemically deposited multilayer stacks as demonstrated in PbS–CuS, ZnS–CuS, and Bi_2S_3–CuS films [241, 269] is very high.

Bismuth trisulfide (Bi_2S_3) in thin-film form is a particularly interesting material for solar cells because of its mid-value bandgap (E_g = 1.7 eV), absorption coefficient of the order of 10^4 to 10^5 cm^{-1}, reasonable conversion efficiency, and stability, together with low cost [270–272]. Nanocrystalline Bi_2S_3 thin films of various thicknesses having grain sizes between 7 and 34 nm have been prepared at 60–70 °C by using CBD onto FTO-coated glass substrates from an aqueous acidic bath (pH 5). Bismuth nitrate and TAM were used as Bi^{3+} and S^{2-} ion sources, respectively. Bi_2S_3/polysulfide junction cells were fabricated and their photoelectrochemical performance was studied. The grain size and thickness of Bi_2S_3 film optimization are key factors for the development of solar cells based on Bi_2S_3 thin films.

Another approach has been reported for optimizing the preparation conditions for Bi_2S_3: deposition temperature from 0 to 80 °C, deposition time from 5 to 30 hours, and pH from 2 to 5 [270]. Using these conditions, Bi_2S_3 thin films were deposited from a solution containing 20 ml of 0.1 M $Bi(NO_3)_3$, 16 ml of 0.1 M EDTA, and 24 ml of 0.1 M TAM onto FTO-coated glass substrates for various deposition time periods to get different thicknesses. A variant of the CBD method is to use microwave radiation to prepare nanorods of B_2S_3. In this case the proposed mechanism of the synthesis of Bi_2S_3 nanorods may be as follows. Firstly, the strong complex action between Bi^{3+} and thiourea leads to the formation of Bi–thiourea complexes in the synthesis chamber, which prevent the production of a large number of free S^{2-} in the solution, thus favoring the formation of nanorods. It has been confirmed that the $Bi(NO_3)_3$ and thiourea can easily dissolve in formaldehyde solution to form yellow Bi–thiourea complexes. Secondly, the Bi–thiourea complexes undergo a decomposition process under microwave irradiation to produce Bi_2S_3 nanorods. During the whole process, HCHO does not take part in the reactions but only acts as a dispersant.

Another possibility similar to the above process consists of the preparation of Bi_2S_3 thin films using 0.2 M solution of disodium salt of EDTA which is added to 0.2 M bismuth nitrate solution, and into which a solution of sodium thiosulfate (0.2 M) is then added. The pH of the resultant solution was between 0.5 and 1.5. The chemical reaction responsible for Bi_2S_3 films from an acidic bath using $Na_2S_2O_3$ as the sulfide ion source [273–277] proceeds according to the following mechanism:

$$Na_2S_2O_3 \rightarrow 2Na^+ + S_2O_3^{2-} \tag{5.101}$$

$Na_2S_2O_3$ is oxidized through the half-cell reaction

$$6S_2O_3^{2-} \rightarrow 3S_4O_6^{2-} + 6e^- \qquad (5.102)$$

Dissociation of $S_2O_3^{2-}$ takes place in the acidic medium as

$$3S_2O_3^{2-} + 3H^+ \rightarrow 3HSO_3^- + 3S \qquad (5.103)$$

The released electrons react with sulfur as:

$$3S + 6e^- \rightarrow 3S^{2-} \qquad (5.104)$$

Bi^{3+} from $Bi(NO_3)_3$ solution or a complex of Bi^{3+} formed by EDTA reacts to give

$$2Bi^{3+} + 3S^{2-} \rightarrow Bi_2S_3 \qquad (5.105)$$

In addition, the CBD of Bi_2S_3 thin films from an alkaline bath (pH 7.3–10.5) at room temperature has been achieved by using disodium salt of EDTA (Na_2EDTA) as a complexing agent and TAM as a sulfide ion source [278]. A typical electrolyte composition of the CBD of Bi_2S_3 thin films was [245, 279]: 10 ml of 0.5 M $Bi(NO_3)_3 \cdot 5H_2O$ solution, 8 ml of 50% TEA, and 4 ml of 1 M TAM, with water making up the remainder. The deposition was carried out at room temperature on the substrate at intermediate durations (5–7 hours), but the films were observed to be adherent when they were allowed to remain in the bath for more than 24 hours.

From these different methods of fabrication, it was found that: (i) the bandgap of Bi_2S_3 can vary from 1.1 to 1.7 eV for the stoichiometric material and from 1.4 to 3.6 eV [274] for the non-stoichiometric Bi_2S_3 when the concentration of Bi increases in the material; (ii) the films are n-type semiconductors; (iii) particle sizes increase as the percentage of Bi decreases; (iv) the films are nanstructured; (v) the particle size increases with percentage of sulfur; and (vi) the variation of bandgap with grain size reveals quantum size effects. These results indicate clearly the importance of optimizing the experimental conditions for the CDB for large-area applications.

The fabrication of Bi_2Se_3 thin films using the CBD method was also carried out [280] for a deposition electrolyte consisting of 7 ml of 0.5 M $Bi(NO_3)_3$ solution, 7 ml of 50% TEA, and 20 ml of 0.07 M N,N-dimethylselenourea solution prepared freshly in 0.01 M Na_2SO_3, with the balance made up of water. Films of 0.2 μm may be deposited in about 9 hours at 40 °C.

SnS thin films can also be fabricated with CBD using the following electrolyte composition [243]: 1 g of $SnCl_2 \cdot 2H_2O$ dissolved in 5 ml of acetone, 12 ml of 50% TEA, 8 ml of 1 M TAM, and 10 ml of 4 M NH_4OH, with water making up the volume to 100 ml. The deposition was carried out at temperatures up to 80 °C to obtain thin films of 0.7 μm thickness.

The CBD method can also be used to fabricate $Sb_{2-x}Bi_xS_3$ (E_g = 1.6–2.4 eV) which is a mixed composition of Sb_2S_3 and Bi_2S_3 compounds, both of which have an orthorhombic crystal lattice. These isomorphous compounds can form solid solutions of a desired mixed composition, giving materials of variable bandgaps [275]. These films were similar to the one described for growth of Sb_2S_3 except bismuth(III) nitrate was also added along with antimony(III) chloride. The value of x in $Sb_{2-x}Bi_xS_3$ can be varied between 0 and 1.90. However, pure films were difficult to obtain, because of poor adherence of the films to the substrate surface. The as-deposited films were amorphous and up to 0.3 μm thick.

Films of Ag_2S ($E_g = 2.2$ eV) and of $AgInS_2$ ($E_g = 1.8$–2.0 eV) have also been fabricated. The precipitation reaction for Ag_2S from thiosulfate solutions is fast even at lower temperatures and thus inconvenient for film growth. Therefore, Ag_2S films were grown in alkaline media. A typical bath for a total volume of 100 ml was prepared as follows. Aqueous ammonia solution (1 volume of ammonia for 3 volumes of water) was gradually added until the forming precipitate of AgOH was just dissolved (complexed by NH_3). The pH of the solution should be 9–11. The solution was then filtered. Then 5.0 ml of 1.0 M aqueous solution of sodium thiosulfate was introduced and distilled water was added to make a total volume of 90–100 ml. Substrates were introduced vertically into the bath and the bath was warmed to 45–55 °C. At this temperature the solution develops a yellow-brown color, after which a brown precipitate fills the beaker. The reaction was completed within 30 minutes after precipitation began, and yellow-brown films of Ag_2S were deposited onto the substrates.

An aqueous method for the deposition of silver indium sulfide ternary semiconductor thin films was demonstrated [281] using analytical-grade silver nitrate ($Ag(NO_3)$), indium nitrate ($In(NO_3)_3 \cdot 5H_2O$), TAM, TEA, sodium citrate ($C_6H_5O_7Na_3 \cdot 2H_2O$), citric acid ($C_6H_8O_7$), sulfuric acid ($H_2SO_4$), and ammonium nitrate (NH_4NO_3) on glass substrates. Aqueous cationic and anionic solutions were prepared separately before deposition. The energy gap, determined from transmission and reflection spectra, was located between 1.8 and 2.0 eV. The thickness of the thin films was in the range 500–700 nm. Electrical resistivity was of the order of $10^4 \Omega$ cm. According to these experimental findings, it was possible to control the compositions of silver indium sulfide thin films from aqueous solutions by suitable control of concentrations of the precursors in the electrolyte and surface functionalities. This technique provides an easy and cost-efficient way to deposit multicomponent semiconductor thin films.

5.2.2.2 Preparation by Electrochemical Deposition

5.2.2.2.1 CdTe Preparation

The cathodic electrodeposition of CdTe films from aqueous electrolytes was first carried out by Panicker *et al.* [63] and Kroger [282], who demonstrated that uniform films with controlled stoichiometry could be obtained using the simple electrodeposition technique. They described the effect of rest potential and the temperature of the bath solution on the properties of the films. Deposition of CdTe took place from an aqueous solution of $CdSO_4$ to which TeO_2 had been added, and electrolysis was carried out using a metal or conducting glass support onto which the CdTe was deposited. The production of a CdTe layer by electrodeposition and its use in the fabrication of PV cells is disclosed in US patent no. 4425194. Various arrangements of electrochemical cell are described; for example, one in which the anode is a tellurium bar, another in which the anode is an inert carbon or stainless steel anode, and another with both a cadmium and a platinum anode, which is described as a neutral anode. US patents 4816120 and 4909857 describe the electrodeposition of CdTe doped with small quantities of Cu, Ag, and Au. It is also possible to use CdTe containing chloride, as described in US patents 4548681 and 4629820. Consequently, the experimental parameters for CdTe deposition are well established in the literature. US patent 4400244 describes the fabrication of homojunction and heterojunction semiconductor materials. Many examples of experimental CdTe deposition parameters (pH, electrodeposition potential, deposition temperature, etc.) were included. It is well established that electrodeposited CdTe films determine the performance of CdS/CdTe thin-film solar cells. Classically, the electodeposition is carried out potentiostatically in a thermoregulated three-electrode cell. The ratio of the concentration of CdS to that of TeO_2 varied between 50 and 1000. The pH of the solution was varied from 1 to 3.5 using sulfuric acid. The dependence of CdTe semiconductor composition on the conditions of electrodeposition has been studied. Several deposition baths were investigated [283–287], including an acidic sulfate bath because of its ability to deposit both n-type and p-type CdTe from the same solution, its lower temperature of operation, and its decreased toxicity. Consequently, several researchers [63, 79, 282, 283, 288] have tried to obtain a suitable range of potential to control the deposition of p-type or n-type CdTe. Panicker et al. and others [63, 282, 284–286] used a quasi-rest potential (QRP) to locate the boundary potential, which is a critical potential separating regions of deposition of p-type and n-type CdTe. They revealed that CdTe obtained at a QRP greater than −0.3 V vs. a saturated calomel electrode (SCE) is p-type and that obtained at a potential of less than −0.3 V/SCE is n-type. It was reported [79] that p-type CdTe is obtained at −0.61 to −0.66 V vs. SCE. Another study [57] found that CdTe film deposited at −0.4 to −0.5 V vs. SCE is p-type and at −0.58 to −0.7 V vs. SCE is n-type. Although using the critical boundary potential to control deposition is not practical because this parameter may depend on deposition parameters such as solute concentration and the temperature and acidity of the electrolyte, the electrodeposition of CdTe is an alloy electrodeposition in which the composition of the alloy is closely related to the semiconductivity type of the CdTe, which, of course, depends on such parameters. It has been shown that, for a deposition at 80 °C, p-CdTe was obtained at −0.5 to −0.56 V vs. SCE and n-CdTe was obtained at −0.58 to −0.64 V vs. SCE [283]. It was shown that the cadmium sulfate concentration has no effect on this boundary potential. If the concentration of [H^+] is increased, the p/n-type boundary potential shifts to more positive value. As the concentration of TeO_2 increases from 0.2 to 0.6 M, the region of n-type increases. But, when the concentration of TeO_2 increases from 0.6 to 1.2 M, the p-type region increases and the n-type region decreases. Since the as-deposited p-type films are of low resistivity depending on the film thickness, such films may indicate a second phase of Te with Cu. However, no efficient device has been demonstrated on as-deposited Cu-doped films. Further, it would be extremely difficult to electroplate a solar cell-grade p-CdTe layer which would yield high-efficiency devices in its as-deposited form. To overcome the problem associated with obtaining a solar cell with generally as-deposited p-CdTe layers, the "type conversion junction formation process" has been developed [284–287, 289] which consists of the electrodeposition of high-resistivity n-CdTe film on a glass/ITO or SnO_2/CdS substrate to a thickness of 1–3 μm. The glass/ITO/CdTe or SnO_2/CdS/CdTe structure is then heat treated in air at around 400 °C for 15 minutes. During the heating process, the n-CdTe layer is converted into solar cell-grade p-CdTe, and the rectifying junction is formed *in situ*, between the CdS film and the converted CdTe layer. This conversion step consists of putting the samples onto a piece of ceramic preheated at 400 °C in air. Experiments carried out in ovens with different atmospheres (H_2, N_2) have shown that the best results were obtained by treatment in an environment containing oxygen. But the factors controlling this type-conversion junction formation are one of the most important points to consider to achieve useful CdS/CdTe thin-film solar cell performance.

5.2.2.2.2 CuInSe$_2$ Fabrication

The direct electrodeposition of $CuInSe_2$ (CIS) thin films was first investigated in acid media (pH < 3) [64, 290–294]. Their electrodeposition from a high pH (>9) selenosulfate solution has also been investigated [295, 296]. It was shown that the deposition process involves the reaction of Cu^+ (CuCl) and In^{3+} ($InCl_3$) with Se^{2-} (Na_2SeSO_3) ions in an aqueous solution. The Cu and In with NH_4OH and citric acid complexes dissociate to give a controlled number of Cu and In ions at the cathode, where they combine with Se^{2-} ions to form CIS films. CIS thin films were also prepared by reducing Cu(II), Se(IV), and In(III) species in an acid solution at a fixed potential between −0.2 and −1.1 V vs. mercury sulfate electrode (MSE) [297]. It was shown that Cu–In–Se alloys can be electrodeposited in a wide range of controlled compositions. The direct electrodeposition provides all the constituents from the same electrolyte in a single step. CIS thin films may also be fabricated by a sequential electroplating process in the first step of which a pure Cu layer is deposited, and, in the second step, precursor films of In–Se, Cu–Se, and Cu–In–Se are deposited using electrodeposition [298, 299]. Direct electrodeposition was accomplished in one step from a plating bath containing calculated proportions of $CuSO_2$, $In_2(SO_4)_3$, and $SeSO_3$ [299–316]. A plating bath consisting of an aqueous solution of 3×10^{-3} M $CuSO_4$, 3×10^{-3} M $In_2(SO_4)_3$, and 5×10^{-3} M SeO_3 with citric acid (0.4 M) as the complexing agent was suggested as an appropriate electrolyte [305, 308]. Molybdenum was used as a substrate in the potentiostatic mode and the plating bath was held at room temperature and not stirred. The current density during the electrodeposition process reached a value between 1 and 3 mA cm^{-2} depending on the applied potential. Other studies have indicated that CIS films may be prepared at a deposition potential of −1.0 V vs. MSE with a plating bath consisting of an aqueous electrolyte of 5×10^{-4} M Cu^{2+} and 6×10^{-4} to 9×10^{-4} M of In^{3+} and Se^{4+} [306–308]. The substrate was SnO_2-coated glass of 6×10^{-4} Ω cm. The solution was stirred and the quantity of electricity needed for electrodeposition was fixed at 2 C cm^{-2}, resulting in films of 1 μm thickness. It was also found [312] that the whole composition of the films depends on the diffusion flux ratios of the species arriving at the surface of the electrode. With an excess of In^{3+} in the solution, the ratio of Se^{4+} and Cu^{2+} fluxes is the key parameter controlling the composition. Lower concentrations of In^{3+} lead to deposition of elemental Se. In this case, the electrodeposition process is limited by the diffusion of the three species. An estimation of the diffusion fluxes is possible in order to predict the chemical composition of the films. Recently, the relation between physical properties and deposition conditions of electrodeposited CIS was determined [309]. It was shown that the as-grown electrodeposited films are polycrystalline and require a thermal treatment to improve their physical properties. This post-treatment, in combination with the electrodeposition process, leads to structural and optical properties that are similar to those reported for thin films prepared by other methods. A structural transition exists for the films from a chalcopyrite to a sphalerite structure as the composition varies from Cu-rich to In-rich. The overpotential and photoelectrochemical properties also depend on the chemical composition. $Cu_{1+x}In_{1+y}Se_{2+z}$ thin films within a wide composition range have been obtained by the one-step method of electrodeposition onto molybdenum substrates [305]. It has been shown that the effective grain size increases as the annealing temperature increases. Chemical etching produces significant changes in film composition, and aqueous KCN solution dissolves secondary phases of copper selenides that are detected in the samples. The characteristics of the films depend, of course, on the sequence of annealing and chemical treatments. In the sequential process, the Cu layer was electrodeposited from a copper sulfate solution followed by an In–Se alloy layer from indium sulfate and selenous acid solutions [305, 309]. It was reported that one-step electrodeposition gives CIS formation at room temperature. Annealing of these samples produces CIS crystallization. For the sequential deposition of In–Se on Cu layers, a potential reaction between the constituents can lead to CIS formation by reaction of binary compounds, mainly $Cu_{2-x}Se$ and In_2Se_3 [317]. It has been shown that cathodic electrodeposition of precursor films of In–Se, Cu–Se, and Cu–In–Se can occur with controlled stoichiometry.

The effect of chemical bath composition, electrodeposition potential, etc., on film composition was determined [317]. The precursor films were loaded in a physical evaporation chamber and additional In or Cu and Se were added to the films to adjust the final composition of CIS. The device fabricated using electrodeposited Cu–In–Se precursor layers had a solar cell efficiency of 9.4%. CIS thin films have also been obtained from different precursors prepared by direct or sequential electrodeposition processes [303]. The results showed that thin crystalline chalcopyrite CIS films with the desired composition can be obtained after annealing, whether directly or sequentially electrodeposited precursors at 400 °C. An improvement in film quality was obtained by using an electrodeposited Cu layer as the growth surface for CIS formation. If elemental Se was also added during the heat treatment, then a higher recrystallization of the films was observed. A new approach for CIS formation by sequential electrodeposition of Cu and In–Se layers and subsequent heat treatment with elemental selenium in $Ar + H_2$ flows has been presented [304]. An increase in the film crystallinity was achieved by maintaining the Se source temperature above 400 °C. It was shown that the introduction of H_2 in the selenizing atmosphere was not suitable because H_2Se formation is more poisonous and less reactive than the elemental selenium vapor. An 8% CIS-based solar cell formed from an electrodeposited precursor film was also achieved [316]. This cell was developed by using an electrodeposited CIS precursor film subjected to post-deposition heat treatment at 550 °C in Se and In atmospheres. The electrodeposition bath used for the co-deposition of Cu–In–Se consisted of 0.025 M $CuCl_2$, 0.025 M $InCl_3$, and 0.025 M H_2SO_3 at pH 1.5. The bath temperature was maintained at 24 °C and the contents were not stirred during film deposition. Co-deposition was achieved using a three-electrode configuration, an SCE as the reference electrode and Pt gauze as the counterelectrode. These various results indicate that the main problem in CIS fabrication is the control of its chemical composition. Thus, further studies must optimize the different compound formation pathways during direct or sequential electrodeposition of CIS. Direct electrodeposition may be considered more favorably because of the low cost of the films fabricated by this method.

5.2.2.3 Preparation by the Sol–Gel Method

The sol–gel method has been successfully applied to the synthesis of CdS and ZnS [102, 318], oxides of Ti, Zr, Al, Zn, Si, V, Ga, In, and B [319], and mixed Ti, Ta oxides [320] as thin films, and mixed Mn-doped ZnS and ZnO on various supports including Si [319].

The synthesis of semiconductor particles by alternate adsorption of anions and cations from aqueous solutions is known as SILAR (successive ionic layer adsorption and reaction) [318]. This technique has been generalized to include molecular precursors, such as metal alkoxides, which can be adsorbed and hydrolyzed as monolayer films [319]. This has been illustrated by the fabrication of patterned TiO_2 films on Si/SiO_2 substrates bearing microcontact-printed lines of an organic polysiloxane [321]. Nanostructures (nanowires and nanotubules) of TiO_2 [318–322], CdS [322], SiO_2 [318, 323], In_2O_3 [324], Ga_2O_3 [325], V_2O_5 [318], MnO_2 [318], WO_3 [318], and many other semiconductor materials have been synthesized using such a sol–gel template synthesis strategy.

Oriented ZnO thin films have been grown by the sol–gel process. Solutions were prepared by dissolving $Zn(CH_3COO) \cdot 2H_2O$ in ethanol and refluxing, after which monoethanolamine (MEA) was added. The ratio $r = [MEA]/[Zn^{2+}]$ was varied between 0.5 and 4. These solutions were aged for variable periods (10–240 hours). Depending on $[Zn^{2+}]$ and r, transparent solutions or colored colloidal dispersions (bluish to yellow) were obtained. The precursor solution was filtered (Millipore cellulose membranes, 0.45 mm) and deposited on glass or quartz substrates by dip- or spin-coating (3000–4500 rpm; 5–30 s); substrates were sonicated or washed with HNO_3 and thoroughly rinsed with water, ethanol, and acetone. As-synthesized films were preheated at 300 °C for 10 minutes after each coating. This procedure was repeated up to six times. The films were subsequently heated up to 550 °C for 2 hours in order to obtain crystallized ZnO. In another procedure, films were dried (100 °C, 5 minutes) between successive spin-coating steps. Subsequently, these films were pretreated at 135 °C for 36–72 hours, and further treated at 450 °C for 2 hours [102].

Another approach for preparing ZnO films [326, 327] begins by refluxing zinc acetate dihydrate in n-propanol after which methanolic tetramethylammonium hydroxide solution is added to the precursor sol of zinc acetate dihydrate in n-propanol. Films were deposited by single dip-coating on various substrates. The withdrawal of the substrate out of the sol was performed at a speed of 20 cm min^{-1}, in an atmosphere with controlled relative humidity: 4% for one of the films deposited on silicon substrate and 30% for all other films. These low and middle values were chosen in order to check on water sorption during dip-coating. The samples were then dried in air at 80° for a minute and annealed at higher temperatures by direct introduction in a hot oven under a controlled atmosphere. Successive annealing steps were performed between 150 and 400 °C for 15 minutes under a flux of pure oxygen or ozone.

The preparation of a TiO$_2$ thin film was investigated using a two-dimensional (2D) sol–gel process [328]. A chloroform solution of monomer or tetramer of tetrabutoxytitanium (TBT) was spread on the surface of an aqueous subphase, hydrolyzed, and polycondensed at the air/water interface to give floating gels. The gels were gathered by 2D compression to yield a uniform gel film on the water surface. The addition of acetic acid (a chelating agent) and acetylacetone to the aqueous subphase or spreading solution was effective, and films were transferred onto the appropriate substrate using Langmuir–Blodgett techniques. Accordingly, n-octadecyl acetoacetate was added in the spreading solution of monomeric Ti butoxide and ethoxide resulting in quantitative transference of the gel films onto substrates without a limitation on the number of depositions. The deposited TiO$_2$-based gel films could be converted into nanometric quantum size TiO$_2$ films by calcination at 500 °C for 30 minutes. The thickness and the density of the nanoparticulate TiO$_2$ films could be controlled by the number of deposited gel film layers and the surface pressure of the gel film layer during deposition.

CdS nanocomposites (CdS cluster sizes less than 5 nm) were prepared via multifunctional inorganic–organic sol–gel processing [329]. In CdS sols as precursors, the CdS clusters can carry inorganic components as stabilizing centers, along with bifunctional ligands acting as inorganic and organic network formers. Hydrolysis and condensation produce an inorganic skeleton yielding viscous liquids useful to prepare films or monoliths. The final organic crosslinking at less than 100 °C results in optically transparent materials. The nature of the stabilizing centers and the starting synthesis conditions influence strongly the resulting electronic properties. Cd^{2+} complexed by mercapto compounds gives smaller cluster sizes with a narrower distribution than Cd^{2+} complexed with amine or carboxylate groups.

5.2.2.4 Thin Films Deposited with Heteropolycompounds
Thin-film semiconductors have been deposited chemically and electrochemically with heteropolyacids for optoelectronic and solar cell devices [33, 192, 330–343].

5.2 Materials and Composite Materials Fabrication | 331

The concept of using heteropolyacids in the chemical bath was developed during the 1990s for deposition of CdSe, Sb_2S_3, CdS, CdTe, etc. The effect of heteropolyacids on the physical, optical, and electrical properties of the films was determined. Their performance in photoelectrochemical, Schottky barrier, and heterojunction solar cells has been investigated.

5.2.2.4.1 CdSe Photoelectrochemical Solar Cells [330, 331]

Deposition of CdSe was carried out by mixing 10 ml of 1 M cadmium acetate, 5 ml of TEA (99%), 10 ml of 30% aqueous ammonia, and 15 ml of 0.45 M sodium selenosulfate. The final pH of the mixed solution was adjusted to 11 [330]. Deposition was conducted at 40 °C by dipping cleaned conducting SnO_2-coated glass plates and holding them vertically on the walls of a 100 ml beaker. The determination of the chemical composition of the films by X-ray diffraction, X-ray photoelectron spectroscopy, and neutron activation has shown the presence of triclinic WO_3 in films deposited with heteropolyacids. It has been shown that the photoelectrochemical solar cell parameters change with the concentration of silicotungstic acid (STA) or phototungstic acid (PTA). The best performances were obtained for electrodes fabricated with 10^{-5} M STA or 10^{-5} M PTA. Without STA or PTA, the CdSe film in 0.45 M $SeSO_3^{2-}$ electrolyte showed an open-circuit voltage (V_{oc}) of 510 mV, a short-circuit photocurrent density (J_{sc}) of 2.7 mA cm^{-2}, a fill factor (FF) of 0.48, and a cell efficiency (η) of 1.6%. The film grown with 10^{-5} M PTA showed the following photoelectrochemical solar cell parameters: $V_{oc} = 570$ mV, $J_{sc} = 7.6$ mA cm^{-2}, FF = 0.55, and $\eta = 5.9\%$. The CdSe film deposited with 10^{-5} M STA showed far superior I–V properties with $V_{oc} = 600$ mV, $J_{sc} = 11.8$ mA cm^{-2}, FF = 0.68, and $\eta = 11.7\%$. The improvement in the photoelectrochemical solar cell properties has been ascribed to the formation of n-CdSe/n-WO_3 heterojunctions, which enhances the charge transfer at the semiconductor/electrolyte interface. These results indicated for the first time the interesting effects of STA and PTA on chemically deposited CdSe films. This opens up a new method for fabricating mixed electrodes with improved physical properties and photoelectrochemical solar cell performances.

5.2.2.4.2 Sb_2S_3 photoelectrochemical Solar Cells [33, 332]

The chemical deposition of Sb_2S_3 thin films was carried out by the reduction of Sb complexed with TEA using thioacetamide as the reducing agent. For a typical film, the chemical bath composition was: 25 ml of 0.1 M potassium antimonyl tartrate, 5 ml of 7.4 M TEA, 3 ml of 17 M ammonia, and various concentrations of STA in a 100 ml beaker. The mixed solution was stirred for 10 minutes to obtain a homogeneous solution. The solution was then diluted to 75 ml and the reduction was initiated by adding 25 ml of 0.1 M thioacetamide with magnetic stirring. Before use, SnO_2-coated and ordinary glass slides were cleaned ultrasonically in isopropyl alcohol and dried in pure nitrogen atmosphere. The cleaned substrates were then clamped vertically in the plating solution at 300 K. During the Sb_2S_3 film formation, the color of the solution changed progressively from a faint to a dark yellow, and then finally to orange-red, at which point a thick deposition of Sb_2S_3 film formed on the substrates. After about 96 hours, the slides were removed and washed thoroughly with water and dried in air. In the case of the Sb_2S_3 films formed with STA, the same procedure was used, but with 5 ml of different concentrations of STA in the chemical bath. Addition of a small amount (10^{-5} M) of STA in the deposition bath enhanced the rate of deposition and significantly improved the photoactivity of the films. A very small (+5%) change in the electrical resistivity was observed for the films in which STA was incorporated. Significant improvement in the film properties was obtained by introducing a small concentration of STA (10^{-5} M) in the deposition bath. The as-deposited films are amorphous in nature, but an annealing treatment (300 °C for an hour in a nitrogen atmosphere) leads to more stoichiometric and polycrystalline films. The optical bandgap of the as-deposited film changed from 1.86 to 1.74 eV after annealing. The films are n-type with resistivity, carrier concentration, and mobility of $5.3 \times 10^6 \Omega$ cm, 1.2×10^{12} cm^{-3}, and 9.8 cm^2 V^{-1} s^{-1}, respectively. WO_3 was detected in the films deposited with STA. The activation energy was found to be 0.53 and 0.56 eV for the films deposited without and with STA, respectively.

5.2.2.4.3 n-Sb_2S_3/p-Ge, n-Sb_2S_3/p-Si, and n-CdSe/p-InP Heterojunctions [192, 332–335]

The chemical composition and structural features of CdSe chemically deposited with different heteropolyacids, and a novel method for fabricating low-cost n-Sb_2S_3/p-Ge heterojunction solar cells by chemical deposition have been reported [333]. The p-Ge substrates used for the fabrication of the heterojunction solar cells were cut from (111) oriented ingots with an acceptor concentration of $N_A = 6 \times 10^{18}$ cm^{-3}. Surfaces were lapped and polished down to 0.25 μm followed by a final polishing carried out mechanochemically using an alkaline silica gel. Then the following cycle was carried out five times: 40% HF dip, ultrapure water rinse, sulfochromic acid mixture dip, water rinse, drying in nitrogen, and a final rinse in 40% HF before n-Sb_2S_3 deposition. It was found that, in the case of n-Sb_2S_3 films chemically deposited with STA on (111) oriented single-crystalline p-Ge and annealed, the heterojunction solar cell properties are considerably improved. Dark current–voltage (I–V) measurements (in the range 298–380 K) for n-Sb_2S_3/p-Ge and STA-deposited n-Sb_2S_3/p-Ge junction devices showed an increase in barrier height (Φ_b) from 0.65 to 0.89 eV, a decrease in ideality factor (n) from 2.21 to 1.38, and a decrease in reverse saturation current density (J_0) from 6.27×10^{-7} to 3.8×10^{-9} A cm^{-2}. Capacitance–voltage (C–V) studies at 1 MHz showed higher values of Φ_o for the improved device. Under AM1 (air mass 1) illumination, the improved junction showed an efficiency (η) of about 7.3% without any antireflection coating, whereas the n-Sb_2S_3 films deposited without STA on p-Ge showed $\eta = 2.4\%$. A significant increase in the normalized spectral response of the n-Sb_2S_3(STA)/p-Ge heterojunction was obtained. This was attributed to a reduction in interface recombination velocity [334]. Analogous results were obtained for n-Sb_2S_3/p-Si junctions where the junction fabricated with Sb_2S_3 containing STA exhibited an efficiency of 7.3% on an active area of 0.05 cm^2

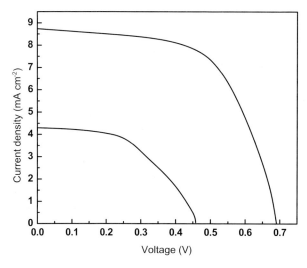

Figure 5.24 Illuminated *I–V* characteristics of n-Sb_2S_3/p-Ge heterojunction devices for Sb_2S_3 films deposited with STA: V_{oc} = 0.69 V, J_{sc} = 8.72 mA cm^{-2}, FF = 62%, η = 7.3%; and for Sb_2S_3 films deposited with STA: V_{oc} = 0.46 V, J_{sc} = 4.32 mA cm^{-2}, FF = 48%, η = 2.4% [334].

(Figure 5.24). Similar results were obtained for n-Sb_2S_3 films deposited with and without STA on p-Si [192] and for n-CdSe/p-InP fabricated with and without STA [335].

5.2.2.4.4 Metal (Pt, Ni, Au)/(CdSe, Sb_2S_3) Thin Films

A novel method for fabricating high-efficiency metal (Pt, Au, Ni)/(CdSe, Sb_2S_3) films for application to Schottky barrier solar cells has been reported [192, 336]. The method is based on the fabrication of n-CdSe or n-Sb_2S_3 thin films chemically deposited with and without STA. The Schottky barrier structures were fabricated on chemically deposited polycrystalline n-CdSe films (about 5 µm) and/or n-Sb_2S_3 films (about 4 µm) on ITO-coated glass (1.5–2.0 Ω□$^{-1}$) [340, 343]. The resulting films were thermally annealed at 350 °C for an hour in a nitrogen atmosphere. After etching, the deposited films in were washed 2% HCl for 5–10 seconds and dried in nitroge. Pt, Ni, or Au of 99.9999% purity was evaporated to form a base for the diodes with different areas in the range 0.04–0.09 cm^2. The Schottky structures were annealed at 100 °C for 5 minutes in a hydrogen atmosphere to improve the reproducibility of the results. Semi-transparent metal (about 130 Å) films served as the rectifying contact, while the ITO-backed contact on the CdSe or Sb_2S_3 acted as an ohmic contact. The performances of the Schottky junctions fabricated with the films deposited with STA, CdSe(STA) or Sb_2S_3(STA) (Figure 5.25), are significantly better than those deposited without STA. Under AM1 illumination, the photovoltaic properties of the improved Pt/CdSe(STA) diode showed V_{oc} = 0.72 V, J_{sc} = 14.1 mA cm^{-2}, FF = 0.70, and η = 5.5%.

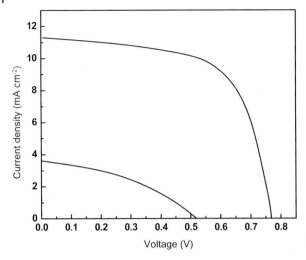

Figure 5.25 Illuminated I–V characteristics of Pt/n-Sb$_2$S$_3$ Schotkky devices for Sb$_2$S$_3$ films deposited with STA: V_{oc} = 0.77 V, J_{sc} = 11.3 mA cm^{-2}, FF = 63%, η = 5.5%; and for Sb$_2$S$_3$ films deposited with STA: V_{oc} = 0.52 V, J_{sc} = 3.6 mA cm^{-2}, FF = 38%, η = 0.7% [336].

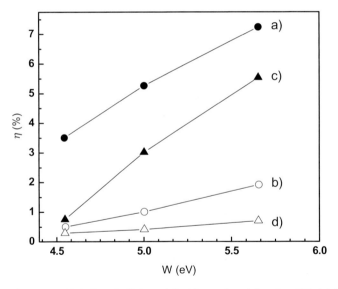

Figure 5.26 Variation of efficiency (η) with metal work function (W): (a) (top level) for metal/n-CdSe(STA); (b) (bottom level) for metal/n-CdSe; (c) (top level) for metal/n-Sb$_2$S$_3$(STA); (d) (bottom level) for metal/n-Sb$_2$S$_3$ [336].

In addition, the ideality factor (n) and saturation current density (J_o) also improved significantly. The variation of the efficiency with the work function of the Schottky metal is important for films fabricated with STA, as shown in Figure 5.26. This variation is very small for films fabricated without STA. This is an indication of Fermi-level pinning for films fabricated without STA. Efficiency measurements at 1 MHz showed that the barrier heights (Φ_o) of the fabricated diodes were 0.62 and 0.59 eV for the Pt/CdSe and Pt/Sb$_2$S$_3$ junctions, respectively, and 0.81 and 0.80 eV for the Pt/CdSe(STA) and Pt/Sb$_2$S$_3$(STA) junctions, respectively. It was also observed that the Φ_o values are independent of the metal work functions, which may be attributed to the Fermi-level pinning of CdSe or Sb$_2$S$_3$ films deposited with and without STA. For Schottky diodes fabricated with CdSe or Sb$_2$S$_3$ deposited without STA, the Fermi levels are pinned at one-third of the bandgap from the valence band edge. For junctions fabricated with CdSe or Sb$_2$S$_3$ deposited with STA, the Fermi levels are pinned at a point slightly below the center of the forbidden gap. The improvement in solar cell properties was attributed to the presence of WO$_3$ in the films deposited with STA. The absorption coefficient is improved due to the presence of WO$_3$ in the films.

5.2.2.4.5 CdS/CdTe Heterojunctions

Various experimental conditions required for CBD on ITO and SnO$_2$ substrates have been identified [337–343]. The chemical bath was based on ammonium acetate (0.5–10^{-3} M), ammonium hydroxide (1–6%), cadmium acetate or cadmium chloride (0.1–10^{-3} M), and thiourea (0.1–10^{-3} M), containing different concentrations and types heteropolyacids (10^{-1}–10^{-8} M). The deposition temperature varied from 50 to 95 °C and the deposition time from 10 minutes to 6 hours. The electrode surface area was 8 cm^2. The annealing temperature was varied from 200 to 500 °C. The effect of the heteropolyacid on the morphology and the chemical composition of the films and on the CdS/CdTe heterojunction parameters (V_{oc}, J_{sc}, FF, η) was determined, as was also the effect of deposition and annealing times. It was found that the presence of heteropolyacids in the deposition bath improved the quality of the films and the solar cell parameters of the CdS/CdTe heterojunction. CdTe was electrodeposited on CdS using a chemical bath composed of HTeO$_2^+$ (10^{-3}–10^{-5} M), CdSO$_4$ (0.1–3 M), and heteropolyacids (10^{-1}–10^{-8} M). The pH of the electrolyte was maintained in the range 1.5–3.5 and the deposition potential was between −0.3 and −0.75 V. A double Ag/AgCl junction was used as the reference electrode and a large Pt grid was used as a counterelectrode. The deposition temperature was in the range 50–95 °C. The deposition time was in the range 1–6 hours. During electrodeposition, the QRP of the electrodeposit was measured at time intervals of 5 seconds in the early stages of the CdTe deposition (i.e., <40 C charge passed) and 20 seconds thereafter. A constant deposition current density was maintained and, after a total plating charge of 20 C had been reached, the plate was removed from the electrolyte, rinsed, and annealed (200–500 °C) for times ranging from 10 minutes to100 hours. The I–V characteristics of the CdS/CdTe-based solar cells were determined under illumination at 100 mW cm^{-2}. The following cell parameters were obtained: V_{oc} = 0.35–0.76 V, J_{sc} = 3–31 mA cm^{-2}, and FF = 25–60%. The reproducibility of the cell parameters was determined and it was found that 80% of the 8 cm^2 area cells exhibited a solar energy conversion efficiency of 6%. This efficiency was obtained for CdS and CdTe preparation and cell fabrication without parameter optimization.

5.3
Systems Development

5.3.1
State-of-the-Art Thin-Film Solar Technology using Chemical, Electrochemical, and/or Sol–Gel Fabrication Methods

The thin-film materials that are discussed in this chapter can be used in depositing one or more thin layers of material on a substrate in the following PV cell technologies: (i) CdTe, (ii) copper indium (gallium) selenide (CIS or CIGS), and (iii) Dye-sensitized solar cell (DSSC). The thickness of such layers can vary from a few nanometers to tens of micrometers.

The CdTe, CIS, and CIGS PV module technologies are in continuous development. Cell efficiency, process optimization, tellurium supply, use and recycling of the extremely toxic cadmium metal, price vulnerability, solar tracking, and market viability are the main issues related to the commercial success of thin-film solar technologies such as CdTe PV solar cells.

The actual module efficiency, which does not exceed 10%, is far from that which CdTe on its own can exhibit. Because of its optimal bandgap for single-junction devices, it may reasonably be expected that efficiencies close to or exceeding 20% (such as already shown by CIS alloys) should be achievable in future practical CdTe cells. Modules of 15% efficiency would then be possible. Improved cells based on a modified CTO/ZTO/CdS/CdTe device structure have been developed at the US National Renewable Energy Laboratory (NREL) which achieved a high FF of 77% and high J_{sc} of nearly $26\,mA\,cm^{-2}$. The CdS/CdTe polycrystalline thin-film solar cell demonstrated (and confirmed by NREL) a total-area efficiency of 16.5% in laboratory studies, the highest efficiency reported for CdTe solar cells [344]. By comparison, CIS and CIGS technologies can exhibit some efficiencies close to 20% for cells with small surface area.

With present technology (2010), if we consider a module efficiency of 11%, we will get an output power of $100\,W\,m^{-2}$. If we assume an actual cost to the customer of \$4 per watt in 2010, we will get $\$400\,m^{-2}$. This corresponds to a panel cost of about \$300. The process optimization of this technology should in the future lead to greater output power at lower cost. Such optimization over a 10- to 20-year period may result in a cost of less than \$1 per watt. For example, if the cost was \$0.5 per watt, the price would drop to $\$50\,m^{-2}$. But if we take into account additional commercialization and installation costs, an installed system might cost \$1.5 per watt or $75\,W\,m^{-2}$. For regions of the world where the sunlight can average $5\,kWh\,m^{-2}\cdot$a day, a price of \$0.03–0.10 per kWh could be achieved.

The availability of Te for use in applications is limited by its abundance (1×10^{-9} kg of Te per kg (or 0.002 ppm) of the Earth's crust). In comparison, the abundance of Cd is 1.5×10^{-7} kg of Cd per kg (or 0.2 ppm) of the Earth's crust [345, 346]. It is a by-product of copper, lead, and gold production. An estimated average of 800–900 metric tons can be produced per year [347], and, in 2007, Te production was 135 metric tons [348]. At the current cell efficiency and thickness, if we take

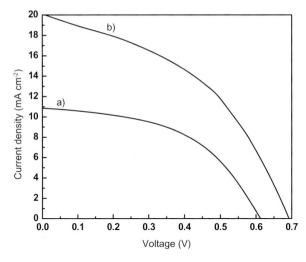

Figure 5.27 Illuminated current-voltage characteristics of the n-CdS/p-CdTe heterojunction devices: (a) for CdS "lm deposited without STA and (b) for CdS "lm deposited with STA.

into account the material density and the mass percentage of Te in CdTe (53%), 1 GW annual production will need approximately 93 metric tons. Because the need for Te is increasing, new suppliers are appearing and its geological exploration may also increase. Development of recycling CdTe PV components may also be an approach that may contribute to more availability of Te. CentennialSolar, a Canadian-based company involved in the PV industry, has a production capacity of CIGS and CdTe of 6 MW.

The abundance of Cd is 100 times higher than that of Te. Cd is produced as a by-product of Zn production. Due to the large quantities of Zn production worldwide, important amounts of Cd are generated from this process. The mass production of CdTe solar modules might therefore not be limited by Cd availability. That is, because Zn is generated in very large quantities, there are substantial amounts of Cd available. Presently when the market does not absorb the Cd, it is cemented and burned, stored for future use, or disposed of as a hazardous material. CdTe PV fabrication may present an interesting and safer option than the current uses of Cd that would be much preferred to its mere disposal [349]. The total production of CdTe/CdS modules of start-up companies manufacturing thin-film solar cells (AVA Solar, PrimeStar, Calyxo/Q-Cells, Antec, Avendi, Abound Solar, and ASP) was 70 MW. At the end of 2009, First Solar, Inc. had the facility to produce 1 GW [350]. The conversion efficiency of the modules is increasing rapidly and First Solar claimed a module efficiency of 12.6% in 2010. This may be compared to Si modules where efficiencies from 13 to 20% are obtained.

New developments related to the improvement in solar cell performance and the influence of the Ga content ($x = Ga/(Ga + In)$) in the absorber for devices using (PVD)In_2S_3-based buffers have been reported [351]. It has been shown that the solar cell efficiency is similar for both buffer layers at lower x values and increases with x only in the case of (CBD)CdS. These trends are in agreement with the occurrence of a conduction band cliff at the CIGSe/(PVD)In$_2$S$_3$ interface [351].

It has also been shown that using window-deposited layers of CdS formed by CBD is convenient for large-area deposition. Morphologies in the absorber layer of CdTe prepared by a close-spaced sublimation technique suited to production have also been demonstrated. A Te-rich layer was produced with chemical etching. The ZnTe/ZnTe:Cu complex layers show superior performance over other back contacts. By using laser scribing and mechanical scribing to fabricate nine minimodules connected in series (54 cm^2), a total-area efficiency of 7.03% was demonstrated [352].

5.3.2
Toxicity and Sustainability Issues

The use and recycling of Cd, which is classified as toxic, might be one of the issues which limits mass production of CdTe PV technology. For example, Cd is one of the six most toxic materials banned by the European Union. Because there are relatively few research results available on the toxicology profile of CdTe, regulatory agencies usually apply Cd criteria as an available approximation [353]. Generally the parameter used to probe the toxicology of a material is the median lethal dose (in mg of substance per kg of organism) of a substance which is defined as the amount of this substance which, when administered (ingested) to a group of experimental animals, will kill 50% of the group in a specific time. From available data in the literature on the median lethal dose of CdS (7080 mg kg^{-1}) [354], Si (3160 mg kg^{-1}) [355], CdTe (2000 mg kg^{-1}) [353], Cd (890 mg kg^{-1}) [356], and CdO (72 mg kg^{-1}) [357], we may conclude that the toxicology of these materials increases as: CdS < Si < CdTe < Cd < CdO. That is, Cd compounds such as CdS and CdTe are less toxic than Cd, while CdO is more toxic than Cd. Accordingly more systematic studies are needed on the toxicology of the materials described in this chapter for thin-film solar cell applications because the increase of their use in PV systems will involve extensive human interfaces. Accordingly, in USA for example, it has been proposed that CdTe be included in the National Toxicology Program [358]. The toxicity of this material may present itself if it is ingested, handled without appropriate gloves, or inhaled as dust [359]. It has been shown that the toxicity is not solely due to the Cd content, but to a highly reactive surface of the CdTe with the interaction of oxygen which can damage the cell membrane and nucleus [360] and to the recrystallization to form toxic cadmium chloride. In contrast it has also been found that the large-scale use of CdTe PV modules does not present any potential health problems and environment issues. Further it was concluded that the environmental risks from CdTe PV modules are minimal because the estimated atmospheric emissions of 0.02 g of Cd per GWh of electricity produced during all the phases of module life are extremely low. Consequently their use is more environmentally acceptable than other uses of Cd including Ni–Cd batteries [349]. By comparison to 10 Ni–Cd batteries, it is found that a CdTe PV module uses Cd about

2500 times more efficiently in producing electricity. A 1 kW CdTe PV system contains less Cd than 10 size-C Ni–Cd batteries. Furthermore, CdTe is more stable and less soluble than the Cd components used in batteries. This is in agreement with the low toxicity of CdTe in comparison to Cd. Moreover, it was shown that the glass plates surrounding CdTe material sandwiched between them (as they are in all commercial modules) seal during a fire and do not allow any Cd release [361].

5.4
Conclusions and Perspectives

During recent decades, the development of PV technologies based on chemically and electrochemically deposited thin films has increased significantly. Growing interest in developing new thin-film materials based on CBD and electrodeposition supports their use in fabrication of low-cost materials. Substantial progress has been made with polycrystalline CdTe. This effort makes possible the fabrication of, for example, Cu–In–Se$_2$ and Cu–In–Ge–Se$_2$ thin films for solar cell technologies. However, technical problems persist and many issues obstruct the successful development of CIS and CdTe thin-film solar cells at low cost. The CdS/CdTe heterojunction still has a fundamental problem of stability due to (i) formation of CdTe$_{1-x}$S$_x$ at the interface, (ii) degradation of the CdTe back contact due to atmospheric interactions and/or corrosion, (iii) impurities in the films, (iv) lack of stability of the films themselves, and (v) lack of control of the chemical composition of CuInSe$_2$ during its direct electrodeposition or sequential deposition. With the increasing use of this technology, health and environmental issues related to human interfaces between the PV materials and the developers and end users must studied in order to determine the level of hazards associated with the use of these materials. That is, the cost of ensuring the safety of these materials must be amortized in the cost of such systems. Also, the availability of the different materials involved in the various compounds must be carefully established. In some cases, *in situ* modification of the original materials, for example with heteropolyacids, has shown significant improvement of cell performance.

The future development of solar energy at low cost (less than $1 per watt) is based on performance which will be obtained from thin-film semiconductors. The elements composing these solar cell systems are in many cases by-products of the production of other elements, and thus production rates cannot be expanded without balancing their cost with additional considerations. A fundamental question therefore arises: will the current generation of PV thin-film materials (CdS, CdTe, CIS, CIGS, Si-based materials) be used to introduce thin-film solar cells into the marketplace? Other direct energy gap materials, for example ZnP$_2$ and Zn$_3$P$_2$, may be considered and developed for solar cell applications. These materials are less environmentally hazardous than CdS, CdTe, CIS, or CIGS. The continuous improvement of the performance of cells and modules using environmentally benign materials is essential for making them available for use in commercial solar systems.

The toxicology of the Cd-based materials increases as: CdS < Si < CdTe < Cd < CdO. That is, Cd compounds such as CdS and CdTe are less toxic than Cd, while CdO is more toxic than Cd. Accordingly more systematic studies are needed on the toxicology of the materials described in this chapter for thin-film solar cell applications because the increase of their use in PV systems will involve extensive human interfaces.

Acknowledgments

We gratefully thank Mr. Oishi and Mrs. Akiko for their generous and great helps in the drawings of the figures.

References

1 Poortmans, J. and Arkhipov, V. (2006) *Thin Film Solar Cells*, John Wiley & Sons, Ltd.
2 Lewis, Nathan S. and Crabtree, George "Basic Research needs for energy utilisation", DOE Report on the Basic Research Workshop on Solar Cell Utilisation, April 18–21, 2005. http://www.sc.doe.gov/bes/reports/files/SEU_rpt.pdf (accessed, January, 10, 2010).
3 Grigas, J., Meshkauskas, J., and Orliukas, A. (1976) *Phys. Stat. Sol. A*, **37**, K39.
4 Anita, T., Hanafusa, A., Kitamura, S., Takakura, H., and Murozono, M. (1991) Proc. 22nd IEEE Photovoltaic Specialists Conference, Las Vegas, NV. IEEE, New York, p. 946.
5 Cattell, A.F. and Cullis, A.G. (1982) *Thin Solid Films*, **92**, 211.
6 Wagner, B.K., Tran, T.K., and Ogle, W. (1996) *J. Cryst. Growth*, **164**, 202.
7 Klechkovskaya, V.V., Maslov, V.N., and Murad, M.B. (1989) *Sov. Phys. Crystallogr.*, **34**, 105.
8 Nicolau, Y.F. and Menand, J.C. (1988) *J. Cryst. Growth*, **92**, 128.
9 Kashani, H. (1996) *Thin Solid Films*, **288**, 50.
10 Lindroos, S., Kanniainen, T., and Leskela, M. (1994) *Appl. Surf. Sci.*, **75**, 70.
11 Mattox, D.M. (1998) *Handbook of Physical Vapor Deposition (PVD) Processing: Film Formation, Adhesion, Surface Preparation and Contamination Control*, Noyes Publications, Westwood, NJ.
12 Peredins, D. and Gauckler, L.J. (2005) *J. Electroceram.*, **14**, 103–111.
13 McCray, W.P. (2007) *Nat. Nanotechnol.*, **2** (5), 2.
14 Morita, Y. and Narusawa, T. (1992) *Jpn. J. Appl. Phys.*, **31**, L1396.
15 Cho, A.Y. and Arthur, J.R., Jr. (1975) Molecular beam epitaxy. *Prog. Solid State Chem.*, **10**, 157.
16 Shirakata, S. and Chichibu, S. (2000) *J. Appl. Phys.*, **87**, 3793.
17 Patkan, H.M. and Lokhandre, C.D. (2004) *Mater. Sci.*, **27** (2), 85.
18 Chichibu, S., Nakanishi, H., and Shirakata, S. (1995) *Appl. Lett.*, **66** (25), 3513.
19 Yanagi, H., Inoue, S.-I., Veda, K., Kawazoe, H., Hosono, H., and Hamada, N. (2000) *J. Appl. Phys.*, **88** (7), 4159.
20 Alonso, M.I., Pascual, J., Garriga, M., Kikuno, Y., Yamamoto, N., and Wakita, K. (2000) *J. Appl. Phys.*, **88** (4), 1923.
21 Tanaka, T., Wakahara, A., Yoshida, A., Oshima, T., Itoz, H., and Okada, S. (2000) *J. Appl. Phys.*, **87** (7), 3283.
22 Hodes, G. (2003) *Chemical Solution Deposition of Semiconductor Films*, Marcel Dekker, New York.
23 Savadogo, O. (1998) *Solar Energy Mater. Solar Cells*, **52**, 361.

24 Nair, P.K., Nair, M.T.S., Garcia, V.M., Arenas, O.L., Pena, Y., Castillo, A., Ayala, I.T., Gomezdaza, O., Sanchez, A., Campos, J., Hu, H., Suarez, R., and Rincon, M.E. (1998) *Solar Energy Mater. Solar Cells*, **52**, 313.

25 Ortega-Borges, R. and Lincot, D. (1993) *J. Electrochem. Soc.*, **140**, 3464.

26 Puscher, C. (1869) *Dingl. J.*, **190**, 421.

27 Wang, Y. and Herron, N. (1987) *J. Phys. Chem.*, **9** (1), 257.

28 Potter, B.G. and Simmons, J.H. (1988) *Phys. Rev. B*, **37**, 10838.

29 Mathieu, H., Richard, T., Allegre, J., Lefebvre, P., and Arnaud, G. (1995) *J. Appl. Phys.*, **77**, 287.

30 Chestnoy, N., Harris, T.D., Hull, R., and Brus, L.E. (1986) *J. Phys. Chem.*, **90**, 3393.

31 Rossetti, R., Hull, R., Gibson, J.M., and Brus, L.E. (1985) *J. Chem. Phys.*, **82**, 552.

32 Chopra, K.L., and Das, S.R. (1983) *Thin Film Solar Cells*, Plenum Press, New York.

33 Savadogo, O. and Mandal, K.C. (1992) *Solar Energy Mater. Solar Cells*, **26**, 117.

34 Mokrushin, S.G. and Yu, D. (1961) Tkachev. *Kolloidn Zh.*, **23**, 438, *Colloid J. USSR* **23** (1961) 366.

35 Kitaev, G.A., Uritskaya, A.A., and Mokrushin, S.G. (1965) *Russian J. Phys. Chem.*, **39**, 1101.

36 Nagao, M., and Watanabe, S. (1968) *Jpn. J. Appl. Phys.*, **7**, 684.

37 Pavaskar, N.R., Menezes, C.A., and Sinha, A.B.P. (1977) *J. Electrochem. Soc.*, **124**, 743.

38 Kaur, I., Pandya, D.K., and Chopra, K.L. (1977) *J. Electrochem. Soc.*, **127**, 943.

39 Call, R.L., Jaber, N.K., Seshan, K., and Whyte, J.R. (1980) *Solar Energy Mater.*, **9**, 373.

40 Danaher, W.J., Lyons, I.E., and Morris, G.C. (1985) *Solar Energy Mater.*, **14**, 137.

41 Mondal, A., Chaudhuri, T.K., and Pramanik, P. (1983) *Solar Energy Mater.*, **7**, 431.

42 Nair, P.K., Nair, M.T.S., Campos, J., and Sansores, L.E. (1987) *Solar Cells*, **22**, 211.

43 Nair, P.K., Nair, M.T.S., and Campos, J. (1987) *Solar Energy Mater.*, **16**, 441.

44 Nair, P.K. and Nair, M.T.S. (1987) *Solar Cells*, **22**, 103.

45 Nair, P.K. and Nair, M.T.S. (1987) *Solar Energy Mater.*, **16**, 431.

46 Lincot, D. and Ortega-Borges, R. (1992) *J. Electrochem. Soc.*, **139**, 1880.

47 Lincot, D., Ortega-Borges, R., and Froment, M. (1993) *Philos. Mag. B*, **68**, 185.

48 Lincot, D., Ortega-Borges, R., and Froment, M. (1993) *Abstract 190, Extended Abstracts*, vol. 93-2, Electrochemical Society, p. 308.

49 Kitaev, G.A., Makurin, Y.N., and Dvoinin, V.I. (1976) *Russian J. Phys. Chem.*, **50**, 1828.

50 Sebastian, P.J. and Nair, P.K. (1992) *Adv. Mater. Opt. Electron.*, **1**, 211.

51 Chu, T.L. (1988) *Solar Cells*, **23**, 131.

52 Dona, J.M. and Herrero, J. (1992) *J. Electrochem. Soc.*, **139**, 2810.

53 Chu, T.L., Chu, S.L., Wang, C., and Wu, C.L. (1992) *J. Electrochem. Soc.*, **139**, 2443.

54 Lincot, D. and Vedel, J. (1991) 10th European Photovoltaic Solar Energy Conference, Lisbon, Portugal.

55 Kartaev, G.A. and Romanov, I.Y. (1974) *Izv. Vyssh. Uchebn. Zaved. Khim. Tekhnol. (Kiev)*, **17**, 1427.

56 Marcotrigiano, D.G., Peyronel, G., and Battistuzzi, R. (1972) *J. Chem. Soc. Perkin Trans. II*, 1539.

57 Fulop, G., Doty, M., Meyer, P., Betz, J., and Liu, C.H. (1982) *Appl. Phys. Lett.*, **40**, 327.

58 (a) Ezema, F.I. (2004) *Acad. Open J. Tech. College Bourgas*, **11**, 1; (b) Okoli, D.N., Ekpunobi, A.J., and Okeke, C.E. (2006) *Acad. Open J. Tech. College Bourgas*, **18**, 1.

59 McNerill, W. (1971) US Patent 3,573,177.

60 Petuenju, E.N. and Savadogo, O. (2008) Characterization of chemically deposited low-cost II-VI thin films solar cells: modifying effects with catalytic silicotungstic acid (STA), poster nmes08-85. Presented at the 7th International Symposium on New Materials for Electrochemical Systems, June 24–27, 2008, Montreal, Canada.

61 Dennison, S. (1993) *Electrochim. Acta*, **38**, 2395.
62 Gobrecht, H., Liess, H.D., and Tausend, A. (1973) *Ber. Dtsch. Bursenges. Phys. Chem.*, **67**, 930.
63 Panicker, M.P.H., Knaster, M., and Kroger, F.A. (1978) *J. Electrochem. Soc.*, **125**, 566.
64 Singh, R.P., Singh, S.L., and Chandra, S. (1986) *J. Phys. D*, **19**, 1299.
65 Lejmi, N. and Savadogo, O. (2001) *Solar Energy Mater. Solar Cells*, **70**, 83.
66 Gernes, D.C. and Montillon, C.H. (1942) *Trans. Electrochem. Soc.*, **81**, 231.
67 Brenner, A. (1963) *Electrodeposition of Alloys*, vol. 2, Academic Press, New York, p. 611.
68 Babu, S.M., Dhanasek, R., and Ramasamy, P. (1991) *Thin Solid Films*, **202**, 67.
69 Engelken, R.D., and Doren, T.P. (1985) *J. Electrochem. Soc.*, **132**, 2904.
70 Kaumpmann, A., Cowache, P., Molsili, B., Ortega-Borges, R., Lincot, D., and Vedel, J. (1989) Proceedings of the 12th European Photovoltaic Solar Energy Conference, Amsterdam, The Netherlands, p. 664.
71 Saraby-Reintges, A., Peter, L.M., Ozsan, M.E., Dennison, S., and Webster, S. (1993) *J. Electrochem. Soc.*, **140**, 2880.
72 Mori, E. and Rajeshwar, K. (1989) *J. Electroanal. Chem.*, **258**, 15.
73 Mishra, K.K. and Rajeskman, K. (1989) *J. Electroanal. Chem.*, **273**, 169.
74 Corwache, P., Lincot, D., and Vedel, J. (1989) *J. Electrochem. Soc.*, **136**, 1646.
75 Danaher, W.J. and Lyons, L.E. (1978) *Nature*, **271**, 139.
76 Danaher, W.J. and Lyons, L.E. (1983) *Aust. J. Chem.*, **36**, 101.
77 Danaher, W.J. and Lyons, L.E. (1984) *Aust. J. Chem.*, **37**, 689.
78 Lyons, L.E., Morris, G.C., Horton, D.H., and Keyes, J.G. (1984) *J. Electroanal. Chem.*, **168**, 101.
79 Llabres, J. (1984) *J. Electrochem. Soc.*, **131**, 464.
80 Bhattacharya, R.N. and Rajeshwar, K. (2002) *J. Electrochem. Soc.*, **131**, 1984.
81 Takahashi, M., Uosaki, K., and Kita, H. (1984) *J. Electrochem. Soc.*, **131**, 2304.
82 Takahashi, M., Uosaki, K., and Kita, H. (1984) *J. Appl. Phys.*, **55**, 3879.
83 Uosaki, K., Takahashi, M., and Kita, H. (1984) *Electrochim. Acta*, **29**, 279.
84 Gerritsen, H.J. (1984) *J. Electrochem. Soc.*, **131**, 136.
85 Carbonelle, P., Labar, C., and Lamberts, L. (1987) *Analysis*, **15**, 286.
86 Sella, C., Boncorps, P., and Vedel, J. (1986) *J. Electrochem. Soc.*, **133**, 2043.
87 Maurin, G., Solorza, O., and Takenouti, H. (1986) *J. Electroanal. Chem.*, **202**, 323.
88 Touskova, J., Kindl, D., and Tousek, J. (1989) *Solar Energy Mater.*, **18**, 377.
89 Sanyal, G.S., Mondal, A., Mandal, K.C., Ghosh, B., Sata, K., and Mukherjee, M.K. (1990) *Solar Energy Mater.*, **20**, 395.
90 Mori, E., Baker, C.K., Reynolds, J.R., and Rajeshwar, K. (1988) *J. Electroanal. Chem.*, **252**, 441.
91 Vittori, O. (1980) *Anal. Chim. Acta*, **121**, 315.
92 Ngac, N.V., Vittori, O., and Quarin, G. (1984) *J. Electroanal. Chem.*, **167**, 227.
93 Kazacos, M.S. and Miller, B. (1980) *J. Electrochem. Soc.*, **127**, 2378.
94 Kazacos, M.S. (1983) *J. Electroanal. Chem.*, **148**, 233.
95 Pottier, D. and Maurin, G. (1989) *J. Appl. Electrochem.*, **19**, 361.
96 (a) Brinker, C.J. and Scherer, G.W. (1990). *Sol-Gel Science: The Physics and Chemistry of Sol-Gel Processing*. Academic Press. ISBN 0121349705. (b) Hench, L.L. and West, J.K. (1990) "The Sol-Gel Process", *Chemical Reviews*, **90**, 33.
97 Brinker, C.J. and Scherer, G.W. (1990) *Sol-Gel Science: The Physics and Chemistry of Sol-Gel Processing*, Academic Press, New York.
98 Lakshmi, B.B., Patrissi, C.J., and Martin, C.R. (1997) *Chem. Mater.*, **9**, 2544.
99 Hulteen, J.C. and Martin, C.R. (1997) *J. Mater. Chem.*, **7**, 1075.
100 Hoyer, P. (1994) *Adv. Mater.*, **8**, 857.
101 Lei, Y., Zhang, L.D., and Meng, G.W. (2001) *Appl. Phys. Lett.*, **78**, 1125.
102 Znaidia, L., Soler Illiab, G.J.A.A., Benyahiaa, S., Sanchezb, C., and Kanaev, A.V. (2003) *Thin Solid Films*, **428**, 257.
103 Peredinis, D. and Gauckler, L.J. (2005) *J. Electroceram.*, **14**, 103.

104 Mochel, J.M. (1951) US Patent 2,564,707.
105 Hill, J.E. and Chamberlin, R.R. (1964) US Patent 3,148,084.
106 Balkenende, A.R., Bogaerts, A., Scholtz, J.J., Tijburg, R.R.M., and Willems, H.X. (1996) *Philips J. Res.*, **50** (3–4), 365.
107 Arya, S.P.S. and Hintermann, H.E. (1990) *Thin Solid Films*, **193** (1–2), 841.
108 Chen, C.H., Kelder, E.M., van der Put, P.J.J.M., and Schoonman, J. (1996) *J. Mater. Chem.*, **6** (5), 765.
109 Viguie, J.C. and Spitz, J. (1975) *J. Electrochem. Soc.*, **122** (4), 585.
110 Choy, K.L. and Su, B. (2001) *Thin Solid Films*, **388** (1–2), 9.
111 Kul, M., Zor, M., Aybek, A.S., Irmak, S., and Turan, E. (2007) *Solar Energy Mater. Solar Cells*, **91**, 882.
112 Pierson, H.O. (1999) *Handbook of Chemical Vapor Deposition: Principles, Technology, and Applications*, 2nd edn, Noyes Publications, Norwich, NY.
113 O. win Bryan, May 7th, 1998, http://www.timedomaincvd.com/CVD_Fundamentals/introduction/introTOC.html (accessed, December 15, 2009).
114 Pich, J. and Koch, W. (1988) *Staub. Reinhaltung der Luft*, **48** (12), 455.
115 Woods, L. and Meyers, P. (2002) ITN Energy Systems Littleton, Colorado. Atmospheric Pressure Chemical Vapor Deposition and Jet Vapor Deposition of CdTe for High Efficiency Thin Film PV Devices. Final technical report, 26 January 2000 to 15 August 2002, NREL/SR-520-32761.
116 Nagayama, H., Honda, H., and Kawahara, H. (1988) *J. Electrochem. Soc.*, **135**, 2013.
117 Kishimoto, H., Takahama, K., Hashimoto, N., Aoi, Y., and Deki, S. (1998) *J. Mater. Chem.*, **8**, 2019.
118 Wang, X.P., Yu, Y., Hu, X.F., and Gao, L. (2000) *Thin Solid Films*, **371**, 148.
119 Yu, J.G., Yu, H.G., Cheng, B., Zhao, X.J., Yu, J.C., and Ho, W.K. (2003) *J. Phys. Chem. B*, **107**, 13871.
120 Dutschke, A., Diegelmann, C., and Lobmann, P. (2003) *J. Mater. Chem.*, **13**, 1058.
121 Deki, S., Iizuka, S., Mizuhata, M., and Kajinami, A. (2005) *J. Electroanal. Chem.*, **584**, 38.
122 (a) Yu, J.G., Yu, H.G., Ao, C.H., Lee, S.C., Yu, J.C., and Ho, W.K. (2006) *Thin Solid Films*, **496**, 273.; (b) Tauste, D., Domenech, X., Casan-Pastor, N., and Ayllon, J.A. (2007) *J. Photochem. Photobiol. A*, **187**, 45.
123 Lee, M.K., Huang, J.J., and Yen, C.F. (2007) *J. Electrochem. Soc.*, **154**, 117.
124 Cui, Y., Du, H., and Wen, L. (2008) *J. Mater. Sci. Technol.*, **24**, 675.
125 Sun, J. and Sun, Y.-C. (2004) *Chin. J. Chem.*, **22**, 661.
126 Houng, M., Wang, Y., Huang, C., Huang, S., and Horng, J. (2000) *Solid-State Electron.*, **44**, 1917.
127 Huang, C., Chen, J., and Huang, S. (2001) *Mater. Chem. Phys.*, **70**, 78.
128 Houng, M., Huang, C., Wang, Y., Wang, N., and Chang, W.J. (1997) *Appl. Phys.*, **82**, 5788.
129 Huang, C., Houng, M., Wang, Y., Wang, N., and Chen, J.J. (1998) *Vac. Sci. Technol. A*, **16**, 2646.
130 Chou, J. and Lee, S. (1994) *Appl. Phys. Lett.*, **64**, 1971.
131 Yeh, C., Chen;, C., Lur, W., and Yen, P. (1995) *Appl. Phys. Lett.*, **66**, 938.
132 Yeh, J., Lin, S., and Hong, T. (1995) *IEEE Electron Device Lett.*, **16**, 316.
133 Yeh, J. and Lee, S. (1999) *IEEE Electron Device Lett.*, **20**, 138.
134 Yeh, K., Jeng, M., and Hwu, J. (1999) *Solid-State Electron.*, **43**, 671.
135 Richardson, T. and Rubin, M. (2001) *Electrochim. Acta*, **46**, 2119.
136 Tsukuma, K., Akiyama, T., Yamada, N., and Imai, H.J. (1998) *J. Non-Cryst. Solids*, **231**, 161.
137 Awazu, K., Kawazoe, H., and Seki, K.J. (1992) *J. Non-Cryst. Solids*, **151**, 102.
138 Deki, S., Aoi, Y., Hiroi, O., and Kajinami, A. (1996) *Chem. Lett.*, 433.
139 Tsukuma, K., Akiyama, T., and Imai, H.J. (1997) *J. Non-Cryst. Solids*, **210**, 48.
140 Ino, J., Hishinuma, A., Nagayama, H., and Kawahara, H. (1989) US Patent 4,882,183; *Chem. Abstr.* (1989) **111**, 62815.
141 Deki, S., Aoi, Y., Miyake, Y., Gotoh, A., and Kajinami, A. (1996) *Mater. Res. Bull.*, **31**, 1399.

142 Najdoski, M., Grozdanov, I., and Minceva-Sukarova, B. (1996) *J. Mater. Chem.*, **6**, 761.
143 Izaki, M. and Omi, T. (1997) *J. Electrochem. Soc.*, **144**, L3.
144 Pramanik, P. and Bhattacharya, S. (1990) *J. Electrochem. Soc.*, **137**, 3869.
145 Deki, S., Aoi, Y., Okibe, J., Yanagimoto, H., Kajinami, A., and Mizuhata, M. (1997) *J. Mater. Chem.*, **7**, 1769.
146 Nicolau, Y.F. (1985) *Appl. Surf. Sci.*, **22–23**, 1061.
147 Ristov, M., Sinadinoovski, G.J., Grozdanov, I., and Mitreski, M. (1987) *Thin Solid Films*, **149**, 65.
148 Ristov, M., Sinadinoovski, G.J., Grozdanov, I., and Mitreski, M. (1985) *Thin Solid Films*, **123**, 63.
149 Tolstoy, V.P. (2006) *Uspekhi Khimii*, **75**, 183.
150 Hodes, G. (2007) *Phys. Chem. Chem. Phys.*, **9**, 2181.
151 Hidber, P.C., Helbig, W., Kim, E., and Whitesides, G.M. (1996) *Langmuir*, **12**, 1375.
152 Warren, S., Reitzle, A., Kazimirov, A., Ziegler, J.C., Bunk, O., Cao, L.X., Renner, F.U., Kolb, D.M., Bedzyk, M.J., and Zegenhagen, J. (2002) *Surf. Sci.*, **496**, 287.
153 Chen, H.-I., Chou, Y.-I., and Chu, C.-Y. (2002) *Sens. Actuators B*, **85**, 10.
154 van der Putten, A.M.T. and de Bakker, J.-W.G.J. (1993) *Electrochem. Soc.*, **140**, 2229.
155 Nakanashi, T., Masuda, Y., and Koumoto, K. (2005) *J. Cryst. Growth*, **284**, 176.
156 Parikh, H. and De Guire, M.R. (2009) *J. Ceram. Soc. Japan*, **117**, 228.
157 Brien, P.O. and Mc Alesse, J. (1998) *J. Mater. Chem.*, **8**, 2309.
158 Breen, M.L., Woodward T.I.V., Schwartz, D.K.J. and Apblett, A.W. (1998) *Chem. Mater.*, **10**, 710.
159 Arias-Carbajal Readigos, A., Garcıa, V.M., Gomezdaza, O., Campos, J., Nair, M.T.S., and Nair, P.K. (2000) *Semicond. Sci. Technol.*, **15**, 1022.
160 Ramanathan, K., Dhere, R.G., Couts, T.J., Chu, T., and Chu, S. (1992) NREL Photovoltaic Advanced Research and Development 11th Review Meeting, Denver, CO, 13–15 May 1992.
161 Chopra, K.L. and Das, S.R. (1983) *Thin Film Solar Cells*, Plenum Press, New York.
162 Nair, P.K., Campos, J., and Nair, M.T.S. (1988) *Semicond. Sci. Technol.*, **3**, 143.
163 Gracia-Jimenez, M., Martinez, G., Martinez, J.L., Gomez, E., and Zehe, A. (1984) *J. Electrochem. Soc.*, **131**, 2974.
164 Sahu, S.N. and Chandra, S. (1987) *Solar Cells*, **22**, 163.
165 Nicolau, Y.P. and Menard, J.C. (1988) *J. Cryst. Growth*, **92**, 128.
166 Padam, G.K., Rao, S.U.M., and Malhotra, G.L. (1988) 20th IEEE Photovoltaic Specialists Conference. IEEE, New York, p. 1591.
167 Nair, M.T.S., Nair, P.K., and Campos, J. (1988) *Thin Solid Films*, **16**, 21.
168 Mondon, F. (1985) *J. Electrochem. Soc.*, **132**, 319.
169 Martinez, G., Martinez, J.L., and Zehe, A. (1982) *Appl. Phys. Lett.*, **40**, 1031.
170 Hall, R.B., Birkmire, R.W., Ester, E., Hench, T.L., and Meakin, J.D. (1980) 14th IEEE Photovoltaic Specialists Conference. IEEE, New York, p. 706.
171 Pawur, S.H., Deshmukh, L.P., and Lokhande, C.D. (1984) *Indian J. Pure Appl. Phys.*, **22**, 315.
172 Dhere, G., Waterhouse, L., Sundaram, B., Melendez, O., Parikh, R., and Patnaik, B.K. (1993) 23rd Photovoltaic Specialists Conference, Louisville, KY.
173 Nair, M.T.S., Nair, P.K., Zingare, R.A., and Meyers, E.A. (1994) *J. Appl. Phys.*, **75**, 1557.
174 Sebastian, P.J. and Hu, H. (1994) *Adv. Mater. Opt. Electron.*, **4**, 407.
175 Baranski, A.S., Fawcett, W.R., and McDonald, A.C. (1984) *J. Electroanal. Chem.*, **160**, 271.
176 Baranski, A.S. and Fawcett, W.R. (1984) *J. Electrochem. Soc.*, **131**, 2509.
177 Preusser, S. and Cocivera, M. (1988) *J. Electroanal. Chem.*, **252**, 139.
178 Fatas, E. and Herrasti, P. (1988) *Electrochim. Acta*, **33**, 959.
179 McCann, J.F. and Kazacos, M.S. (1981) *J. Electroanal. Chem.*, **119**, 409.
180 Power, G.P., Peggs, D.R., and Parker, A.J. (1981) *Electrochim. Acta*, **26**, 681.

181 Fatas, E., Duo, R., Herrasti, P., Arjona, F., and Garcia-Camarero, E. (1984) *J. Electrochem. Soc.*, **131**, 2243.

182 Fatas, E., Herrasti, P., Arjona, F., Garcia-Camarero, E., and Leon, M. (1986) *J. Mater. Sci. Lett.*, **5**, 583.

183 Jackowska, K. and Skompska, M. (1986) *Polish J. Chem.*, **60**, 551.

184 Das, S.K. and Morris, G.C. (1992) *J. Appl. Phys.*, **72**, 4940.

185 Morris, G.C., Tanner, P.G., and Tottszer, A. (1991) *Mater. Forum*, **15**, 21.

186 Němec, P., Šimurda, M., Němec, I., Formánek, P., Němcová, Y., Sprinzl, D., Trojánek, F., and Mal, P. (2008) *Phys. Stat. Sol. (a)*, **205** (10), 2324.

187 Němec, P., Němec, I., Nahálková, P., Němcová, Y., Trojánek, F., and Malý, P. (2002) *Thin Solid Films*, **403/404**, 9.

188 Němec, P., Němec, I., Nahálková, P., Knížek, K., and Malý, P. (2002) *J. Cryst. Growth*, **240**, 484.

189 Trojánek, F., Cingolani, R., Cannoletta, D., Mikeš, D., Němec, P., Uhlířová, E., Rohovec, J., and Malý, P. (2000) *J. Cryst. Growth*, **209**, 695.

190 Němec, P., Mikeš, D., Rohovec, J., Uhlířová, E., Trojánek, F., and Malý, P. (2000) *Mater. Sci. Eng. B*, **69/70**, 500.

191 Němec, P., Šimurda, M., Němec, I., Sprinzl, D., Formánek, P., and Malý, P. (2006) *J. Cryst. Growth*, **292**, 78.

192 Savadogo, O. and Mandal, K.C. (1993) *Appl. Phys. Lett.*, **63**, 12.

193 Nair, M.T.S., Nair, P.K., Pathirana, H.M.K.K., Zingaro, R.A., and Meyers, E.A. (1993) *J. Electrochem. Soc.*, **140**, 2987.

194 García, V.M., Nair, M.T.S., Nair, P.K., and Zingaro, R.A. (1996) *Semicond. Sci. Technol.*, **11**, 427.

195 Chopra, K.L. and Das, S.R. (1983) *Thin Film Solar Cells*, Plenum, New York; Nair, M.T.S. and Nair, P.K. (1989) *Semicond. Sci. Technol.*, **4**, 191.

196 Grozdanov, I. and Najdoski, M. (1995) *J. Solid State Chem.*, **114**, 469.

197 Nair, M.T.S. and Nair, P.K. (1989) *Semicond. Sci. Technol.*, **4**, 189.

198 Nair, P.K. and Nair, M.T.S. (1987) *Solar Cells*, **22**, 103.

199 Stanley, A.G. (1975) Cadmium sulfide solar cells, in *Applied Solid State Sciences*, vol. 5 (ed. R. Wolfe), Academic Press, New York, p. 251.

200 Cotton, F.A. and Wilkinson, G. (1980) *Advanced Inorganic Chemistry*, John Wiley & Sons, Inc., New York, p. 80.

201 Griffith, E.H., Hunt, G.W., and Amma, E.L. (1976) *J. Chem. Soc. Chem. Commun.*, 432.

202 Pathana, H.M., Desaib, J.D., and Lokhande, C.D. (2002) *Appl. Surf. Sci.*, **202**, 47.

203 Pathan, H.M., Lokhande, C.D., Amalnerkar, D.P., and Seth, T. (2003) *Appl. Surf. Sci.*, **211**, 48.

204 Estrada, C.A., Nair, P.K., Nair, M.T.S., Zingaro, R.A., and Meyers, E.A. (1994) *J. Electrochem. Soc.*, **141**, 802.

205 Shafizade, R.B., Ivanova, I.V., and Kazinets, M.M. (1978) *Thin Solid Films*, **55**, 211.

206 Levy-Clement, C., Neumann-Spallart, M., Haram, S.K., and Santhanam, K.S.V. (1997) *Thin Solid Films*, **302**, 12.

207 Garg, J.C., Garg, P., Sharma, R.P., and Sharma, K.C. (1990) *J. Semicond. Mater. Devices*, **2**, 9.

208 Padam, G.K. (1987) *Thin Solid Films*, **150**, L-89.

209 Lakshmi, M., Bindu, K., Bini, S., Vijayakumar, K.P., Sudha, C., Kartha, C.S., Abe, T., and Kashiwaba, Y. (2001) *Thin Solid Films*, **386**, 127.

210 Pejova, B. and Grozdanov, I. (2001) *J. Solid State Chem.*, **158**, 59.

211 Nair, P.K., García, V.M., Fernandez, A.M., Ruiz, H.S., and Nair, M.T.S. (1991) *J. Phys. D*, **24**, 441.

212 Nair, P.K. and Nair, M.T.S. (1992) *Semicond. Sci. Technol.*, **7**, 239.

213 Nair, P.K., Nair, M.T.S., and Campos, J. (1987) *Proc. SPIE*, **823**, 256.

214 Padam, G.K. and Rao, S.U.M. (1986) *Solar Energy Mater.*, **13**, 297.

215 Cahen, D. (1987) *Solar Energy Mater.*, **15**, 225.

216 Padam, G.K. and Rao, S.U.M. (1987) *Solar Energy Mater.*, **15**, 227.

217 Arjona, F., Elizalde, E., Garcia-Camerero, E., Feu, A., Lacal, B., Leon, M., Llabres, J., and Rueda, F. (1979) *Solar Energy Mater.*, **1**, 379.

218. Mondal, A. and Pramanik, P. (1983) *J. Solid State Chem.*, **47**, 81.
219. Mondal, A. and Pramanik, P. (1983) *J. Solid State Chem.*, **55**, 116.
220. Hathaway, B.J. and Tomlinson, A.A.G. (1970) *Coordin. Chem. Rev.*, **5**, 1.
221. Pavaskar, N.R., Menezes, C.A., and Sinha, A.B.P. (1977) *J. Electrochem. Soc.*, **124**, 743.
222. Boudreau, R.A. and Rauh, R.D. (1983) *J. Electrochem. Soc.*, **130**, 513.
223. Ubale, U., Sangawar, V.S., and Kulkarni, D.K. (2007) *Bull. Mater. Sci.*, **30** (2), 147.
224. Hankare, P.P., Bhuse, V.M., Garadkar, K.M., Delekar, S.D., and Mulla, I.S. (2004) *Semicond. Sci. Technol.*, **19**, 70.
225. Hankare, P.P., Bhuse, V.M., Garadkar, K.M., Delekar, S.D., and Bhagat, P.R. (2004) *Semicond. Sci. Technol.*, **19**, 277.
226. Patil, V.B., More, P.D., Sutrave, D.S., Shahane, G.S., Mulik, R.N., and Deshmukh, L.P. (2000) *Mater. Chem. Phys.*, **65**, 282.
227. Schredre, B., Schmidt, T., Patschek, V., Winkler, U., Materny, A., Umbach, E., Lerch, M., Muller, G., Kiefer, W., and Spanhel, L. (2000) *J. Phys. Chem. B*, **104**, 1677.
228. Li, G.M., Zingaro, R.A., Segi, M., Reibenspies, J.H., and Nakajima, T. (1997) *Organometallics*, **16**, 756.
229. Li, G.M. and Zingaro, R.A. (1998) *J. Chem. Soc. Perkin Trans.*, **1**, 647.
230. Sotelo-Lerma, M., Zingaro, R.A., and Castillo, S.J. (2001) *J. Organometallic Chem.*, **623**, 81.
231. Mane, R.S. and Lokhande, C.D. (2000) *Thin Solid Films*, **65**, 1.
232. Grozdanov, I. (1994) *Semicond. Sci. Technol.*, **9**, 1234.
233. Schlamp, M.C., Peng, X.G., and Alivisators, A.P. (1997) *J. Appl. Phys.*, **82**, 5837.
234. Mattoussi, H., Radzilowski, L.H., Dabbousi, B.O., Thomas, E.L., Bawendi, M.G., and Rubner, M.F. (1998) *J. Appl. Phys.*, **83**, 7965.
235. Huynh, W., Peng, X., and Alivisatos, A.P. (1999) *Adv. Mater.*, **11**, 123.
236. Mach, R. and Muller, G.O. (1992) *Phys. Stat. Sol. A*, **59**, 11.
237. Mika, P.V. and Lindroos, S. (1988) *Appl. Surf. Sci.*, **136**, 131.
238. Fukarova, M., Juruskovska, M., Ristov, M., and Andonow, A. (1997) *Thin Solid Films*, **299**, 149.
239. Asikainen, T., Ritola, M., and Leskela, M. (1994) *Appl. Surf. Sci.*, **82**, 122.
240. Naghavi, N., Spiering, S., Powalla, M., Cavana, B., and Lincot, D. (2003) *Prog. Photovoltaics*, **11**, 437.
241. Nair, P.K., Huang, L., Nair, M.T.S., Hu, H., Meyers, E.A., and Zingaro, R.A. (1997) *J. Mater. Res.*, **12**, 651.
242. Moskovits, M. (1990) *Chemical Physics of Atomic and Molecular Clusters* (ed. G. Soles), North Holland, Amsterdam, p. 397.
243. Nair, M.T.S. and Nair, P.K. (1991) *Semicond. Sci. Technol.*, **6**, 132.
244. Nair, P.K., Ocampo, M., Fernandez, A., and Nair, M.T.S. (1990) *Energy Mater.*, **20**, 235; García, V.M., Nair, M.T.S., and Nair, P.K. (1991) *Solar Energy Mater.*, **23**, 47.
245. Nair, M.T.S. and Nair, P.K. (1990) *Semicond. Sci. Technol.*, **5** (125), 1225.
246. Varkey, A.J. (1991) *Solar Energy Mater.*, **21**, 291.
247. Engelken, R.D., McCloud, H.E., Lee, C., Slayton, M., and Ghreishi, A. (1987) *J. Electrochem. Soc.*, **134**, 2696.
248. Hariskos, D., Spiering, S., and Powalla, M. (2005) *Thin Solid Films*, **480–481**, 99.
249. Busev, A.I. (1962) *The Analytical Chemistry of Indium*, Pergamon Press.
250. Swiftand, E.H. and Butler, E.A. (1956) *Anal. Chem.*, **28**, 146.
251. Govender, K., Smyth-Boyle, D., and O'Brien, P. (2002) *MRS Symp. Proc.*, **692**, H9.33.
252. Bayón, R. and Herrero, J. (1999) Proceedings of the 11th Workshop on Quantum Solar Energy Conversion (QUANTSOL'98), March 14–19, 1999, Wildhaus, Switzerland.
253. Barreau, N., Bernède, J.C., El Maliki, H., Marsillac, S., Castel, X., and Pinel, J. (2002) *Solid State Commun.*, **122** (7–8), 445.
254. Yi, S.-H., Choi, S.-K., Jang, J.-M., Kim, J.-A., and Jung, W.-G. (2007) *J. Colloid Interface Sci.*, **313**, 705.
255. Cui, J.B., Daghlian, C.P., Gibson, U.J., Püsche, R., Geithner, P., and Ley, L. (2005) *J. Appl. Phys.*, **97**, 044315.

256 Hsu, J.W.P., Tian, Z.R., Simmons, N.C., Matzke, C.M., Voigt, J.A., and Liu, J. (2005) *Nano Lett.*, **5**, 83.

257 Drici, A., Djeteli, G., Tchangbedji, G., Derouiche, H., Jondo, K., Napo, K., Bernède, J.C., Ouro-Djobo, S., and Gbagba, M. (2004) *Phys. Stat. Sol. A*, **201**, 1528.

258 Borges, R.O., Lincot, D., and Vedel, J. (1992) Proceedings of the 11th European Photovoltaic Solar Energy Conference, Montreux, Switzerland, p. 862.

259 Vidal, J., Vigil, O., DeMelo, O., Lopez, N., and Zelaya-Angel, O. (1999) *Mater. Chem. Phys.*, **61**, 139.

260 Hariskos, D., Spiering, S., and Powalla, M. (2005) *Thin Solid Films*, **99**, 480.

261 Ennaoui, A., Bär, M., Klaer, J., Kropp, T., Sáez-Araoz, R., and Lux-Steiner, M.Ch. (2006) *Prog. Photovolt. Res. Appl.*, **14**, 499.

262 O'Brien, P., Otway, D.J., and Smyth-Boyle, D. (2000) *Thin Solid Films*, **17**, 361.

263 Govender, K., Boyle, D.S., Kenway, P.B., and O'Brien, P. (2004) *J. Mat. Chem.*, **14**, 2575.

264 Bohnsack, G. (1988) *Ber. Bunsen-Ges. Phys. Chem.*, **92**, 203.

265 Chu, J.B., Huang, S.M., Zhang, D.W., Bian, Z.Q., Li, X.D., Sun, Z., and Yin, X.J. (2009) *Appl. Phys. A*, **95**, 849.

266 Estrada, C.A., Nair, P.K., Nair, M.T.S., Zingaro, R.A., and Meyers, E.A. (1994) *J. Electrochem. Soc.*, **141**, 802.

267 Huang, L., Nair, P.K., Nair, M.T.S., Zingaro, R.A., and Meyers, E.A. (1994) *J. Electrochem. Soc.*, **141**, 2536.

268 Bode, D.E. (1966) *Physics of Thin Films* (eds G. Hass and R.E. Thun), vol. 3, Academic Press, New York, p. 275.

269 Stanley, A.G. (1975) Cadmium sulfide solar cells, in *Applied Solid State Sciences*, vol. 5 (ed. R. Wolf), Academic Press, New York, p. 251.

270 Mane, R.S., Sankapal, B.R., and Lokhande, C.D. (1999) *Mater. Chem. Phys.*, **60**, 196.

271 Bhattacharya, R.N. and Pramanik, P.P. (1983) *J. Electrochem. Soc.*, **128**, 332.

272 Lokhande, C.D. (1991) *Mater. Chem. Phys.*, **29**, 1.

273 Ubale, A.U., Daryapurkar, A.S., Mankar, R.B., Raut, R.R., Sangawar, V.S., and Bhosale, C.H. (2008) *Mater. Chem. Phys.*, **110**, 180.

274 Sonawane, P.S. and Patil, L.A. (2007) *Mater. Chem. Phys.*, **105**, 157.

275 Lokhande, C.D. (1991) *Mater. Chem. Phys.*, **28**, 145.

276 Desai, J.D. and Lokhande, C.D. (1993) *Indian J. Pure Appl. Phys.*, **31**, 152.

277 Sonawane, P.S., Wani, P.A., Patil, L.A., and Seth, T. (2004) *Mater. Chem. Phys.*, **84**, 221.

278 Desai, D. and Lokhande, C.D. (1994) *Indian J. Pure Appl. Phys.*, **32**, 964.

279 Nair, P.K., Campos, J., Sanchez, A., Banos, L., and Nair, M.T.S. (1991) *Semicond. Sci. Technol.*, **6**, 393.

280 Garcia, V.M., Nair, M.T.S., Nair, P.K., and Zingaro, R.A. (1997) *Semicond. Sci. Technol.*, **12**, 645.

281 Lin, L.-H., Wu, C.-C., Lai, C.-H., and Lee, T.-C. (2008) *Chem. Mater.*, **20**, 4475.

282 Kroger, F.A. (1978) *J. Electrochem. Soc.*, **125**, 2028.

283 Chen, J.H. and Wan, C.C. (1994) *J. Electroanal. Chem.*, **365**, 87.

284 Basol, B.M., Tseng, E.S., and Rod, R.L. (1983) US Patent 4,388,483.

285 Bassol, B.M. and Tseng, E.S.F. (1984) EP0118579.

286 Basol, B.M. (1984) *J. Appl. Phys.*, **55**, 601.

287 Basol, B.M., On, S.S., and Stupsudd, O.M. (1985) *J. Appl. Phys.*, **58**, 3809.

288 Volvoda, V., Touskova, J., and Kindle, D. (1986) *Cryst. Res. Technol.*, **21**, 975.

289 Basol, B.M. (1988) *Solar Cells*, **23**, 69.

290 Bhattacharya, R.N. (1983) *J. Electrochem. Soc.*, **130**, 2040.

291 Pern, F.J., Goral, J., Matson, R.J., Gessert, T.A., and Noufi, R.N. (1988) *Solar Cells*, **24**, 81.

292 Hodes, G. and Cahen, D. (1986) *Solar Cells*, **16**, 245.

293 Khare, N., Razzini, G., and Bicelli, L.P. (1990) *Thin Solid Films*, **186**, 113.

294 Sahu, S.N., Kristensen, R.D.L., and Haneman, D. (1989) *Solar Energy Mater.*, **18**, 385.

295 Garg, P., Garg, J.C., and Rastogi, A.C. (1990) Proceedings of the 21st IEEE

296 Garg, P., Garg, A., Restogi, A.C., and Garg, J.C. (1991) *J. Phys. D*, **24**, 2026.

297 Lincot, D., Guillemeles, J.F., Covale, P., Massaccesi, J., Thouin, L., Fezza, K., Boisinon, F., and Vedel, J. (1994) Proceedigs of the 1st World Conference on Photovoltaic Energy Conversion, Waikoloa, HI, 5–9 December 1994.

298 Kapur, V.K., Basol, B.M., and Tseng, E.S. (1987) *Solar Cells*, **21**, 65.

299 Chu, T.L., Chu, S.S., Lin, S.C., and Yue, J. (1984) *J. Electrochem. Soc.*, **131**, 2182.

300 Thouin, L. and Vedel, J. (1995) *J. Electrochem. Soc.*, **142**, 2996.

301 Herrero, J. and Ortega, J. (1987) *Solar Energy Mater.*, **16**, 477.

302 Guillen, C. and Herrero, J. (1994) Proceedings of the 12th European Photovoltaic Solar Energy Conference, Amsterdam, p. 593.

303 Guillen, C. and Herrero, J. (1995) *J. Electrochem. Soc.*, **142**, 1834.

304 Guillen, G. and Herrero, J. (1995) *J. Electrochem. Soc.*, **143**, 493.

305 Guillen, C. and Herrero, J. (1994) *J. Electrochem. Soc.*, **141**, 225.

306 Rockett, A., Abou-Elfotouh, F., Albin, D., Bode, M., Ermer, J., Klenk, R., Lommasson, T., Russel, T.W.F., Tomlinson, R.D., Tuttle, J., Stolt, L., Walter, T., and Peterson, T.M. (1994) *Thin Solid Films*, **237**, 1.

307 Dimmler, B., Grunwald, F., Schmid, D., and Schock, H.W. (1991) Proceedings of the 22nd IEEE Photovoltaic Specialists Conference, Las Vegas, NV. IEEE, New York, p. 951.

308 Noufi, R. and Dick, J. (1985) *J. Appl. Phys.*, **58**, 3884.

309 Guillen, C., Galiano, E., and Herrero, J. (1991) *Thin Solid Films*, **195**, 137.

310 Herrero, J. and Ortega, J. (1990) *Solar Energy Mater.*, **20**, 53.

311 Pern, F.J., Noufi, R., Matson, A., and Franz, A. (1991) *Solar Energy Mater.*, **21**, 299.

312 Qiu, C.X. and Shih, I. (1987) *Solar Energy Mater.*, **16**, 219.

313 Thouin, L., Massaccesi, S., Sanchez, S., and Vedel, J. (1994) *J. Electroanal. Chem.*, **374**, 81.

314 Vedel, J., Thouin, L., and Lincot, D. (1996) *J. Electrochem. Soc.*, **143**, 2173.

315 Thouin, L., Sanchez, S., and Vedel, J. (1993) *Electrochim. Acta*, **38**, 2387.

316 Fernandez, A.M., Sebastian, P.J., Bhattacharya, R.N., Noufi, R., Contreras, M., and Hermann, A.M. (1996) *Semicond. Sci. Technol.*, **11**, 1.

317 Bhattacharya, R.N., Fernandez, A.M., Contreras, M.A., Keane, J., Tennant, A.L., Ramanathan, K., Tuttle, J.R., Noufi, R.N., and Herman, A.M. (1996) *J. Electrochem. Soc.*, **143**, 854.

318 Lakshmi, B.B., Patrissi, C.J., and Martin, C.R. (1997) *Chem. Mater.*, **544**, 2544.

319 Hoyer, P. (1994) *Adv. Mater.*, **8**, 857.

320 Kovtyukhova, N.I., Buzaneva, E.V., Waraksa, C.C., Martin, B.R., and Mallouk, T.E. (2000) *Chem. Mater.*, **12**, 383.

321 Lei, Y., Zhang, L.D., Meng, G.W., et al. (2001) *Appl. Phys. Lett.*, **78**, 1125.

322 Cao, H.Q., Xu, Y., Hong, J.M., Liu, H.B., Yin, G., Li, B.L., Tie, C.Y., and Xu, Z. (2001) *Adv. Mater.*, **13**, 1393.

323 Nakamura, H. and Matsui, Y. (1995) *J. Am. Chem. Soc.*, **117**, 2651.

324 Cheng, B. and Samulski, E.T. (2001) *J. Mater. Chem.*, **11**, 2901.

325 Limmer, S.J., Seraji, S., Forbess, M.J., Wu, Y., Chou, T.P., Nguyen, C., and Cao, G.Z. (2001) *Adv. Mater.*, **13**, 1269.

326 Hilgendorff, M., Spanhel, L., Rothenhausler, C.h., and Muller, G. (1998) *J. Electrochem. Soc.*, **145**, 3632.

327 Brenier, R. and Ortega, L. (2004) *J. Sol-Gel Sci. Technol.*, **29**, 137.

328 Moriguchi, I., Maeda, H., Teraoka, Y., and Kagawa, S. (1997) *Chem. Mater.*, **9**, 1050.

329 Spanhell, L., Arpac, E., and Schmidth, H. (1992) *J. Non-Cryst. Solids*, **147–148**, 657.

330 Savadogo, O. and Mandal, K.C. (1992) *Mater. Chem. Phys.*, **31**, 301.

331 Savadogo, O. and Frechette, M. (1997) Poster 35, Presented at the Second International Symposium on New Materials for Elctrochemical Systems, July 9–13, Montreal, Canada.

332 Savadogo, O. and Mandal, K.C. (1992) *J. Electrochem. Soc.*, **139**, L16.

333 Savadogo, O., Boutin, E., and Frechette, M. (1997) Poster 37, Presented at the Second International Symposium on New Materials for Elctrochemical Systems, July 9–13, Montreal, Canada.
334 Savadogo, O. and Mandal, K.C. (1992) *J. Phys. D*, **27**, 1070.
335 Savadogo, O. and Boutin, E. (1997) Poster 41, Presented at the Second International Symposium on New Materials for Elctrochemical Systems, July 9–13, Montreal, Canada.
336 Savadogo, O. and Mandal, K.C. (1994) *J. Electrochem. Soc.*, **141**, 2871.
337 Savadogo, O. (1994) Progress Report, Ecole Polytechnique de Montreal, October 1994, 32 pp.
338 Savadogo, O. (1995) CDT Contract, P.1937, Progress Report, Ecole Polytechnique de Montreal, July 1995, 32 pp.
339 Savadogo, O. (1995) CDT Contract, P.1937, Progress Report, Ecole Polytechnique de Montreal, December 1995, 25 pp.
340 Savadogo, O. (1996) Final Report, Ecole Polytechnique de Montreal, March 1996, 45 pp.
341 Savadogo, O., Ndzebet, E., Levesque, S., and Lacroix, M. (1998) Final Report, for MERNQ and 5N plus, 55 pages.
342 Savadogo, O. and Lacroix, M. (1999) Poster 45, Presented at the third International Symposium on New Materials for Elctrochemical Systems, July 6–10, Montreal, Canada.
343 Savadogo, O. (1999) Poster 49, Presented the third International Symposium on New Materials for Elctrochemical Systems, July 6–10, Montreal, Canada.
344 Wu, X., Dhere, R.G., Albin, D.S., Gessert, T.A., DeHart, C., Keane, J.C., Duda, A., Coutts, T.J., Asher, S., Levi, D.H., Moutinho, H.R., Yan, Y., Moriarty, T., Johnston, S., Emery, K., and Sheldon, P. (2001) NREL/CP-520-3102. Présented at the NCPV Program Review Meeting, Lakewood, Colorado 14–17 October 2001.
345 16 January 2010 m (51,804 bytes) (sortable tables) http://en.wikipedia.org/wiki/Abundances_of_the_elements_(data_page) (accessed, February 05, 2010).
346 Haxel, Gordon B., Hedrick, James B., and Orris, Greta J. Created in November 20, 2002, Last modified May 17, 2005. http://pubs.usgs.gov/fs/2002/fs087-02/ (accessed, December 11, 2009).
347 29 December 2009 MrOllie (talk | contribs) (31,165 bytes) (Reverted 1 edit by Donwikiworker; Rv daystar spammer. (TW)) (undo) http://www.nrel.gov/pv/thin_film/docs/telluriumworldindustrialminerals2000.doc (accessed Fabruary 05, 2010). Assessment of critical thin film resources.
348 George, Micheal W. (2009) Tellurium. Mineral Commodity Summaries. United States Geological Survey. http://minerals.usgs.gov/minerals/pubs/commodity/selenium/mcs-2009-tellu.pdf (accessed Janvier 12, 2010).
349 Fthenakis, V.M. (2004) *Renewable Sustainable Energy Rev.*, **8**, 303.
350 (2010) VolkovBot (discuter | contributions), http://fr.wikipedia.org/wiki/First_Solar. (11 127 octets) (robot Ajoute: es:First Solar) (accessed January 20, 2010).
351 Jacob, F., Barrea, N., Gall, S., and Kessler, J. (2007) *Thin Solid films*, **515**, 552.
352 Wei, L., Lianghuan, F., Jingquan, Z., Lili, W., Yaping, C., Jiagui, Z., Wei, C., Bing, L., and Zhi, L. (2008) *Sci. China Ser. E-Tech. Sci.*, **51**, 33.
353 Zayed, J. and Philippe, S. (2009) *Int. J. Toxicol.*, **28** (4), 259.
354 (23 January 2010) Yilloslime (talk | contribs) (14,574 bytes) http://www.alfa.com/content/msds/german/A14544.pdf (accessed January 29, 2009).
355 (1998–2009) Lennetech, http://www.lenntech.fr/periodique/elements/si.htm (accessed, December 10, 2009).
356 June 11 2008 Sciencelab.com, Inc. http://www.sciencelab.com/xMSDS-Cadmium-9923223 (Decemeber 10, 2009).
357 (July 2010), Merck, http://assets.chemportals.merck.de/documents/sds/

358 United States Department of Health and Human Services (2003) Nomination of Cadmium Telluride to the National Toxicology Program. 2003-04-11.

359 (8 September 2009), Ablazev, http://en.wikipedia.org/wiki/Cadmium_telluride (accessed December 22, 2009).

360 (2005) Unmodified Cadmium Telluride Quantum Dots Prove Toxic. Nano News (National Cancer Institute). 2005-12-12. http://nano.cancer.gov/news_center/nanotech_news_2005-12-12c.asp (accessed January 25, 2010)

361 Fthenakis, V., Fuhrmann, M., Heiser, J., and Wang, W. (2004) (free download pdf). Experimental Investigation of Emissions and Redistribution of Elements in CdTe PV Modules during Fires. Paris, France: 19th European PV Solar Energy Conference, p. 5BV.1.32. http://www.nrel.gov/pv/thin_film/docs/fthenakis_2004_cdte_fires_paris_preprint.pdf.

362 Madelung, O. (1992) *Semiconductors Other Than Group IV Elements, III-V Compounds*, Springer, Berlin.

[preceding text: emd/deu/de/1020/102015.pdf (accessed February 18, 2010).]

Index

a

absorber
– layer 4f., 11
– thickness 71, 149
absorption
– coefficient 46, 49, 65, 222
– core–shell 93
– edge 222
– length 71
– site-specific 121
– SJW-E1 dye 201
– UV-visible 239
acceptor, see doping
accumulation layer 119f., 158
adhesion 301f.
annealing 5, 10f., 47, 51f., 280
aqueous chemical growth 226
atom–by–atom deposition 282
atomic force microscopy (AFM) 83, 86, 166
– contact mode (CM-AFM) 83f., 105, 108f., 116f., 151
– Derjaguin–Muller–Toporov model 100
– InP 167f.
– Kelvin probe 119f.
– tapping mode (TM-AFM) 99f., 137, 167f.
atomic sensitivity factor 136
Auger electron emission (AEE) 90f., 141
autocatalytic redox process 306

b

backbond splitting 160
band
– conduction 63, 77, 104
– energy 64, 73
– illumination 239f.
– valence 63, 69, 77, 104
bandgap
– absorber materials 49
– CdS 44
– CdTe 45f.
– CZTS 30
– energy 49, 51ff.
– porous wide- 222ff.
– Sb_2S_3 322f.
– TiO_2 223
blisters 12, 21
Boltzmann approximation 62
Brewster angle analysis (BAS) 84f., 126, 129
broadening parameter 51f.
buffer layer 198f., 320f., 337

c

capping layer 21
carrier type
– majority 42
– minority 65f., 68, 71, 81, 89f.
– profiles 65ff.
carrier surface concentration 82
CdTe film 4f., 7, 10, 14f., 17ff.
chalcopyrite 1, 11
– crystal structures 4
– $Cu(In,Ga)(S,Se)_2$ (CIGS) 2f., 5f., 9, 12, 19ff.
– $CuInS_2$ (CIS) 1, 12, 42, 52f., 138ff.
– interface conditioning 137ff.
– polycrystalline 143ff.
– single-crystal 53
– ternary 138, 145f.
charge
– collection 67
– injection 77, 231
charge transfer
– light-induced 241
– rate 64, 68, 81f., 90, 155
– velocity 83
chemical bath deposition (CBD) 4, 44, 278ff.
– Ag_2S 325

- Bi_2S_3 323ff.
- CdS layer   32, 48, 282ff.
- induction   282, 286
- linear growth   282, 286
- porous films   280
- Sb_2S_3 322ff.
- SnS   324
- termination   282, 286
- ZnO   320f.
- ZnS   320f.

chemical solution deposition (CSD)   278, 285

chemical vapor deposition (CVD)   4, 94, 226, 278, 302ff.
- atmospheric pressure (APCVD)   304

chemical wet deposition (WD)   278
Cherenkov radiation   93
chlorophyll derivates   207f.
chronoamperometry
- conditioning   144
- coulometry   105
- profiles   82, 166

coating
- antireflection   332
- low-temperature   184ff.
- paste   191
- SnO_2-   285
- ZnO-based   252

co-deposited
- $CuInSe_2$ (CIS)   6f., 19, 22f., 329
- metal   31, 295
- simultaneous   27
- ZnO–ZnS–Zn(OH)   321

colloidal
- particles   279, 283
- systems   280

complexing agents   30, 281ff.
- metal–ligand   307
- methods   283ff.
- non-interfering   281f., 289

composite materials fabrication   279ff.
condenstaion   296ff.
conditioning, see interface
contacts
- back   294, 304, 333
- electrolyte   40f., 49
- liquid   41, 140
- metal   41
- semiconductor–redox electrolyte   62, 64, 150

conversion
- down-   222
- internal   236f.

- interphase   313
- up-   222

copper indium dichalcogenides ($CuInSe_2$), see chalcopyrite

core-level   155f., 158
Cottrell equation   150
cracks   12, 49, 197
current
- burst   121
- dark   43, 62, 131, 147
- density, see photocurrent
- diffusion   63f.
- light-induced   63

d

DC
- parallel connection   216
- series connection   213f.

Debye length   68
decomposition
- anodic   14, 73
- cathodic   14, 73, 134
- energy   75
- hydrolytic   318
- level   72, 133
- reaction rate   76f.

defect
- blue   48
- compositional   50
- density   283
- interface   226
- structural   50

dehydration/condensation reaction   186
density of states (DOS)   75, 96f., 102f., 159
depletion layer   64f., 81
diffusion
- coefficient   23, 29, 67, 150
- constant   233f.
- controlled   27
- length   49, 65f., 68, 70f., 148f., 152, 197
- region   66
- surface   10

dissociation
- constant   282
- first-order   307
- light-induced water   78ff.
- reaction   79

dissolution reaction   74, 281
- silicon   108ff.

Doniach–Sunjic function   84
doping   13
- acceptor density   50
- CdTe   326
- concentration   136

– density 44
– donor 285
– level 133, 285
– profile 166
dye
– multilayers 246
– ZnO/dye materials 244ff.

e
efficiency
– cell 20f.
– charge collection 67f.
– complexation 316
– conversion 19, 22, 38, 146, 152f., 162, 187, 189, 196f., 243
– external quantum (EQE) 42ff.
– incident photon conversion (IPCE) 42, 242
– internal quatum (IQE) 44, 82
– light-harvesting 252
– module 336
– plating 20
Einstein relation 67
electrochemical atomic layer epitaxy (ECALE) 6
– liquid-phase 278
electrochemical deposition (ED) 4, 6
– alloy 19
– and annealing (EDA) 7, 10f., 37
– anodic 289
– cathodic 14, 17, 289
– CdTe 14f., 290, 293f., 325ff.
– Cu(In,Ga)(S,Se$_2$) (CIGS) 19f., 294f.
– hybrid ZnO/organic thin films 244ff.
– one-step 22
– Pt 154f.
– pulsed 257f.
– single-element 19
– stoichiometry 5
– two-step 5
– ZnO thin films 226f., 251
electrochemical phase diagram 28
electrode
– back 226, 229, 238
– counter 78f., 186, 190ff.
– dye-sensitized ITO–PEN 186ff.
– fabrication 185
– mercury sulfate (MSE) 23f., 28f.
– passivated 24
– plastic 190ff.
– rotating disk (RDE) 9, 16, 20f., 24, 26, 195
– semi-transparent plastic 191f.
– SHE 23
– SJW-E1 sensitized plastic 197ff.
– standard (saturated) calomel (SCE) 21, 30, 134, 166, 168, 326
– working 25, 30, 78f., 196
electrodeposition, see electrochemical deposition (ED)
electroless deposition 306f.
electrolyte contacts, *see* contacts
electrolyte electroabsorbance (EEA) 50ff.
electrolyte electroreflectance (EER) spectroscopy 49ff.
electron energy loss spectroscopy (EELS) 232
– high-resolution (HREELS) 90, 99, 104
electron
– affinity 133f., 138, 154, 165f.
– back transfer 198f., 229, 231, 255f.
– diffusivity 188f.
– -donating groups 232
– –hole pair 44
– injection 108, 114, 206
– lifetime 188, 228
– light-induced defect 111
– mobility 89
– transfer rate 22
electrophoretic deposition 185, 188
electropolishing 129, 132
ellipsometry 84
energy
– band diagram 80f.
– binding 116, 118, 133
energy dispersive X-ray (EDX) 9, 32, 48
– Cu 11
– Cz/Sn/Cu/Zn 35
– In$_2$Se$_3$/Cu–Se 26
– stacked precursor film 35
energy distribution curve (EDC) 97f.
epitaxial layers 17, 162, 225, 306
– homo- 165f., 226
EQE, *see* efficiency
etching
– bromine/ethanol 294
– chemical 105ff.
– electrochemical 104
– indium phosphite (InP) 131
– KCN 143, 146
– oxide 129
– photoelectrochemical 142
– silicon 111ff.
excess carrier concentration 66f., 70
excess minority carrier profiles 65f., 68, 71, 81, 89f.
excess microwave conductivity 65f.

excess microwave reflectivity
– profiles   67, 82
– signal   67, 81f.
– stationary   81, 87
excitation energy transfer   80, 171
extinction coefficients
– polymethine dyes   203
– sensitizer molecules   229
– SJW-E1   197ff.

f

Fermi level   62, 70, 335
– bulk   134
– gradients   5
– pinning   102
– quasi-   69ff.
Feynman path integrals   123
fill factor (FF)   189, 193, 199, 206, 211, 251
fluorescence emission   236f.
flux ratio   23, 28
Förster and Dexter transfer   171, 207
Fourier transform infrared (FTIR) spectroscopy   121
frontier orbital positions
– highest occupied molecular orbital (HOMO)   231f., 235, 237ff.
– lowest unoccupied molecular orbital (LUMO)   231f., 235, 237ff.
– second highest occupied molecular orbital (SHOMO)   239
FTO (fluorine-doped tin oxide)   3, 17, 185, 190, 322f.
– -coated glass   17, 44, 46, 186, 212

g

Gärtner equation   49, 67
Gerischer model   110
Gibbs
– energy of formation   14, 280
– free energy   26, 293
– reaction enthalpy   73
Grotthuß-like transport mechanism   228

h

Hamaker constant   100
Hammett coefficients   233
Helmholtz layer   63, 79, 89
heterojunction   13, 19, 44, 222, 224, 332ff.
heteropolycompounds   330
hole injection   150, 155, 160
hydrodynamically controlled   23, 26f.
hydrogen
– aggregation   200

– diffusion   150
– evolution   21, 34, 165, 168
hydrolysis   296f., 315

i

ideality factor   354
impedance spectroscopy   29, 209f., 309
impurity   5, 294
incident photon-to-current quantum efficiency (IPCE)   42, 242ff.
incident photon-to-electron conversion quantum efficiency   187f.
indium phosphite (InP)
– (111) A-face   130f., 147f.
– conditioning, see interface
– metal–interphase–semiconductor   165ff.
– photocurrent–voltage characteristic   132
– wafer substrate   131
indium tin oxide (ITO)   184, 285, 309
indoline dyes   204
inductive coupled plasma mass spectrometry (ICP-MS)   20, 27, 31
inorganic crystal structure database (ICSD)   37
intensity-modulated photocurrent spectroscopy (IMPS)   246, 248
intensity-modulated photovoltage spectroscopy (IMVS)   248
interaction
– chromophore   242, 246
– dye–dye   246
– electron bunch–laser   93
– intermolecular   234
– sensitizer–surface   230
– substrate–adsorbate   15
intercalation
– charge-balancing counterions   232, 234
– proton   233
interface
– CdTe–electrolyte   46
– charge transfer   228
– conditioning   104ff.
– electrolyte–oxide–Si   129
– interpenetrating   222
– ITO–electrolyte   215
– ITO–TiO$_2$ 215
– Pc–electrolyte   237
– recombination velocity   332
– semiconductor–electrolyte   72
– Si–oxide   104
– solid–liquid   61
– substrate–ion   282
– support–semiconductor   307
– tailoring   61

interfacial film
- amorphous   167
- formation   77, 131, 134, 136f., 165
- oxide   132, 161
- passivating   143
- stress   123, 125
internal quantum conversion efficiency   206
intersystem crossing   236
inversion layer extension   159
ion–by–ion condensation   282ff.

j

jet vapor deposition (JET)   304
junction
- metal–oxide–semiconductor   160
- sample–electrode   42
- Schottky   42, 62, 333
- semiconductor– redox electrolyte   41f., 62f., 78

k

kesterite   3 , 5, 11
- crystal structures   4
- Cu_2ZnSnS_4 (CZTS)   1, 4, 30ff.
- synthethis   12
kinetic   9
- CBD   282
- control   20
- decomposition   74
- deposition   312
- electrode   234, 249
- electrodeposition   9f.
- electron transfer   12
- energy   133, 156f., 232
- molecular   304
- stability   72, 143f.
- surface parameter   68
kink site   160
Koutecky–Levich plots   195

l

Langmuir adsorption isotherm.
Lawson–Woodward theorem   93
layer-by-layer growth   6, 279
lifetimes   70, 72
lift-off techniques   165
ligand   30, 281
- -centered second reduction   235
- electron-withdrawing   232, 234, 238
- exchange equilibrium reaction   304
- substitution   232ff.
light-emitting diode (LED)   42

liquid-phase atomic layer epitaxy, *see* electrochemical atomic layer epitaxy (ECALE)
liquid-phase deposition (LPD)   304f.
low-pressure metal organic vapor-phase epitaxy (LPMOVE)   278
lock-in amplifier   51

m

Madelung term   95
Marcus–Gerischer distribution   62
mass transfer-controlled   20
mass transport   9, 11
mesopores
- silicon   218
- TiO_2 184f.
metal–ligand charge transfer mechanism   203
metal organic vapor-phase deposition (MOVPD)   4
metal organic vapor-phase epitaxy (MOVPE)   131, 136, 165
modulation   50f.
molecular beam epitaxy (MBE)   4, 94, 242, 278
Mott–Schottky plot   149
multilayer films   246, 280, 300, 319

n

nanocomposites   330
nanocrystalline   187, 280
nanoemitter structures   147ff.
- p-Si   162ff.
nanopores   121f., 197
- etching   126
- oxide   122, 127, 147, 162
- wide-bandgap semiconductors   222ff.
nanotopographies
- oxide-related   121
- silicon   107ff.
National Renewable Energy Laboratory (NREL)   336
Nernst equation   294f.
nonresonant X-ray emission spectroscopy (NXES)   142
nucleation
- CdTe   16
- center   316
- compound   281
- density   5, 33
- heterogeneous   282f., 289
- homogeneous   282
- metal layers   33
- secondary   5

o

operation mode
– photoelectrochemical solar cells (PECS) 77f.
– photovoltaic (PECS) 77f.
oscillators
– thickness 123ff.

p

passivating layer 148, 256
PbS thin film 279
phase
– intermediate 10
– secondary 5
– transformation 12
phosphorescence emission 236f.
photocorrosion 73, 75, 110, 143, 241, 256
– anodic 131
– cathodic 131
– fractal 126
photocurrent 21, 42, 46, 62, 64, 83, 147, 242
– density 78, 128f., 186, 189, 196, 202, 256f., 295, 331, 334f,
– oscillation 151
– spectroscopy 65
– switchable 240
photoelectron emission microscopy (PEEM) 90
photon
– energy 51, 118, 133
– flux 49, 66
photosensitization
– anodic 241
– phthalcocyanines 235ff.
– porphyrins 235ff.
– ZnO 227f.
photosynthesis 79, 92
photovoltammetry 42, 72
phthalcocyanines 232ff.
– epitaxial films 242
– photoelectrochemistry 236
push–pull arrangement 243
physical vapor deposition (PVD) 19, 94, 278, 303
pinhole 12, 21, 49
– formation 5
– -free films 6
plating
– electro- 22, 225
– ionic 185, 303
– pulse 27, 30
polarization
– anodic 121, 123
– cathodic 295
– concentration 295
– cyclic 136f.
– energy 232
– potentiostatic 240
– spectroscopy 29
Poisson equation 68
pore deepening 149ff.
porosity 186, 195f.
porphyrins 230f., 235
potential
– boundary 326
– cathodic 14
– chemical 69, 78
– contact 164ff.
– decomposition 73
– deposition 10
– doping 159
– equilibrium 14
– flatband 41, 43, 106, 135, 149, 154, 159
– Galvani 69, 119f.
– open circuit 114, 118, 158f.
– over- 74, 79f., 150, 295
– quasi-rest (QRP) 326
– reduction 8f., 23ff.
– reference electrode 41, 295
– standard reduction 19
Pourbaix diagram 290ff.
precursor
– Cz/Sn/Cu/Zn stacks 35f.
– film 6ff.
– metallic 8f.
– routes 226, 296
– stacked film 35
– stoichiometry 30
– types 11f.
printable materials 183ff.
– PEDOT-based paste 196
pulsed laser-induced current transients 202
pulsed layer deposition 278

q

quadrupole mass spectroscopy (QMS) 94
quantum
– confinement 222
– efficiency 241f., 248, 294
quantum yield 81f.
quarz crystal microbalance technique (QCM) 284, 309
quenching 207

r

radiofrequency diode sputtering (RDS) 278
reactive magnetron sputtering 143

recombination   5, 43, 66, 230
– bulk   67
– catalysis   231
– losses   5, 48
– minority carrier   164
– surface/interface   135
– velocity   66, 68, 332
recrystallization   26, 51
redox
– electrolyte   147, 223f., 228
– energy   62
– level   78
– reaction   76
– species   63, 65
reduction
– cathodic   17, 294
– four-electron   15f., 293
– six-electron   293
– two-electron   16
reflectance   51, 128f.
refractive index   217f.
relaxation
– energy   95
– process   15f., 90
reorganization energy   62f.
Richardson constant   62
roll-to-roll process   20, 185
room temperature ionic liquids (RTILs)   39

s
scanning electrochemical microscopy (SECM)   162
scanning electron microscopy (SEM)   6, 33
– CuInSe$_2$ film   6
– Cz/Sn/Cu/Zn precursor stack   37
– high-resolution (HRSEM)   83
– In$_2$Se$_3$ 26f.
– ZnO/dye hybrid thin films   245
scanning tunneling spectroscopy (STS)   99ff.
– distance (DTS)   102
– Pt islands   161
Schottky
– barrier   41, 333
– diodes   335
self-assembly   279
self-limiting reaction   279
sensitization, see photosensitization
sensitizer molecules   227ff.
– aggregation   232
– organic dyes   244
– panchromatic   229
– Ru complexes   228, 230f., 243, 255
skew lines   156f.

semiconductor thin film
– crystallinity   6f., 10, 21
– indirect-gap   75
– morphology   8, 31
– n-type   41f., 70, 78
– optoelectronic properties   6f.
– polycrystalline   5, 306
– preparation   307ff.
– p-type   5f., 13ff.
– single-crystal   49
– stoichiometric   5, 7f., 10
– thickness   9, 26, 77, 286, 311ff.
– porous wide-bandgap   222ff.
Shockley relation   62
Shockley–Queisser limit   222
Showa–Shell–Siemens industrial process   19
silicon
– amorphous   222
– crystalline module   217
– current–voltage characteristics   107, 109, 126, 128
– divalent dissolution   108, 111, 113, 128
– nanotopographies   107ff.
site-selective deposition   152
solar cells
– bifacial-type module   217
– CdS/CdTe   13, 44, 335ff.
– CdS/CIS   19, 286
– CdSe   331
– CIGS/CdS/ZnO   19ff.
– configurations   3ff.
– CTO/ZTO/CdS/CdTe   336
– CZTS   37ff.
– durability   208ff.
– dye-sensitized (DSSC)   77, 143, 183ff.
– excitonic   78
– fabrication   2, 30f., 279ff.
– glass-based DSSCs   208f.
– heterojunction   13, 19, 44, 222, 224, 332ff.
– high-voltage DSSC modules   212
– large-area plastic DSSCs   212ff.
– lifetime   211
– liquid junction   145
– maximum power point   78, 80f., 134f.
– metal thin films   333f.
– nanoemitter   107, 147ff.
– performance   199, 201
– plastic DSSCs   191f., 252f.
– photoelectrocatalytic   61, 77f., 153
– photoelectrochemical (PECS)   13, 61ff.
– photovoltaic (PV)   1, 7, 61, 183ff.
– p/n-type   222
– power density   251

– regenerative   223
– Sb_2S_3 331f.
– solid-state photochemical   13, 143, 146, 280
– stability   224, 228, 252
– tandem   80, 222
– textile-based   253ff.
– thin-film   1, 4
– ultrathin   234
sol–gel technique   224, 242, 279, 295ff.
– 2D (two-dimensional)   330
– multifunctional inorganic–organic   330
solid–liquid analysis system (SoLiAS)   94f., 154
solid-state
– diode   62
– p–n heterojunction device   140, 222
– reaction   10, 15, 293f., 319
solubility product   280f., 284
solvent breakdown   22
Soret band maximum   243
space charge layer   64, 66f., 81, 148f.
spray pyrolysis   278, 299ff.
– electrostatic spray-assisted vapor-deposition (ESAVP)   301
– ultrasonic   302
sputtering   226, 303
stabilization
– chemical   75f.
– kinetic   75f., 113
– physical   75, 77, 252
stacked elemental approach (SEL)   30ff.
stacked layered approach   19
– binary layers   23
– Cu/In/Ga   20f.
– In_2Se_3/Cu–Se precursor   25
stacking faults   17f.
Stokes shift   207
structure-directing agent (SDA)   248
substrate
– degradation   10
– electrode   8
– InP/CdS   18
– single-crystal   6, 225
successive ionic layer adsorption and reaction (SILAR)   278, 306, 329
sulfurization   31, 34, 43, 48
surface area
– CdTe film   17
– internal   250, 252
surface
– blocking agents   228
– bond-terminating   75
– core-level shift   109, 111, 141
– passivation   33, 76, 83
– recombination   64, 68f., 71, 73, 76, 164
– recombination rate   81f.
– recombination velocity   83, 90
– roughness   20
– step bunched   111ff.
synchronization function   125f.
synchrotron radiation photoelectron spectroscopy (SRPES)   90, 94, 97ff.
– photocorrosion   110
– silicon dissolution   109

t
tautomerization reaction   286
thermal stability   252
thermodynamic
– analysis   14
– decomposition   74
– electrodeposition   10
– equilibrium   280
– potentials   73
– stability   61, 72f.
thin film, see semiconductor
third-generation concept   240
Thomas–Fermi screening approximation   119
TiO_2
– bandgap   223
– ITO–TiO_2 215
– liquid-phase deposition (LPD)   305
– mesopores   184f.
– porous   254
toxicity   140, 145, 318, 326, 338f.
transitions
– multiple efficient   222
– optical   239
transmission coefficient   62
transmission electron microscope (TEM)
– electrodeposited CdTe   18
– high resolution (HRTEM)   18
– InP   167ff.
– Pt islands   161
transmission losses   222
transmittance   51
transparent conducting oxide (TCO)   8, 153, 184, 252f., 258
traveling heater   278
tunneling thickness   77, 145, 230
type conversion   10, 46

u
ultraviolet photoelectron spectroscopy (UPS)   90, 94, 132ff.
underpotential deposition (UPD)   16, 33f.
– Kroger   23
UV–visible absorption spectroscopy   246

v

vacuum
– evaporation 278
– sputtering 4, 22
voltammogram
– cyclic 14f., 193f., 234
– linear 23
– reverse cycling 20

w

wafer 221
wet chemical methods 277ff.

x

X-ray diffraction (XRD) 6, 29, 93
– $CuInSe_2$ 6, 22, 27
– Cz/Sn/Cu/Zn precursor 36ff.
X-ray emission spectroscopy (XES) 90ff.
X-ray fluorescence energy 91
X-ray photoelectron spectroscopy (XPS) 29, 96f., 119, 121, 134f.

z

ZnO
– crystal growth 246, 248
– /dye materials 244ff.
– electrodeposition 226ff.
– EY/ZnO 248ff.
– hybrid thin films 244ff.
– internal surface area 250, 252
– porous networks 249, 255
– sensitization 227f.
– sol–gel technique 299, 329